ORPHAN CROPS FOR SUSTAINABLE FOOD AND NUTRITION SECURITY

Orphan Crops for Sustainable Food and Nutrition Security discusses the issues, challenges, needs and opportunities related to the promotion of orphan crops, known also as neglected and underutilized species (NUS).

The book is structured into six parts, covering the following themes: introduction to NUS, approaches, methods and tools for the use enhancement of NUS, integrated conservation and use of minor millets, nutritional and food security roles of minor millets, stakeholders and global champions, and, building an enabling environment. Presenting a number of case studies at the regional and country levels, the chapters cover different but highly interlinked aspects along the value chains, from acquisition and characterization of genetic diversity, cultivation and harvesting to value addition, marketing, consumption and policy for mainstreaming. Cross-cutting issues like gender, capacity building and empowerment of vulnerable groups are also addressed by authors. Representatives from communities, research for development agencies and the private sector also share their reflections on the needs for the use enhancement of NUS from their own perspectives.

This book will be of great interest to students and scholars of food security, sustainable agriculture, nutrition and health and development, as well as practitioners and policymakers involved in building more resilient food and production systems.

Stefano Padulosi is a Senior Scientist at the Alliance of Bioversity International and CIAT, Via di San Dominico, 1, 00153, Rome, Italy.

E.D. Israel Oliver King is Principal Scientist at the M.S. Swaminathan Research Foundation, Chennai, India.

Danny Hunter is a Senior Scientist at the Alliance of Bioversity International and CIAT, Via di San Dominico, 1, 00153, Rome, Italy.

M.S. Swaminathan is the Founder of the M.S. Swaminathan Research Foundation, Chennai, India.

Issues in Agricultural Biodiversity
Series editors: Michael Halewood and Danny Hunter

This series of books is published by Earthscan in association with the Alliance of Bioversity International and CIAT. The aim of the series is to review the current state of knowledge in topical issues associated with agricultural biodiversity, to identify gaps in our knowledge base, to synthesize lessons learned and to propose future research and development actions. The overall objective is to increase the sustainable use of biodiversity in improving people's well-being and food and nutrition security. The series' scope is all aspects of agricultural biodiversity, ranging from conservation biology of genetic resources through social sciences to policy and legal aspects. It also covers the fields of research, education, communication and coordination, information management and knowledge sharing.

Farmers and Plant Breeding
Current Approaches and Perspectives
Edited by Ola T. Westengen and Tone Winge

Agrobiodiversity, School Gardens and Healthy Diets
Promoting Biodiversity, Food and Sustainable Nutrition
Edited by Danny Hunter, Emilita Monville Ora, Bessie Burgos, Carmen Nyhria Roel, Blesilda M. Calub, Julian Gonsalves and Nina Lauridsen

Biodiversity, Food and Nutrition
A New Agenda for Sustainable Food Systems
Edited by Danny Hunter, Teresa Borelli and Eliot Gee

Orphan Crops for Sustainable Food and Nutrition Security
Promoting Neglected and Underutilized Species
Edited by Stefano Padulosi, E.D. Israel Oliver King, Danny Hunter and M.S. Swaminathan

For more information about this series, please visit: www.routledge.com/Issues-in-Agricultural-Biodiversity/book-series/ECIAB

ORPHAN CROPS FOR SUSTAINABLE FOOD AND NUTRITION SECURITY

Promoting Neglected and Underutilized Species

Edited by
Stefano Padulosi, E.D. Israel Oliver King, Danny Hunter and M.S. Swaminathan

First published 2022
by Routledge
2 Park Square, Milton Park, Abingdon, Oxon OX14 4RN

and by Routledge
605 Third Avenue, New York, NY 10158

Routledge is an imprint of the Taylor & Francis Group, an informa business

© 2022 selection and editorial matter, Stefano Padulosi, E.D. Israel Oliver King, Danny Hunter and M.S. Swaminathan; individual chapters, the contributors

The right of Stefano Padulosi, E.D. Israel Oliver King, Danny Hunter and M.S. Swaminathan to be identified as the authors of the editorial material, and of the authors for their individual chapters, has been asserted in accordance with sections 77 and 78 of the Copyright, Designs and Patents Act 1988.

All rights reserved. No part of this book may be reprinted or reproduced or utilised in any form or by any electronic, mechanical, or other means, now known or hereafter invented, including photocopying and recording, or in any information storage or retrieval system, without permission in writing from the publishers.

Trademark notice: Product or corporate names may be trademarks or registered trademarks, and are used only for identification and explanation without intent to infringe.

British Library Cataloguing-in-Publication Data
A catalogue record for this book is available from the British Library

Library of Congress Cataloging-in-Publication Data
Names: Padulosi, S. (Stefano), editor. | King, E. D. I. Oliver
(E. D. Israel Oliver), editor. | Hunter, Danny, editor. |
Swaminathan, M. S. (Monkombu Sambasivan), editor.
Title: Orphan crops for sustainable food and nutrition security :
promoting neglected and underutilized species / edited by
Stefano Padulosi, E.D.Israel Oliver King, Danny Hunter, and
M.S. Swaminathan.
Description: Abingdon, Oxon ; New York, NY : Routledge, 2022. |
Includes bibliographical references and index. |
Identifiers: LCCN 2021014641 (print) | LCCN 2021014642 (ebook) |
ISBN 9780367491581 (hardback) | ISBN 9780367902827 (paperback) |
ISBN 9781003044802 (ebook)
Subjects: LCSH: Food crops. | Crops—Germplasm resources. |
Germplasm resources conservation. | Millets—Breeding—Case
studies. | Agrobiodiversity. | Agrobiodiversity conservation. | Plant
species diversity. | Plant diversity conservation. | Food security. |
Sustainable agriculture.
Classification: LCC SB175 .O77 2021 (print) | LCC SB175 (ebook) |
DDC 333.95/34—dc23
LC record available at https://lccn.loc.gov/2021014641
LC ebook record available at https://lccn.loc.gov/2021014642

ISBN: 978-0-367-49158-1 (hbk)
ISBN: 978-0-367-90282-7 (pbk)
ISBN: 978-1-003-04480-2 (ebk)

DOI: 10.4324/9781003044802

Typeset in Bembo
by codeMantra

CONTENTS

List of contributors   *xi*
Preface and acknowledgments   *xxiii*

PART I
Introduction to Neglected and Underutilized
Species (NUS)   **1**

1 NUS: what they are and why we need them more than ever   3
   Stefano Padulosi, Gennifer Meldrum, E.D. Israel Oliver King
   and Danny Hunter

2 Landmark NUS events and key publications   19
   Stefano Padulosi and Danny Hunter

3 Mainstreaming NUS for nutrition-sensitive agriculture:
   a holistic approach   43
   Stefano Padulosi, Gennifer Meldrum, Shantanu Mathur
   and Danny Hunter

4 NUS and peoples sense of dietary diversity and well-being   58
   Gigi Manicad

5 Diversity and small millets in India   68
   Annadana Seetharam and Prabhakar Bhat

vi Contents

6 From discovery to food system diversification with African
   neglected and underutilized species   78
   Céline Termote, Stepha McMullin and Hendre Prasad

7 Barriers to adopting a diversity of NUS fruit trees in Latin
   American food systems   88
   *Robin Van Loon, Elisabeth Lagneaux, Gabriela Wiederkehr
   Guerra, Fidel Chiriboga-Arroyo, Evert Thomas, Bruno Gamarra,
   Maarten van Zonneveld and Chris Kettle*

8 Promoting buckwheat in China   109
   Zongwen Zhang

9 Neglected and underutilized species of Mexico: status and
   priorities for their conservation and sustainable use   118
   *Tiziana Ulian, Hugh W. Pritchard, Alejandro Casas,
   Efisio Mattana, Udayangani Liu, Elena Castillo-Lorenzo,
   Michael Way, Patricia Dávila Aranda and Rafael Lira*

10 Neglected and underutilized species in Brazil: from
   obscurity to non-conventional edible plants   128
   *Nuno Rodrigo Madeira, Valdely Ferreira Kinupp
   and Lidio Coradin*

11 Underutilized genetic resources and crop diversification in
   Europe   138
   Ambrogio Costanzo

12 Neglected and underutilized species and indigenous
   foodways of Oceania   147
   *Danny Hunter, Nick Roskruge, Simon Apang Semese,
   Philip Clarke and Gerry Turpin*

PART II
Approaches, methods and tools for the use
enhancement of NUS   **161**

13 The BFN Mainstreaming Toolkit. A roadmap to using
   neglected and underutilized species for food system change   163
   *Teresa Borelli, Daniela Beltrame, Victor W. Wasike,
   Gamini Samarasinghe, Ayfer Tan and Danny Hunter*

Contents **vii**

14 Development of seasonal calendars for sustainable
diets – experiences from Guatemala, Mali and India 174
Gaia Lochetti

15 Enhancing the use of underutilized food crops: partnerships
in a success story of a pop cereal business in Kenya 182
*Yasuyuki Morimoto, Patrick Maundu, Elizaphan Gichangi
Mahinda and Daniel Kirori*

16 Agritourism and conservation of neglected and
underutilized native Andean crops in Santiago
de Okola, Bolivia 198
Stephen R. Taranto, Eliseo Mamani Alvarez and Wilfredo Rojas

17 Mainstreaming African vegetables to improve diets
and livelihoods 208
*Sognigbe N'Danikou, Maarten van Zonneveld, Fekadu Fufa
Dinssa, Roland Schafleitner, Jody Harris, Pepijn Schreinemachers
and Srinivasan Ramasamy*

18 Slow Food and NUS: protecting and promoting
endangered food products 216
Charles Barstow, Edie Mukiibi and Dauro Mattia Zocchi

PART III
Integrated conservation and use of minor millets **225**

19 Conservation and utilization of small millets genetic
resources: global and Indian perspectives 227
Kuldeep Singh, Nikhil Malhotra and Mohar Singh

20 Challenges to conserving millets in
ecologically sensitive areas 237
Amit Mitra

21 Community-centred value-chain development of
nutri-millets: challenges and best practices in India 245
*E.D. Israel Oliver King, Karthikeyan Muniappan, Prashant
K Parida, Sharad Mishra, Kumar Natarajan, Melari Shisha
Nongrum, Manjula Chinnadurai, Nirmalakumari, Carl O.
Rangad, Somnath Roy, Gennifer Meldrum and Stefano Padulosi*

viii Contents

22 Taking millets to the millions: experiences from
government-driven value chains 263
Dinesh Balam, Saurabh Garg, Srijit Mishra, Bhagyalaxmi,
Mallo Indra and Jayshree Kiyawat

23 Millet-based intercropping systems facilitated by beneficial
microbes for climate-resilient, sustainable farming in tropics 273
Natarajan Mathimaran, Devesh Singh, Rengalakshmi Raj,
Thimmegowda Matadadoddi Nanjundegowda, Prabavathy
Vaiyapuri Ramalingam, Jegan Sekar, Yuvaraj Periyasamy,
E. D. Israel Oliver King, Bagyaraj Davis Joseph, Thomas Boller,
Ansgar Kahmen and Paul Mäder

24 Status of minor millets processing technologies in India:
an overview 281
B. Dayakar Rao and Vilas A Tonapi

PART IV
Nutritional and food security roles of minor millets **293**

25 Modelling the food security role of millets under climate
change in eastern Madhya Pradesh 295
Gennifer Meldrum, Victoria Rose, Somnath Roy and
Ashis Mondal

26 Germplasm characterization and novel technologies to
unleash the nutritional potential of small millets 309
Vasudevan Sudha, Nagappa Gurusiddappa Malleshi, Chamarthy
Venkata Ratnavathi, Shanmugam Shobana, Mani Vetriventhan,
Krishna Hariprasanna, Bakshi Priyanka, Viswanathan Mohan
and Kamala Krishnaswamy

27 Millets in farming systems in support of nutrition and
social safety net programmes 319
Priya Rampal, Aliza Pradhan, Akshaya Kumar Panda,
Sathanandham Raju and R.V. Bhavani

28 The smart food approach: the importance of the triple
bottom line and diversifying staples 327
Joanna Kane-Potaka, Nigel Poole, Agathe Diama,
Parkavi Kumar, Seetha Anitha and Oseyemi Akinbamijo

Contents **ix**

PART V
Stakeholders and global champions 335

29 Voices from the communities: the custodians of neglected
and underutilized species 337
*Aiti Devi, Bibiana Ranee, Biswa Sankar Das, Girigan Gopi,
Indra Bai, Kamla Devi, Loichan Sukia, Loknath Naure,
Maganbhai Ahir, Manikandan, Maruthan Ganeshan, Maruthi,
Meera Bai, Melari Shisha Nongrum, Nanchiyamma, Phool Bai,
Prashant Kumar Parida, Rakesh Kumar, Ramesh Makavana,
Rami, Ramkali Bai, Ridian Syiem, Sangeeta Devi, Shalini
Devi, Sharad Mishra, Sunadei Pitia, Sunamani Muduli,
Tukuna Burudi and Usha Devi*

30 Voices from the private sector engaged in the use
enhancement of NUS 344
*Ana Luiza Vergueiro, Daniel Kirori, Jacqueline Damon, Jose
Alfredo Lopez, Leon Kenya, Mahesh Sharma, Ram Bahadur
Rana, Margaret Komen, Meghana Narayan, Michael Ngugi,
Rohan Karawita, Sergio Vergueiro, Serkan Eser, Shauravi Malik,
Simon Nderitu, Sohini Dey, Sridhar Murthy Iriventi and
Vikram Sankaranarayanan*

31 Voices from the agencies: research for development (R4D)
agencies supporting efforts to conserve and promote the
sustainable use of NUS 365
*Claudio Bogliotti, Generosa Calabrese, Hamid El Bilali,
Isabella Rae, Marco Platzer,Mario Marino, Mary Jane Ramos
de la Cruz, Paul Wagsta, Pietro Pipi, Shantanu Mathur,
Stineke Oenema and Tobias Kiene*

PART VI
Building an enabling environment 385

32 Equity, gender and millets in India: implications for policy 387
Nitya Rao, Amit Mitra and Raj Rengalakshmi

33 What is going on around the world: major NUS actors and
ongoing efforts 395
*Stefano Padulosi, Gennifer Meldrum, E. D. Israel Oliver King
and Danny Hunter*

x Contents

34 In a well-nourished world, underutilized crops will be on
the table 416
Marco Antonio Rondon and Renaud DePlaen

Index 427

CONTRIBUTORS

Maganbhai Ahir, Farmer, Ningal, Anjar Block, Kachchh, Gujarat, India.

Aiti Devi, Farmer, Mirag Village, Chamoli District, Uttarakhand, India.

Oseyemi Akinbamijo, Executive Director, Forum for Agricultural Research in Africa.

Eliseo Mamani Alvarez, Fundación para la Promoción e Investigación de Productos Andinos – PROINPA, 538 Calle Americo Vespucio, Piso 3, La Paz, Bolivia.

Seetha Anitha, International Crops Research Institute for the Semi-Arid Tropics (ICRISAT), Hyderabad, India.

Patricia Dávila Aranda, Facultad de Estudios Superiores Iztacala, Universidad Nacional Autónoma de México, Mexico.

Dinesh Balam, Watershed Support Services and Activities Network, Nabakrushna Choudhury Centre for Development Studies, Po-RRL, Institutional Area, Gajapati Nagar, Bhubaneswar.750041.

Charles Barstow, Slow Food International, Piazza XX Settembre 5, 12042, Bra, Italy.

Daniela Beltrame, Biodiversity for Food and Nutrition Project, Ministry of the Environment, Brasília-DF, Brazil.

xii Contributors

Bhagyalaxmi, Watershed Support Services and Activities Network (WASSAN), Plot. No. 685, Road. No. 12 Narasimha Swamy Colony, Nagole, Hyderabad 500 068.

Prabhakar Bhat, Former Project Coordinator (Small Millets), No. 15, 12th A Cross, 2nd Main Road, Sir MV Layout, Thindlu, Bengaluru, Karnataka, India.

R.V. Bhavani, M.S. Swaminathan Research Foundation, Chennai, India.

Claudio Bogliotti, International Centre for Advanced Mediterranean Agronomic Studies (CIHEAM-Bari), Via Ceglie 9, 70010 Valenzano (Bari), Italy.

Thomas Boller, Department of Environmental Sciences – Botany, University of Basel, Schönbeinstrasse 6, 4056-Basel, Switzerland.

Teresa Borelli, Alliance of Bioversity International and CIAT, Via di San Domenico, 1, 00153, Rome, Italy.

Generosa Calabrese, International Centre for Advanced Mediterranean Agronomic Studies (CIHEAM-Bari), Via Ceglie 9, 70010 Valenzano (Bari), Italy.

Alejandro Casas, Facultad de Estudios Superiores Iztacala, Universidad Nacional Autónoma de México, Mexico.

Elena Castillo-Lorenzo, Royal Botanic Gardens Kew, Wellcome Trust Millennium Building, Wakehurst Place, Ardingly, West Sussex RH17 6TN, United Kingdom.

Celine Termote, Alliance Bioversity International and CIAT, c/o ICIPE Duduville Campus, Off Kasarani Road, PO box 823-00621, Nairobi, Kenya.

Manjula Chinnadurai, M.S. Swaminathan Research Foundation, Chennai, India.

Fidel Chiriboga-Arroyo, Ecosystem Management. Department of Environmental System Science, ETH Zürich, Zurich, Switzerland.

Philip Clarke, South Australian Museum, Adelaide, South Australia, Australia.

Lidio Coradin, National Coordinator, Plants for the Future Initiative and National Project Director, Biodiversity for Food and Nutrition, Brasília DF, Brazil.

Ambrogio Costanzo, The Organic Research Centre, Trent Lodge, Stroud Road, Cirencester, GL7 6JN, United Kingdom.

Contributors **xiii**

Renaud DePlaen, Climate Resilient Food Systems Program, IDRC Headquarters, 150 Kent St, Ottawa, K1P 0B2, Canada.

Agathe Diama, International Crops Research Institute for the Semi-Arid Tropics (ICRISAT), Bamako, Mali.

Fekadu Fufa Dinssa, World Vegetable Center, P.O. Box 10, Duluti, Arusha, Tanzania.

Hamid El Bilali, International Centre for Advanced Mediterranean Agronomic Studies (CIHEAM-Bari), Via Ceglie 9, 70010 Valenzano (Bari), Italy.

Serkan Eser, Bilgi Sistem Yönetimi – İthalat & İhracat, Serik Antalya, Turkey.

Bruno Gamarra, Valuable Forests GmbH, Rötelstrasse 121, CH-8037 Zürich, Switzerland.

Maruthan Ganeshan, Farmer, Nakkupathi Village, Agali Panchayat, Attappadi Block, Kerala, India.

Saurabh Garg, Krushi Bhawan 3rd Floor, Gopabandhu Marg, Keshari Nagar, Bhubaneswar – 751001, India.

Girigan Gopi, MSSRF, Wayanad, Kerala, India.

Gabriela Wiederkehr Guerra, Alliance of Bioversity International and CIAT, Via di San Domenico, 1, 00153, Rome, Italy.

Krishna Hariprasanna, ICAR-Indian Institute of Millet Research, Rajendranagar, Hyderabad – 500030, India.

Jody Harris, World Vegetable Center, P.O. Box 1010 (Kasetsart University), Bangkok 10903, Thailand.

Hendre Prasad, World Agroforestry (CIFOR-ICRAF), AOCC Genomics Laboratory, World Agroforestry (CIFOR-ICRAF), United Nations Avenue, Gigiri, Nairobi, 00100, KENYA.

Danny Hunter, Alliance of Bioversity International and CIAT, Via di San Domenico, 1, 00153, Rome, Italy.

Mallo Indra, Mahila Arthik Vikas Mahamandal (MAVIM) (Under Government of Maharashtra) Griha Nirman Bhavan (MHADA), Mezzanine floor, Kalanagar, Bandra (E), Mumbai – 400051, India.

xiv Contributors

Indra Bai, Farmer, Dhiravan Village, Dindori District, Madhya Pradesh, India.

Sridhar Murthy Iriventi, GoBhaarati Agro Industries and Services Pvt. Ltd., Hyderabad, Telangana, India.

Isabella Rae, Evaluation Consultant, FAO.

Bagyaraj Davis Joseph, Centre for Natural Biological Resources and Community Development, 41, RBI Colony, Anand Nagar, Bangalore – 560024, Karnataka, India.

Ansgar Kahmen, Department of Environmental Sciences - Botany, University of Basel, Schönbeinstrasse 6, 4056-Basel, Switzerland.

Kamla Devi, Farmer, Paini Village, Chamoli District, Uttarakhand, India.

Joanna Kane-Potaka, International Crops Research Institute for the Semi-Arid Tropics (ICRISAT), Hyderabad, India.

Rohan Karawita, National Food Promotion Board, Ministry of Agriculture, Colombo, Sri Lanka.

Karthikeyan Muniappan, DHAN Foundation, Madurai, India.

Chris Kettle, Alliance of Bioversity International and CIAT, Via di San Domenico, 1, 00153, Rome, Italy, and Ecosystem Management, Department of Environmental System Science, ETH Zürich, Zurich, Switzerland.

E.D. Israel Oliver King, M.S. Swaminathan Research Foundation, Chennai, India.

Valdely Ferreira Kinupp, Professor/Researcher, Inst. Fed. Amazonas (IFAM), Av. Cosme Ferreira, 8045, 69085-015, Manaus-AM, Brazil.

Daniel Kirori, D. K. Engineering Company Ltd., Food Processing Equipment Sales & Services, Nairobi, Kenya.

Jayshree Kiyawat, Directorate of Women and Child Development Department, Vijayaraje Vatsalya Bhawan, Plot No 28A, Arera Hills, Bhopal, Madhya Pradesh – 462011.

Margaret Komen, MACE FOODS Ltd., Eldoret, Kenya.

Kamala Krishnaswamy, Madras Diabetes Research Foundation, Dr. Mohan's Diabetes Specialties Centre, WHO Collaborating Centre for Non-Communicable Diseases, Gopalapuram, Chennai, India.

Contributors **xv**

Parkavi Kumar, International Crops Research Institute for the Semi-Arid Tropics (ICRISAT), Hyderabad, India.

Rakesh Kumar, HESCO, Dehradun, Uttarakhand, India.

Elisabeth Lagneaux, Plant Production Systems Group, Wageningen University, 6700 AK, Box 430, Wageningen, The Netherlands, and Institute for Environmental Sciences, Group of Environmental Economics, University of Koblenz-Landau, Fortstraße 7, 76829 Landau, Germany.

Rafael Lira, Facultad de Estudios Superiores Iztacala, Universidad Nacional Autónoma de México, Mexico.

Udayangani Liu, Royal Botanic Gardens Kew, Wellcome Trust Millennium Building, Wakehurst Place, Ardingly, West Sussex RH17 6TN, United Kingdom.

Gaia Lochetti, Consultant, Alliance of Bioversity International and CIAT, Via di San Domenico, 1, 00153, Rome, Italy.

Ing. Jose Alfredo Lopez L., EUROTROPIC, S.A., Guatemala, Central America.

Nuno Rodrigo Madeira, Researcher, Embrapa Vegetables, P.O. Box 218, Zip Code 70250-970, Brasília-DF, Brazil.

Paul Mäder, Department of Soil Sciences, Research Institute of Organic Agriculture (FiBL), Ackerstrasse 113, CH 5070 Frick, Switzerland.

Elizaphan Gichangi Mahinda, Kieru Ltd, P.O. Box 1378–60100 Embu, Kenya.

Ramesh Makavana, Satvik, Gujarat, India.

Nikhil Malhotra, ICAR-National Bureau of Plant Genetic Resources Regional Station, Shimla, India.

Nagappa Gurusiddappa Malleshi, Madras Diabetes Research Foundation, Dr. Mohan's Diabetes Specialties Centre, WHO Collaborating Centre for Non-Communicable Diseases, Gopalapuram, Chennai, India.

Gigi Manicad, Independent senior consultant. Eindstede 34, The Hague 2543 BL, The Netherlands.

Manikandan, Farmer, Kallakkara Village, Attappadi Block, Kerala, India.

Maruthi, Farmer, Marrappalam Village, Sholayur Panchayath, Attappadi Block, Kerala, India.

xvi Contributors

Natarajan Mathimaran, M.S. Swaminathan Research Foundation, Chennai, India.

Shantanu Mathur, International Fund for Agricultural Development (IFAD), Rome, Italy.

Efisio Mattana, Royal Botanic Gardens Kew, Wellcome Trust Millennium Building, Wakehurst Place, Ardingly, West Sussex RH17 6TN, United Kingdom.

Mario Marino, Mary Jane Ramos de la Cruz and Tobias Kiene, ITPGRFA, Rome, Italy.

Patrick Maundu, Alliance of Bioversity International and CIAT, Africa Hub, Kenya Regional Office, c/o ICIPE. P.O. Box 823-00621 Duduville Campus, Nairobi, Kenya, and National Museum of Kenya. Museum Hill P.O. Box 40658-00100, Nairobi, Kenya.

Shantanu Mathur, Lead Adviser, Global and Multilateral Engagement, IFAD, Rome, Italy.

Stepha McMullin, World Agroforestry (ICRAF), United Nations Avenue, Gigiri, Nairobi, 00100, Kenya.

Gennifer Meldrum, Alliance of Bioversity International and CIAT, Via di San Domenico, 1, 00153, Rome, Italy.

Meera Bai, Farmer, Magartagar Village, Dindori District, Madhya Pradesh, India.

Meghana Narayan, Shauravi Malik and Sohini Dey, Slurrp Farm, New Delhi, India.

Sharad Mishra, Action for Social Advancement, Bhopal, India.

Srijit Mishra, Indira Gandhi Institute of Development Research, Gen. A.K. Vaidya Marg Goregaon East, Mumbai – 400065.

Amit Mitra, Independent Researcher, E 170 Sarita Vihar, New Delhi – 110076, India.

Viswanathan Mohan, Madras Diabetes Research Foundation, Dr Mohan's Diabetes Specialties Centre, WHO Collaborating Centre for Non-Communicable Diseases, Gopalapuram, Chennai, India.

Contributors **xvii**

Ashis Mondal, Action for Social Advancement, Bhopal, India.

Yasuyuki Morimoto, Alliance of Bioversity International and CIAT, Africa Hub, Kenya Regional Office, c/o ICIPE. P.O. Box 823-00621, Duduville Campus, Nairobi, Kenya.

Sunamani Muduli, Farmer, Janiguda Village, Koraput District, Odisha, India.

Edie Mukiibi, Slow Food International, Piazza XX Settembre 5, 12042, Bra, Italy.

Sognigbe N'Danikou, World Vegetable Center, P.O. Box 10, Duluti, Arusha, Tanzania.

Nanchiyamma, Farmer, Kallakkara Village, Attappadi Block, Kerala, India.

Thimmegowda Matadadoddi Nanjundegowda, University of Agricultural Sciences, GKVK, Bangalore −560065, Karnataka, India.

Kumar Natarajan, Sathyabama Institute of Science and Technology, Chennai, India.

Loknath Naure, Farmer, Kearandiguda, Bisamacataka, Rayagada, Odisha, India.

Simon Nderitu, Leon Kenya and Jacqueline Damon, African Forest, Nairobi, Kenya.

Michael Ngugi, Simlaw Seeds, Nairobi, Kenya.

Nirmalakumari, Tamil Nadu Agricultural University, Coimbatore, India.

Melari Shisha Nongrum, North East Slow Food and Agrobiodiversity Society, Shillong, India.

Stineke Oenema, Coordinator, UNSCN, Rome, Italy.

Stefano Padulosi, Alliance of Bioversity International and CIAT, Via di San Domenico, 1, 00153, Rome, Italy.

Akshaya Kumar Panda, M.S. Swaminathan Research Foundation, Chennai, India.

Prashant K Parida, M.S. Swaminathan Research Foundation, Chennai, India.

xviii Contributors

Yuvaraj Periyasamy, M.S. Swaminathan Research Foundation, Chennai, India.

Pietro Pipi, Head, Agriculture and Rural Development Department (AICS), Italy.

Sunadei Pitia, Farmer, Machhara Village, Koraput District, Odisha, India.

Phool Bai, Farmer, Bhilai Village, Dindori District, Madhya Pradesh, India.

Nigel Poole, Former Board Chair, ICRISAT governing board, ICRISAT Ambassador of Goodwill, Crowthorne, United Kingdom.

Marco Platzer, Senior Specialist, Agriculture and Rural Development Department, Italian Agency for Development Cooperation (AICS), Italy.

Aliza Pradhan, ICAR-National Institute of Abiotic Stress Management, Baramati, Pune, India.

Hugh W. Pritchard, Royal Botanic Gardens Kew, Wellcome Trust Millennium Building, Wakehurst Place, Ardingly, West Sussex RH17 6TN, United Kingdom.

Bakshi Priyanka, Madras Diabetes Research Foundation, Dr. Mohan's Diabetes Specialties Centre, WHO Collaborating Centre for Non-Communicable Diseases, Gopalapuram, Chennai, India.

Sathanandham Raju, M.S. Swaminathan Research Foundation, Chennai, India.

Prabavathy Vaiyapuri Ramalingam, M.S. Swaminathan Research Foundation, Chennai, India.

Srinivasan Ramasamy, World Vegetable Center, P.O. Box 42, Shanhua, Tainan 74199, Taiwan.

Ram Bahadur Rana, Chief Executive Officer, Anamol Biu Pvt. Ltd., Chitwan, Nepal.

Rami, Farmer, Marrappalam Village, Sholayur Panchayath, Attappadi Block, Kerala, India.

Ramkali Bai, Farmer, Magartagar Village, Dindori District, Madhya Pradesh, India.

Priya Rampal, Indian Council for Research in International Economic Relations, New Delhi, India.

Contributors **xix**

Bibiana Ranee, Farmer, Nongtraw, Meghalaya, India.

Carl O. Rangad, North East Slow Food and Agrobiodiversity Society, Shillong, India.

Dayakar Rao Benhur, ICAR, Indian Institute of Millets Research, Hyderabad, India.

Nitya Rao, School of International Development, University of East Anglia, Norwich NR47TJ, UK.

Chamarthy Venkata Ratnavathi, ICAR-Indian Institute of Millet Research, Rajendranagar, Hyderabad – 500030, India.

Raj Rengalakshmi, M.S. Swaminathan Research Foundation, Chennai, India.

Wilfredo Rojas, Fundación para la Promoción e Investigación de Productos Andinos – PROINPA, 538 Calle Americo Vespucio, Piso 3, La Paz, Bolivia.

Marco Antonio Rondon, Climate Resilient Food Systems Program, IDRC Asia Regional Office, 208 Jor Bagh, New Delhi – 110003 India.

Victoria Rose, Alliance of Bioversity International and CIAT, Via di San Domenico, 1, 00153, Rome, Italy.

Nick Roskruge, School of Agriculture and Environment, Massey University, Palmerston North, New Zealand.

Somnath Roy, Action for Social Advancement, Bhopal, India.

Gamini Samarasinghe, Ministry of Mahaweli, Agriculture, Irrigation and Rural Development, 80/5, Govijana Mandiraya, Rajamalwatta Lane, Battaramulla, Sri Lanka.

Sangeeta Devi, Farmer, Saloor Village, Chamoli District, Uttarakhand, India.

Biswa Sankar Das, WASSAN, Odisha, India.

Vikram Sankaranarayanan, Director, San Lak Agro-Industries Pvt. Ltd. Coimbatore, India.

Roland Schafleitner, World Vegetable Center, P.O. Box 42, Shanhua, Tainan 74199, Taiwan.

xx Contributors

Pepijn Schreinemachers, World Vegetable Center, P.O. Box 1010 (Kasetsart University), Bangkok 10903, Thailand.

Annadana Seetharam, Emeritus Professor and Former Project Coordinator (Millets), No. 20233, Tower 20, Prestige Ferns Residency, Haralur Road, Bengaluru, Karnataka, India.

Jegan Sekar, M.S. Swaminathan Research Foundation, Chennai, India.

Simon Apang Semese, Maori and Pasifika Directorate, Massey University, Palmerston North, New Zealand.

Mahesh Sharma, Chief Executive Officer, Anamol Biu Pvt. Ltd., Chitwan, Nepal.

Shalini Devi, Farmer, Poona Village, Chamoli District, Uttarakhand, India.

Shanmugam Shobana, Madras Diabetes Research Foundation, Dr. Mohan's Diabetes Specialties Centre, WHO Collaborating Centre for Non-Communicable Diseases, Gopalapuram, Chennai, India.

Devesh Singh, Yale-NUS College, 16, #01–220, College Ave West, Singapore – 138527.

Kuldeep Singh, ICAR-National Bureau of Plant Genetic Resources, Pusa Campus, New Delhi, India.

Mohar Singh, ICAR-National Bureau of Plant Genetic Resources Regional Station, Shimla, India.

Vasudevan Sudha, Madras Diabetes Research Foundation, Dr. Mohan's Diabetes Specialties Centre, WHO Collaborating Centre for Non-Communicable Diseases, G–opalapuram, Chennai, India.

Loichan Sukia, Farmer, Machhara Village, Koraput District, Odisha, India.

Ridian Syiem, Farmer, Khweng, Meghalaya, India.

Tukuna Burudi, Farmer, Khiloput, Koraput District, Odisha, India.

Ayfer Tan, Retired, Aegean Agricultural Research Institute, Izmir, Turkey.

Stephen R. Taranto, Sendas Altas – Operadores en Turismo, Edificio California, 2022 Avenida Ecuador, Piso 7, La Paz, Bolivia.

Evert Thomas, Alliance of Bioversity International and CIAT, c/o CIP Avenida La Molina 1895, La Molina, Lima 12, Peru.

Vilas A Tonapi, ICAR, Indian Institute of Millets Research, Hyderabad, India.

Gerry Turpin, Queensland Herbarium, Department of Environment and Science, Mount Coot-tha Botanical Gardens, Mount Coot-tha Road, Toowong, QLD 4066, Australia, and Tropical Indigenous Ethnobotany Centre, Australian Tropical Herbarium, James Cook University, McGregor Road, Smithfield, QLD 4879, Australia.

Tiziana Ulian, Royal Botanic Gardens Kew, Wellcome Trust Millennium Building, Wakehurst Place, Ardingly, West Sussex RH17 6TN, United Kingdom.

Usha Devi, Farmer, Saldhar Village, Chamoli District, Uttarakhand, India.

Robin Van Loon, Camino Verde, Of. Serpost Casilla Postal 120, Puerto Maldonado, Madre de Dios 17001, Peru.

Maarten van Zonneveld, Genetic Resources and Seed Unit, World Vegetable Center, 74151, Shanhua, Taiwan.

Ana Luiza Vergueiro, ECONUT Comércio de Produtos Naturais Ltda., Brazil.

Sergio Vergueiro, ECONUT Comércio de Produtos Naturais Ltda., Brazil.

Mani Vetriventhan, International Crops Research Institute for the Semi-Arid Tropics (ICRISAT), Patancheru, Telangana, India.

Paul Wagsta, Senior Agriculture Advisor, Self Help Africa.

Victor W. Wasike, Genetic Resources Research Centre, P.O. Box 30148-00100, Nairobi, Kenya.

Michael Way, Royal Botanic Gardens Kew, Wellcome Trust Millennium Building, Wakehurst Place, Ardingly, West Sussex RH17 6TN, United Kingdom.

Zongwen Zhang, Alliance of Bioversity International and CIAT, c/o Chinese Academy of Agricultural Sciences, Beijing 100081, China.

Dauro Mattia Zocchi, University of Gastronomic Sciences of Pollenzo, Piazza Vittoria Emanuele II 9, 12042, Pollenzo, Italy.

PREFACE AND ACKNOWLEDGMENTS

Today we are confronted with two major paradoxes globally – the persistence of hunger in the midst of an impressive technological capacity to grow more food; and the narrowing of crop diversity within global food systems in the face of a fast-growing world population. Both conditions call for urgent actions to ensure a world without hunger.

Agriculture is the mother of nutrition security. It is believed that agriculture or settled cultivation began over 12,000 years ago with women growing crops of importance to life on earth. Yet, over time, the importance of genetic variability was not given adequate recognition, leading to an over-reliance on a few crops such as rice, wheat and maize. Not only has this resulted in a large number of plants becoming extinct, but it has also put the food and nutrition security of the poor and marginalized, dependent on a range of crop species, at risk. With a focus on the standardization of production systems, not only are species lost, but so are markets – national and international – as is research interest in improving the productivity of these crops. This is why planned initiatives for the conservation, cultivation, consumption and commerce of genetic resources are important.

Maintaining the genetic diversity of crops is even more critical in today's context of climate change. It is, in fact, the dynamic maintenance of agrobiodiversity operated by farmers in situ/on farm that could help make our food systems more resilient. Whereas the world can feel relatively comforted by the 1,450 gene banks that have been built to safeguard crop diversity – including the Svalbard Vault in Norway – much more needs to be done to map, collect, characterize, document and evaluate the thousands of orphan and underutilized crops, today just marginally conserved in gene banks, but whose survival is thanks to the laborious work of millions of farming communities around the world, a service done for the public good – but at their own personal cost! In that regard, the work pursued by the M.S. Swaminathan Research Foundation and other NGOs in building capacities of farmers in conserving local crop needs should be encouraged and supported.

Given the scarcity of land, we must recognize that ensuring food security cannot be addressed by expanding land available for agricultural activities. We need to embrace a different paradigm that, while using less land, can provide more food from crops that are better adapted to climate change. At the same time, we need to do more to safeguard biodiversity so as to kee foodscapes healthy and productive; reduce the depletion of finite resources like water and soil; promote equitable, gender-inclusive food systems and foster greater synergy between scientific and indigenous knowledge. The role of women in feeding the planet needs to be better recognized and supported.

One of the great achievements of science was the development of our ability to describe the genomic structure of major crop species and their wild relatives, which has provided a wealth of information useful for increasing both crop production and productivity. However, what we are witnessing now is that yields of major crops are reaching a plateau that may not be easy to overcome. More innovative solutions are needed to address the yield bottleneck. Leveraging agrobiodiversity to grow more and diverse nutritious food in difficult areas with poor soil and challenging climatic conditions should receive greater attention. This is not new to us, since risk-aversion practices have always guided generations of farmers who have been growing different crop and varietal mixes to buffer against shocks. Farming families have often been motivated by a desire to minimize risks, not just maximize profits; hence, the wisdom underlying the decisions to balance subsistence and market motivations needs to be recovered.

In fact, significant progress in agronomic research and the adoption of systems approaches have shown the many benefits that biodiversity-based practices can bring about in challenging conditions, not just to the environment, but equally to ensuring the food and nutrition security of local communities. Crops considered 'orphan, underutilized and neglected' are all extremely valuable as not only are they often more nutritious than the major cereals, but also tend to need less water and are more tolerant of high temperatures and grow better in difficult climatic conditions.

I have stressed, on various occasions, the disparities in access to technology that we are witnessing today. Despite the many exciting developments – be it in digital and precision agriculture, biotechnologies or ecotechnologies – we still register what I describe as a 'technological apartheid', which is contributing to 'orphan crops remaining orphans' in relation to the choice of research areas for their use enhancement. A good example is the case of technology for processing minor millets, which until recently was hardly accessible for rural households, discouraging the consumption of these nutritious foods where they are most needed.

Back in 1968, I warned that if all locally adapted crop varieties were replaced with one or two high-yielding strains, it could lead to serious damage from pests, pathogens and weeds, contributing to the making of major agricultural and ecological disasters. Fifty years later, I believe we still face such a risk, and the diversification of crops and varieties in production systems has become even more critical to safeguarding our future. This is one reason why I emphasized the need for a new "ever-green revolution" guided by diverse nutritious crops to strengthen our diets, ones that would require less water and fewer chemicals and would be able to grow in periods of change. Orphan crops will be a big part of this revolution I am advocating. In October 2018, I proposed to the UN's Committee on Food Security that we have a year devoted to promoting these crops, to try and revive them by revitalizing markets, research and the culinary tradition that used them.

Very important here is the need to ensure farmers' engagement and, indeed, their food sovereignty. Unfortunately, while the concept of farmers' rights is widely discussed in national and international fora, its practical application remains largely inadequate. The international community should be made more aware of the fact that the loss of every gene and species limits our options for a secure future, particularly in the context of climate change and related unforeseen shocks.

This book covers many of the issues hindering the successful promotion of orphan crops in India and around the world. Minor millets are the leitmotif of this publication and are leveraged as an example to describe the needs, challenges and opportunities in bringing to scale the cultivations of orphan crops. More than 20 years ago, at an FAO conference, I pleaded that we should refer to millets as 'nutri-rich and climate smart' food grains. I felt that the change in the terminology being used to refer to these crops was much needed. It is heartening to acknowledge that the Indian government has taken a great step in including millets in the public distribution system under the 2013 National Food Security Act – a testimony to the fact that millets are no longer perceived as inferior foods.

The book covers numerous aspects related to the promotion of underutilized crops, including participatory approaches, methods and tools for their use enhancement, conservation methods, market analyses and promotional strategies and policy needs for their mainstreaming. Cross-cutting issues like gender, capacity-building and empowerment of vulnerable groups are also covered. Representatives from various local communities and the private sector also share their reflections and perspectives on the wider use of these crops, as do several international agencies. I would like to acknowledge and thank them for supporting numerous projects to advance the agenda of 'NUS' around the world.

Several chapters focusing on millets have been developed from talks delivered at the international conference on this theme that took place in April 2018 in Chennai, India. These have been complemented by additional studies focusing on other representative orphan crops from around the world. I believe that, together, these contributions will enrich the socio-cultural perspectives and the R&D outlook for the future of these crops to help build more inclusive and sustainable food systems.

xxvi Preface and acknowledgments

The book is the result of a successful close cooperation between the M.S. Swaminathan Research Foundation (MSSRF) and Bioversity International (now The Alliance of Bioversity International and CIAT), an organization whose establishment I had the honor of contributing to some 46 years ago and whose mission I still see as being of high strategic value to the world.

We trust that such a publication will be of great use to students and scholars, practitioners and stakeholders, including policymakers involved in building more resilient food and production systems. I also believe that some of the insights in this book could contribute towards building farming systems for nutrition, a concept that signifies bringing agriculture, nutrition and health into a sustainable public-health management system.

I wish to thank Stefano Padulosi, Israel Oliver King and Danny Hunter for their invaluable ideas in conceiving and bringing this book to reality. Their highly committed compiling and editing efforts are much appreciated. I join them in expressing a special thanks to the Indo-Swiss Collaboration in Biotechnology – ISCB, Switzerland, for their financial contribution in support of this publication. Many are those organizations that have been supporting NUS projects around the world, leading to invaluable discoveries and lessons, now shared through this book. While acknowledging their support in each respective chapter, the other editors and I would like to convey our sincere gratitude to them for championing the NUS agenda at the national and international levels. I hope this book will be an inspiration to many others in further strengthening ongoing efforts for bringing NUS back to full fruition, for the benefit of current and future generations.

Prof. M.S Swaminathan

PART I

Introduction to Neglected and Underutilized Species (NUS)

1

NUS

What they are and why we need them more than ever

Stefano Padulosi, Gennifer Meldrum,
E.D. Israel Oliver King and Danny Hunter

Introduction

Agricultural biodiversity represents a strategic resource in ensuring food and nutrition security for humankind (Thrupp, 2000; Frison et al., 2006; Bioversity International, 2017). It keeps us healthy, as diets with poor biodiversity often lack crucial vitamins and micronutrients and are associated with diet-related non-communicable diseases (diabetes, heart attacks, overweight, obesity and cancer), which are a leading cause of death at the global level (Branca et al., 2019). Diversity on our plate is of paramount importance to everybody, and making sure this diversity is safeguarded and promoted should receive the utmost attention of decision makers, both at the national and the international level.

The period of intense agricultural growth from the early 1960s to mid-1980s, known as the Green Revolution,[1] was characterized by an unprecedented expansion in the production of staple crops through the development of high-yielding varieties (HYV). The Green Revolution contributed to a reduction of poverty. An increase of approximately 15% in per capita GDP as a result of a 10% use increase of the HYV in the period 1960–2000 was observed, with an associated reduction of food insecurity for billions of people, and an estimated 18–27 million hectares of natural ecosystems safeguarded from being converted to agricultural land (Hazell, 2003; FAO, 2011; Pingali, 2012; Stevenson et al., 2013; Gollin et al., 2018). Among the key players of the Green Revolution were the international agricultural research centers of the CGIAR Consortium, responsible for the development of HYVs of major staples—mostly cereals—whose production more than doubled in developing nations between the years 1961 and 1985 (Tribe, 1994; Conway, 1998). Unfortunately, this success came with a heavy cost to the environment (e.g., in the loss of wild and cultivated biodiversity, water scarcity, increased crop vulnerability to pests and diseases and loss of soil fertility), and caused a deterioration in human nutrition

DOI: 10.4324/9781003044802-2

4 Stefano Padulosi et al.

(e.g., with essential amino acid deficiencies and a general lack of balanced essential fatty acids, vitamins and minerals from cereal-dominated diets), as well as increased health hazards from the widespread use of pesticides. This period also brought about socio-political instabilities (e.g., rural–urban migration of farmers unable to afford introduced technologies, as well as social conflicts and marginalization) (Jennings, 1988; Fowler and Mooney, 1990; Pingali, 2012).

The greater availability of calories resulting from the Green Revolution's efforts has not represented an escape from hunger for millions of people, as HYV have been scarcely adopted in Africa, for instance, and the narrow crop diversity has led to food systems simplification, reducing options for healthy diets (Mooney and Fowler, 1990; Vanhaute, 2011; Willett et al., 2019). Nearly one in three people globally are still afflicted by malnutrition—a situation predicted to worsen in coming years based on current trends (FAO et al., 2018; UNSCN, 2018).

The diversity reductionist approach followed by the Green Revolution has influenced the way agricultural development strategies and programs have been developed in every country for decades. This approach continues to have an influence today at different levels; however, it is increasingly questioned and less accepted, with urgent calls for a global transformation of the food system growing (Schutter and Vanloueren, 2011).

As we tackle the ambitious Sustainable Development Goal 2 (SDG2) of achieving 'zero hunger' by 2030, we are faced with the disturbing paradox that is hindering our efforts: of the 5,000 food crops estimated to exist worldwide, global food systems are dominated by just three (rice, wheat and maize), which provide half the world's plant-derived calories (FAO, 2015; Willis, 2017). An incredible wealth of nutritious crops, and other wild edible food plants, is largely overlooked in our battle to produce food and tackle malnutrition, and this is happening now when the need for diversification of production and food systems has never been greater. Crop uniformity and standardization of agricultural fields are causing food systems to be enormously vulnerable to climate change, reducing farmers' capacity to absorb shocks and leaving consumers with fewer choices for nutritious and healthy diets (Padulosi et al., 2019). Those thousands of nutritious plant species left behind by the Green Revolution, which we call 'neglected and underutilized species' (NUS), represent a unique treasure for humanity that must be recovered from their state of neglect and must be mobilized to fuel a truly evergreen agricultural revolution (Swaminathan, 1996, 2020).

NUS: beyond a definition

Neglected and underutilized species have become a popular topic lately, both in public debates and scientific papers. Different terminologies (orphan, forgotten, lost, alternative, minor, novel, local, traditional, etc.) are interchangeably used to refer to these species, creating confusion among researchers and development practitioners. It is one of the aims of this book to dispel this confusion. NUS is a terminology subject to different interpretations, reflecting people's own cultural background and sensitivity, and we hope to put everybody on the same level of understanding through the following observations.

In simple words, NUS are plant species that—although appreciated at local level—are forgotten, abandoned or rarely explored by researchers and other agriculture and food systems R&D actors, for various reasons (e.g., low economic competitiveness; lack of improved seed, adequate cultivation practices, or processing technologies and reduced consumer appeal). They include wild, semi or fully domesticated plants from different food plant groups (including cereals, vegetables, legumes, roots and tubers, fruits, nuts, spices, etc.), diverse growth forms (herbs, shrubs, vines, trees, etc.) and life cycles (annual, biennial, perennial) (Padulosi et al., 2018). Although this book focuses solely on food plants, we should point out that NUS as a concept may well refer to other species too (plants or animals), as exemplified by the growing interest in the use of insects as a source of sustainable and cheap proteins for food and feed (van Huis et al., 2013; Dickie et al., 2019).

The reduced use of NUS over time has led to the loss of both their genetic diversity and a wealth of traditional practices and associated knowledge that was developed by generations of traditional farmers for managing sustainable harvests (if occurring wild), cultivation, processing and preparation. Also contributing to this marginalization is the widespread perception, registered particularly among younger consumers, that NUS are the legacy of backwardness and hardship of traditional rural societies, the food of the poor, and should therefore be abandoned (Durst and Nomindelger, 2014; Padulosi et al., 2019).

The term 'neglected and underutilized species' was first conceived by IPGRI (the predecessor of Bioversity) in the late 1990s (Eyzaguirre et al., 1999). The reason why this term was chosen relates to its usefulness in conveying two key messages upfront: firstly, the status of neglect by research and development efforts of these traditional resources; and secondly, the status of underuse in relation to the multiple benefits they can bring to improving nutrition and health, livelihoods, the environment and biodiversity conservation, if better harnessed by society. Furthermore, the use of the word 'species' instead of 'crops', was also preferred in reminding people that NUS do encompass both naturally occurring species that are harvested in the wild and domesticated species. Lastly, the word 'neglected' is useful for evoking in our narratives those millions of 'neglected' people (vulnerable groups, marginalized members of society), who rely on NUS for their livelihood and for whom the improvement of these resources, to which they are culturally connected, represents an opportunity of economic growth, empowerment and reaffirmation of identity.

The term 'orphan crops', present also in the title of this book, is often used interchangeably with NUS, though the latter is broader in scope as it includes both wild and cultivated species that may not immediately convey the messages with the same intensity described in Box 1.1. For more reflections on the semantics of the term NUS, the reader can refer to Padulosi et al. (2004, 2008).

6 Stefano Padulosi et al.

Box 1.1 Short description of NUS and other similar terms commonly found in literature

Term	Description	Source
Neglected crops	*"Neglected crops are those grown primarily in their centres of origin by traditional farmers, where they are still important for the subsistence of local communities. Some species may be widely distributed around the world but tend to occupy special niches in the local ecology and in local production and consumption systems. While these crops continue to be maintained by sociocultural preferences and the ways they are used, they remain inadequately documented and neglected by formal research and conservation".*	IPGRI, 2002
NUS	*"Acronym standing for Neglected and Underutilized Species and applied to useful plant species which are marginalized, if not entirely ignored, by researchers, breeders and policy makers; they belong to a large, biodiverse group of thousands of domesticated, semi-domesticated or wild species; they may be locally adapted minor crops as well as non-timber forest species. The 'NUS' term is a fluid one, as when a crop is simultaneously a well-established major crop in one country and a neglected minor crop in another. NUS tend to be managed with traditional systems, use informal seed sources and involve a strong gender element".* In a wider sense, the term NUS also could be used to refer to animal species.	Padulosi et al., 2013
Orphan crops	*"Orphan crops are defined as crops that have either originated in a geographic location or those that have become 'indigenized' over many years (> 10 decades) of cultivation as well as natural and farmer selection (Dawson et al., 2007). The term 'orphan' has often been used to refer to crops that may have originated elsewhere, but have undergone extensive domestication locally, thus giving rise to local variations, i.e., 'naturalized/indigenized crops' (Mabhaudhi et al., 2017)".*	Mabhaudhi et al., 2019.
Underutilized crops	*"Underutilized crops were once grown more widely or intensively but are falling into disuse for a variety of agronomic, genetic, economic and cultural reasons. Farmers and consumers are using these crops less because they are in some way not competitive with other species in the same agricultural environment. The decline of these crops may erode the genetic base and prevent distinctive and valuable traits being used in crop adaptation and improvement".*	IPGRI, 2002

Beyond the terminology used, the best way to describe NUS is to refer to their key characteristics, which can be helpful in identifying the common limitations behind their marginalization as well as the multiple useful traits that can be leveraged for their promotion. Box 1.2 provides a list of these recurrent features that we have been documenting around the world, during our work focusing on the use enhancement of NUS. In fact, such a list of features could be well considered a 'terms of reference' for a shared comprehension of NUS by workers.

Referring to these key features of NUS is useful in making a clear distinction between NUS and landraces of major crops (e.g., local varieties of wheat, rice or beans), whose status of neglect and underuse is also recorded in many countries. Our position is that these varieties should not be considered NUS, because their status can be improved by leveraging the vast network of research focusing on major crops that *already exists in most countries*. The growing interest on NUS by the research community and development practitioners (and hopefully accompanied by much needed funding!) should, therefore, strategically be used to tackle the promotion of NUS '*sensu stricto*' instead of being used for underutilized varieties of major crops for which the R&D infrastructure already exists.

Box 1.2 Key features of neglected and underutilized species (from Padulosi et al., 2019—modified)

- *Low competitiveness:* little R&D investment has left NUS behind in terms of advances in their conservation, cultivation, harvest, postharvest, marketability, nutritional profiles, gender, policies and legal frameworks.
- *Relevant only to local consumption and production systems:* being intimately linked to local food cultures, NUS are used in traditional food preparations and are associated with social and religious ceremonies and rituals.
- *Adapted to agroecological niches and marginal areas:* NUS often demonstrate comparative advantages over commercial crops due to natural selection or selection carried out by local growers against biotic and abiotic stresses, which makes them perform comparatively better under low input and biological agriculture techniques.
- *Resilient to climate change:* compared with commodity crops, NUS are perceived by local growers as highly adapted to biotic and abiotic stresses related to climate change, something that is being increasingly confirmed by scientific research.
- *Represented by ecotypes or landraces:* owing to the poor attention received from breeders, they are often represented by germplasm material that is not performing so well, requiring some degree of genetic improvement—a fact that hinders their competitiveness in production systems.

8 Stefano Padulosi et al.

- *Rich in traditional knowledge:* in view of the ongoing cultural erosion affecting traditional societies, associated knowledge on NUS is being rapidly lost, which, in turn, leads to the loss of genetic diversity and continued opportunities for appreciation by consumers, especially the younger generation.
- *Poorly represented in ex situ gene banks:* their genetic diversity is maintained largely in situ and on-farm and (possibly) in private seed collections (e.g., the one maintained by the Baker Creek Heirloom Seeds Company[2]) or community seed banks (such as those maintained by MSSRF in south India[3]).
- *Characterized by poorly developed or non-existent seed supply systems:* inferior quality of seed that has a negative impact on their performance in cultivation.
- *Highly relevant in Indigenous Peoples' societies:* for Indigenous communities, NUS are the result of sophisticated trials and accumulation of experience over many centuries and generations: they are a manifestation of a systematic process that involved intricate ways of learning and accumulating experience (Padulosi et al., 2019).
- *Multi-functionality and multiple benefits:* they are often able to provide people with not just nutritious food, but also valuable non-food products and ecosystem services. Excellent examples of such multi-functionality can be found in Bambara groundnut, chaya or minor millets that are presented in the chapters in this book.

With regard to how many NUS exist at the global level, we can safely say that taking into account both wild and cultivated species, the number is in the order of thousands (Padulosi et al., 2018). In fact, most of the estimated 5,000 cultivated food crops recently recorded as existing (RBG Kew, 2016; Ulian et al., 2020) are in some state of marginalization. For instance, in the case of cultivated vegetables, within the family Leguminosae, there are an estimated 23,000 species (Plant List, 2020), of which some 653 have been cultivated (Khoshbakht and Hammer, 2008). Taking as an example the 127 species of cultivated vegetables belonging to this family, the majority have been found to be neglected, based on three key indicators viz. (a) number of records in Google Scholar as indicator of research effort devoted to the species, (b) number of accessions maintained in ex situ germplasm collections worldwide and (c) production data from the FAO's FAOSTAT, an indicator of knowledge on species distribution and production levels (Meldrum et al., 2018). Similarly, in the family Compositae the numbers are as follows: of its 27,000 species, 284 are cultivated plants, of which 85 vegetables are also largely neglected.

Further complications in a shared understanding of NUS may arise also from the fact that some species that are clearly underutilized in one country are, on the contrary, very popular in another. This is the case for the cereal tef, (*Eragrostis tef*), a staple food in Ethiopia, that is hardly consumed outside the country, except in small amounts primarily by the Ethiopian diaspora. The opposite case is that of quinoa (*Chenopodium quinoa)*, which was an underutilized crop until a few years ago, when it started to spread from the Andean region across all continents to become a global commodity (Bazile et al., 2016). The spicy rocket (collective name for *Eruca sativa* and *Diplotaxis* spp. originating from the Mediterranean/Central Asia) was another typical underutilized vegetable until the early 1990s (Padulosi et al., 2008); yet, today it is being cultivated on all continents, with 400,000 tons/year produced in a small area near Salerno ("*Piana del Sele*") in south Italy alone, representing 73% of the entire Italian production (Pignataro, 2019).

The status of NUS nowadays changes rapidly (for better or for worse) and it becomes hard to keep track of these changes in a consistent manner. What we see today, is very often a 'mosaic' situation, with a crop being popular in one country (or region) and highly marginalized in others. Such a scenario poses great challenges to workers engaged in setting priorities for choosing the 'best' NUS to support agricultural development programs. In our opinion, more than a silver bullet, what the world really needs today are portfolios of '*silver baskets*' of NUS diversity to meet the various needs of consumers from different socioeconomic and cultural backgrounds across regions and countries. The reader may find more on priority-setting approaches for NUS in Chapters 2, 9, 13 of this volume, and in the wider scientific literature (Padulosi, 1999a,b; Hunter et al., 2019, 2020; Ulian et al., 2020, and references therein).

Importance of NUS

The chapters in Part I provide several examples of benefits associated with the use enhancement of emblematic NUS from India and other regions of the world. An overview of livelihood benefits arising from the use enhancement of NUS can be also found in the works of Ravi et al. (2010), Padulosi et al. (2011, 2013), Massawe et al. (2015), Mabhaudhi et al. (2019) and Raneri et al. (2019).

In Table 1.1, we have summarized the wide range of the benefits associated with NUS and compare these to similar ones that can be obtained from commodity crops. As can be appreciated, benefits from the wider cultivation and use of NUS far outweigh those associated with commodity crops. Research investment is urgently needed to translate these potential benefits into concrete realities by providing high-quality seed, refined cultivation practices, more efficient processing technologies, marketing that is more efficient, etc. It is worth stressing—as mentioned by Professor Swaminathan in his foreword to this book—that our societies are well equipped to address the bottlenecks hindering the use of NUS; what is most needed is the political will to leverage the enormous technology and expertise accumulated from working on commodity crops, to benefit of the NUS cause (see also Chapters 3, 13 and 17 for more on mainstreaming needs and opportunities).

TABLE 1.1 Comparing the benefits from NUS and commodity crops (adapted from DIVERSIFOOD[4])

Type of benefit		NUS	Commodity crops
Agroecological	Adaptability and resilience	High: great source of adaptability across wide inter- and intra-specific diversity	Low: large genetic uniformity present in cultivations limits adaptation
	Soil stability and fertility enhancement	High: many species available able to strengthen these roles, especially through the use of perennial and tree species	Low: less diversity of options available, mostly in the form of annual crops, limiting these benefits
Nutritional	Food and nutrition security and healthier diets	High: excellent source of micronutrients in numerous species, enhancing these benefits	Low: deployment of calorie-rich but micro-nutrient-poor staples limit nutrition security
	Strengthening local food systems and self-reliance	High: many locally available species helpful in strengthening local food systems and buffering populations from shocks and periods of adversity	Medium-low: limited contributions possible due to limited inter-specific diversity available in major crops feeding the world
Economic	Empowerment of value chain actors	High: great opportunities for local actors, women and vulnerable groups participating in value chains	Medium: limited contributions due to limited inter- and intra-specific diversity present in markets
	Income opportunities in marginal areas	High: many opportunities for NUS as a result of their remarkably well-adapted diversity occurring in marginal areas	Medium-low: limited contributions, but opportunities may exist, especially in landraces
	Reduced dependence on food importation	High: highly relevant impact, especially in countries where food import is remarkably high	Medium-low: limited contributions, but opportunities may exist especially in the use enhancement of landraces
	Income stability and reduced risks to farmers/ market actors	High: marketing of NUS diversity to act as robust and sustainable buffer against economic risks and shocks	Low: high dependence on imports makes markets highly vulnerable to shocks caused by climate change or other factors
	Contribution to agri-tourism activities	High: NUS richness represents an important element in support of this sector useful to also reinforce territory identity and local pride	Medium-low: limited diversity available, with missed opportunities in the use of landraces

Social	Healthier 'foodscapes'	High: presence of many resilient, nutritious and soil-enhancing species	Low: limited contributions due to limited diversity available and deployment of poorly sustainable environmental practices
	Conservation and plant genetic resources for food and nutrition security	High: the NUS basket contains more than 5,000 cultivated food species—their conservation would safeguard food and nutrition security for current and future generations	Medium-low: staple crops' diversity already well conserved—more attention on conservation of crop wild relatives (CWR) both ex situ and in situ
	Safeguard of food culture and identity	High: great contribution in keeping alive relevant food traditions, strengthening territory identity and pride of local populations	Medium-low: valorization of landraces could be leveraged to better capture these benefits
	Fun and pleasure for farmers and for consumers (exploring and experimenting new ways of cooking and eating)	High: the large NUS portfolio offer tremendous opportunities in terms of an 'adventure' experience in national and international cuisines	Low: comparatively fewer opportunities owing to limited diversity available
	Cultural exchange and learning across generations	High: ample opportunities to use rich knowledge associated with NUS as a lever for keeping culture and traditions alive	Medium-low: fewer opportunities compared to NUS, owing to limited diversity available in these crops
	Aesthetics of diversified landscapes, shapes and colors	High: the universe of NUS can offer huge opportunities to combine healthy food with attractive diets	Medium-low: fewer opportunities compared to NUS, owing to limited diversity available in these crops

12 Stefano Padulosi et al.

Consistent with this long list of benefits is that NUS can be powerful levers in support of important activities framed in the context of International Treaties and Agreements, such as those of the UN's Sustainable Development Goals (for example SDGs 2, 7, 12, 13, 15 and 17, the Aichi Biodiversity Target 13,[5] Activity 11 [and others as well] of the FAO Second Global Plan of Action on PGRFA,[6] the IFAD Action Plans on Mainstreaming Nutrition[7] 2019–2025, the IFAD Rural Youth Action Plan 2019–2021[8] and IFAD Strategy and Action Plan on Environment and Climate Change 2019–2025[9]). Interesting to note is that in 2014, at the Second International Conference on Nutrition (ICN2), member states acknowledged the key role played by diversified and sustainable diets, including traditional foods such as NUS, in reducing malnutrition[10] (see the UNSCN's commentary provided in Chapter 31).

With regard to climate change, one of the most interesting examples of the drought-escaping capacity of NUS in stress-avoiding strategies comes from minor millets and in particular from barnyard millet (*Echinochloa colona*), whose grains can reach maturity just 45 days after sowing (Hulse et al., 1980), and hence represent ideal crops in areas with very short rainy seasons. Minor millets can produce a reasonable harvest with only about 10% of the water required to grow rice, which makes them an ideal (and cheap) alternative source of carbohydrates and other important micronutrients for areas like South Asia, which are suffering from climate change and where the cultivation of rice is increasingly challenged (Padulosi et al., 2009). Similarly, the highly nutritious tepary bean (*Phaseolus acutifolius*)—which completes its reproductive cycle rapidly and thus avoids subsequent drought (Nabhan, 1990)—is able to produce mature beans in just 60 days (Wolf, 2018). This legume is a precious source of protein, especially in Central America, where the crop used to be popular in the past and where it could help counteract the decline in common bean production as a result of recurrent droughts (Bioversity International, 2018). Cañahua (*Chenopodium pallidicaule*), a close relative of quinoa, is a very nutritious and hardy crop but its fate has been very different from that of the popularity of quinoa. In fact, cañahua is capable of withstanding cold stresses, a trait not present in quinoa and one that makes it much desired by local farmers who are experiencing more frequent morning frosts around Lake Titicaca in Bolivia and Peru. Unfortunately, the abandonment of this crop by farmers, who replaced it with the more remunerative quinoa, has led to the disappearance of its germplasm, making its re-introduction to farming communities hard to realize (Bioversity International, 2017).

Supporting evidence also exists in the literature for the risk-avoidance practices adopted by small-scale farmers in areas where the crop diversity of NUS is particularly high (Altieri, 1987; Holt-Gimenez, 2006). An example of mixed production systems in which farmers can harness the multiple and complementary benefits (agronomic, economic, social) from NUS to mitigate harvest failures due to climate change (Mijatovic et al., 2019) is that of the 'baranaja' mixed cropping system. This farming practice, used in the Garhwal region of the state of Uttarakhand in India, involves sowing 12 or more crops on the same

NUS: what they are and why we need them **13**

plot, including various types of beans, grains and millets, and harvesting them at different times (Ghosh and Dhyani, 2004). Several of these crops are NUS and such practices help farmers make the best use of limited land, save on the cost of chemical fertilizers through the N-fixing capacity of some species and make use of enhancing soil capacities and different maturity times of crops to help secure a regular and nutritious supply of foods to households.

Other NUS that are often mentioned in the literature for their drought tolerance/adaptation include the baobab (*Adansonia digitata*), pigeon pea (*Cajanus cajan*), jute mallow (*Corchorus olitorius*), chaya (*Cnidoscolus aconitifolius*), fonio (*Digitaria exilis*), moringa (*Moringa oleifera*), tepary bean (*Phaseolus acutifolius*), Bambara groundnut (*Vigna subterranea*), jujube (*Ziziphus mauritania*), strawberry tree (*Arbutus* spp.), marula (*Sclerocarya birrea*), amaranth (*Amaranthus* spp.) and grass pea (*Lathyrus sativus*) (Padulosi et al., 2011).

From a nutritional point of view, it is very important to highlight that NUS have similar or sometimes remarkably higher nutritional profiles than those found in commodity crops (Hunter et al., 2019). Considering that the 51 essential nutrients needed for sustaining human life (Graham et al., 2007) cannot be found easily in the few staple crops feeding the world today, we can make a strong case for the theory that the vast NUS portfolio that fortunately still exists at local level is a strategic source of vital nutrients in the fight against malnutrition (Raneri et al., 2019). The diversity of NUS can be a source of essential nutrients year round and, at the same time, can make peoples' meals more delicious, tastier and more enjoyable. More NUS diversity on the plate helps people of different age groups better meet their own nutritional needs, and can allow the fulfillment of individual health requirements and food preferences. Interestingly, 10 of the 17 essential micronutrients and 13 (out of 13) vitamins can be found in fruits and vegetables, a category of crops represented by thousands of species that are today largely underutilized.

Furthermore, whereas affluent people nowadays have the privilege of accessing food originating from all parts of the world on supermarket shelves at any time of the year, for the majority of the world's population such a privilege is largely denied. It is a fact that for most people living in vulnerable areas of sub-Saharan Africa or Asia, food security is largely dependent on crops grown locally. Due to the shrinking of the food basket that these areas are seeing, local populations are more vulnerable to the so-called *lean season*, the period preceding the harvest of main staples during which households have often exhausted their household food reserves. In this regard, our research in Mali, India and Guatemala has discovered that several local NUS (wild and cultivated) that are currently scarcely used due to low awareness of their nutritional values, lack of seed or other constraints can be revitalized and become helpful allies in fighting food insecurity during the lean season (Lochetti et al., 2020). More on the seasonal calendars developed to guide local populations in managing this untapped diversity is provided in Chapter 14, whereas a more extensive articulation of NUS and nutrition-sensitive agriculture is provided in Chapter 3.

As stressed previously, the promotion of NUS can be leveraged to support the empowerment of local communities, because it is in those areas—inhabited especially by Indigenous Peoples—that a great diversity of NUS is found. Agricultural systems in Indigenous Peoples' territories have not been subjected to intensive cultivation practices and local communities living there have been safeguarding a great amount of diversity compared to other societies (IFAD, 2019; Singh and Rana, 2020). The intimate link between NUS and local communities, whose own identity is quite dependent on these foods, offers an opportunity for the revitalization of the gastronomic culture through, for instance, agritourism approaches as practiced in depressed areas of Lake Titicaca by Bioversity and its partners in the mid-2000s (Taranto and Padulosi, 2009).

Conclusion

As noted in previous publications (Eyzaguirre et al., 1999; Padulosi et al., 2013; Hunter et al., 2019; Borelli et al., 2020; Hunter et al., 2020), NUS are often called 'minor' crops in view of the fact that their production volume is much lower compared to that of 'major' crops (commodities or staples). In terms of their multiple livelihood and other benefits, NUS are not minor at all and, hence, call for more robust and consistent actions by decision makers for their recognition and promotion.

Tapping the portfolio of an estimated 653 leguminous NUS (Khoshbakht and Hammer, 2008) to provide proteins to a hungry world in ways that are more sustainable to the environment, using less water and fewer chemical fertilizers as compared to livestock farms, is, for example, an important contribution to safeguarding our planet and is consistent with the objective of the so-called 'sustainable intensification' of agriculture that is advocated by many (Cassman and Grassini, 2020).

Dramatic climate change scenarios predicted in the not-too-distant future warn that a rise in the sea level is expected to submerge many coastal areas around the world, and most affected would be the small islands of the Pacific and other such low-lying regions (Nurse et al., 2014). The NUS basket can certainly be of great help to those areas, for providing plants adapted to cope with increasing levels of salinity and water logging associated with such changes; thus, this approach deserves the urgent attention of governments, researchers and other stakeholders.

The intense urbanization all over the world is also leading to greater demand for more nutritious food, which—by encouraging the practice of growing locally and leveraging the diversity of well-adapted (smart') NUS—could help mitigate the impact that increased food production would have on the environment. The potential of tapping this diversity to provide fresh, healthy, safe and tasty food to millions of people living in large cities through vertical farming is also another opportunity worth seizing.

The dramatic COVID-19 pandemic that the world has been experiencing lately is also reminding us of the need to strengthen localized sourcing of food in order to help communities cope with shocks of this nature.

Ultimately, consumers play a fundamental role in changing the paradigm of uniformity and standardization that currently affects our food systems and that makes the world more vulnerable. Their attitude towards NUS needs to change, so that these crops are no longer perceived as the food of the poor, but, rather, as the food of the future.

Notes

1 The term 'Green Revolution' was coined by William Gaud late in its unfolding, at a meeting of the Society for International Development in DC in 1968 (reported in Patel 2013).
2 https://www.rareseeds.com/
3 https://www.mssrf.org/content/promoting-community-seed-grain-gene-banks
4 Costanzo A. Introduction: A Working Definition of Underutilized Crops: Proceedings of Conference 'DIVERSIFOOD Embedding crop diversity and networking for local high quality food systems' (http://www.diversifood.eu/wp-content/uploads/2017/11/DIVERSIFOOD-WP21_Inventory-of-Underutilized-crops.pdf) accessed on 10 July 2020.
5 https://www.cbd.int/sp/targets/rationale/target-13/ To note that in the post 2020 Biodiversity Framework the Aichi Targets will be replaced by other indicators.
6 http://www.fao.org/3/i2650e/i2650e.pdf
7 https://webapps.ifad.org/members/eb/126/docs/EB-2019-126-INF-5.pdf
8 https://webapps.ifad.org/members/eb/125/docs/EB-2018-125-R-11.pdf
9 https://webapps.ifad.org/members/eb/125/docs/EB-2018-125-R-12.pdf
10 The ICN2 Framework for Action can be found at http://www.fao.org/3/a-mm215e.pdf

References

Altieri, M. (1987) 'The significance of diversity in the maintenance of the sustainability of traditional agro-ecosystems. *ILEIA Newsletter*, 3: 2, July 1987.

Bazile, D., Jacobsen, S.E. and Verniau, A. (2016) The global expansion of Quinoa: Trends and limits. *Frontiers in Plant Science*, 7: 622. doi: 10.3389/fpls.2016.00622

Bioversity International (2017) *Mainstreaming agrobiodiversity in sustainable food systems: Scientific foundations for an agrobiodiversity index*. Rome: Bioversity International, 180 p. ISBN: 978-92-9255-070-7

Bioversity International (2018) Tale of the Tepary. Retrieved on 20 May 2020. https://www.bioversityinternational.org/news/detail/untold-tale-of-the-tepary/

Borelli, T., et al. (2020) 'Local solutions for sustainable food systems: The contribution of orphan crops and wild edible species', *Agronomy*, 10(2): 231.

Branca, F., Lartey, A., Oenema, S., Aguayo, V., Stordalen, G. and Richardson, R. (2019) Transforming the food system to fight non-communicable diseases. *BMJ* 364: 1296.

Cassman, K. and Grassini, P. (2020) A global perspective on sustainable intensification research. *Nature Sustainability* 3: 262–268. doi: 10.1038/s41893-020-0507-8

Conway, G. (1998) *The doubly green revolution: Food for all in the twenty-first century*. Ithaca: Comstock Pub. ISBN 978-0-8014-8610-4

Dawson, I., Guarino, L. and Jaenicke, H. (2007) Underutilized plant species: Impacts of promotion on biodiversity. ICUC Position Pap. 23. ISBN: 978-955-1560-05-9

Dickie, F. Miyamoto, M. and Collins, C.M. (2019) The potential of insect farming to increase food security. doi: 10.5772/intechopen.88106

16 Stefano Padulosi et al.

Durst, P. and Bayasgalanbat, N. (2014) *Promotion of underutilized indigenous food resources for food and nutrition in Asia and in the Pacific.* RAP Publication 2014/07. Food and Agriculture Organization of the United Nations, Regional Office for Asia and the Pacific. Bangkok, Thailand.

Eyzaguirre, P., Padulosi, S. and Hodgkin, T. (1999) IPGRI's strategy for neglected and underutilized species and the human dimension of agrobiodiversity. In Padulosi, S. (ed.) *Priority setting for underutilized and neglected plant species of the Mediterranean region.* Report of the IPGRI Conference, 9–11 February 1998, ICARDA, Aleppo. Syria. International Plant Genetic Resources Institute, Rome, Italy.

FAO (2011) *Save and Grow, A. Policymaker's guide to the sustainable intensification of smallholder crop production.* Rome: FAO. 112 pp.

FAO (2015) *The second report on the state of the world's plant genetic resources for food and agriculture.* Rome: FAO.

FAO, IFAD, UNICEF, WFP, WHO (2018) The state of food security and nutrition in the world 2018. Retrieved on 10 May 2020. www.fao.org/3/I9553EN/i9553en.pdf

Fowler C. and Mooney, P. (1990) *The threatened gene: Food, politics and the loss of genetic diversity.* Lutterworth Press, Cambridge.

Frison, E.A., Smith, I.F., Johns, T., Cherfas, J. and Eyzaguirre, P.B. (2006) Agricultural biodiversity, nutrition, and health: Making a difference to hunger and nutrition in the developing world. *Food and Nutrition Bulletin* 27(2): 167–179.

Ghosh, P. and Dhyani, P.P. (2004) Baranaaja: The traditional mixed cropping system of the central Himalaya. *Outlook on Agriculture,* 33(4): 261–266.

Gollin, D., Worm Hansen, C. and Wingender, A. (2018) Two blades of grass: The impact of the green revolution. doi: 10.3386/w24744

Graham, R.D., et al. (2007) Nutritious subsistence food systems. *Advances in Agronomy* 92: 1–74.

Hazell, P.B.R. (2003) The green revolution. In Mokyr, J. (eds.) *Oxford encyclopedia of economic history.* Oxford: Oxford University Press, pp. 478–480.

Holt Gimenez, E. (2006) *Campesino a Campesino: Voices from the farmer-to-farmer movement for sustainable agriculture in Latin America.* Oakland: Food First, 300 pp.

Hulse, J.H., Laing, E.M. and Peason, O.E. (1980) *Sorghum and millets: Their composition and nutritive value.* London: Academic Press.

Hunter, D., Borelli, T., Beltrame, D., Oliveira, C., Coradin, L., Wasike, V., Mwai, J., Manjella, A., Samarasinghe, G., Madhujith, T., Nadeeshani, H., Tan, A., Tuğrul Ay, S., Güzelsoy, N., Lauridsen, N., Gee, E. and Tartanac, F. (2019) The potential of neglected and underutilized species for improving diets and nutrition. *Planta* 250(3): 709–729. doi: 10.1007/s00425-019-03169-4

Hunter, D., Borelli, T. and Gee, E. (2020) *Biodiversity, food and nutrition: A new agenda for sustainable food systems. Issues in agricultural biodiversity.* Earthscan/Routledge, London.

IPGRI (2002) *Neglected and underutilized plant species: Strategic action plan of the international plant genetic resources institute.* Rome: International Plant Genetic Resources Institute.

Jennings, B.H. (1988) *Foundations of international agricultural research: Science and politics in Mexican agriculture.* Boulder: Westview Press, p. 51.

Khoshbakht, K. and Hammer, K. (2008) Species richness in relation to the presence of crop plants in families of higher plants. *Journal of Agriculture and Rural Development in the Tropics and Subtropics* 109(2): 181–190.

Lochetti, G., Meldrum, G., Kennedy, G. and Termote, C. (2020) *Seasonal food availability calendar for improved diet quality and nutrition: Methodology guide.* Rome: Bioversity International, ISBN 978-92-9255-184-1

Mabhaudhi, T., Chimonyo, V.G.P., Chibarabada, T.P. and Modi, A.T. (2017) Developing a roadmap for improving neglected and underutilized crops: A case study of South Africa. *Frontiers in Plant Science* 8: 8. doi:10.3389/fpls.2017.02143

Mabhaudhi, T., Chimonyo, V.G.P., Hlahla, S., et al. (2019) Prospects of orphan crops in climate change. *Planta* 250: 695–708. doi:10.1007/s00425-019-03129

Massawe, F.S., Mayes, A., Cheng, H., Chai, P., Cleasby, R.C., Symonds, W., Ho, A., Siise, Q., Wong, P., Kendabie, Y., Yanusa, N., Jamalluddin, A., Singh, R., Azman, S.N. and Azam-Ali (2015) The potential for Underutilized crops to improve food security in the face of climate change. *Environmental Science Proceedings* 29: 140–141.

Meldrum, G., Padulosi, S., Lochetti, G., Robitaille, R. and Diulgheroff, S. (2018) Issues and prospects for the sustainable use and conservation of cultivated vegetable diversity for more nutrition-sensitive agriculture. *Agriculture* 8: 112.

Mijatović, D., Meldrum, G. and Robitaille, R. (2019) Diversification for Climate Change Resilience: Participatory Assessment of Opportunities for Diversifying Agroecosystems. Rome: Bioversity International and the Platform for Agrobiodiversity Research

Nabhan, G.P. (1990) *Gathering the desert*. Tucson: The University of Arizona Press

Nurse, L.A., McLean, R.F., Agard, J., Briguglio, L.P., Duvat-Magnan, V., Pelesikoti, N., Tompkins, E. and Webb, A. (2014) Small islands. In Barros, V.R., Field, C.B., Dokken, D.J., Mastrandrea, M.D., Mach, K.J., Bilir, T.E., Chatterjee, M., Ebi, K.L., Estrada, Y.O., Genova, R.C., Girma, B., Kissel, E.S., Levy, A.N., MacCracken, S., Mastrandrea, P.R. and White, L.L. (eds.) *Climate change 2014: Impacts, adaptation, and vulnerability. Part B: Regional aspects*. Contribution of working group II to the fifth assessment report of the intergovernmental panel on climate change. Cambridge and New York: Cambridge University Press, pp. 1613–1654.

Padulosi, S. (1999a) Criteria for priority setting in initiatives dealing with underutilized crops in Europe. In Gass, T., Frese, F. Begemann, F. and Lipman, E. (compilers) *"Implementation of the global plan of action in Europe – conservation and sustainable utilization of plant genetic resources for food and agriculture. Proceedings of the European Symposium*, 30 June–3 July 1998. Braunschweig, Germany: International Plant Genetic Resources Institute, Rome.

Padulosi, S. (1999b) *Priority Setting for underutilized and neglected plant species of the Mediterranean region*. Report of the IPGRI Conference, 9–11 February 1998, ICARDA, Aleppo. Syria. International Plant Genetic Resources Institute, Rome, Italy.

Padulosi, S., Heywood, V., Hunter, D. and Jarvis, A. (2011) Underutilized species and climate change: Current status and outlook. *Crop adaptation to climate change*. Oxford: John Wiley-Blackwell, pp. 507–521.

Padulosi, S. and Hoeschle-Zeledon, I. (2004) Underutilized plant species: What are they? In: Agriculture Network (ed.) *LEISA* 20(1): 5–6.

Padulosi, S., Mal, B., Ravi, B., Gowda, J., Gowda, K.T.K., Shanthakumar, G., Yenagi, N. and Dutta, M. (2009) Food security and climate change: Role of plant genetic resources of minor millets. *Indian Journal of Plant Genetic Resources* 22(1): 1–16.

Padulosi, S., Mareè, D., Cawthorn, C., Meldrum, G., Flore, R., Halloran, A. and Mattei, F. (2018) *Leveraging neglected and underutilized plant, fungi, and animal species for more nutrition sensitive and sustainable food systems, reference module in food science*. Elsevier. doi:10.1016/B978-0-08-100596-5.21552-7

Padulosi, S., Hoeschle-Zeledon, I. and Bordoni, P. (2008) Minor crops and underutilized species: Lessons and prospects. In Maxted, N., Ford-Lloyd, B.V., Kell, S.P., Iriondo, J.M., Dulloo, M.E. and Turok, J. (eds.) *Crop wild relative conservation and use*. Wallingford: CAB International, pp. 605–624. ISBN 978-1-84593-099-9. http://bit.ly/1XcIA2r

Padulosi, S., Thompson, J. and Rudebjer, P. (2013) *Fighting poverty, hunger and malnutrition with neglected and underutilized species: Needs, challenges and the way forward: Neglected and underutilized species*. Rome: Bioversity International.

Padulosi, S., Roy, P. and Rosado-May, F.J. (2019) *Supporting nutrition sensitive agriculture through neglected and underutilized species – operational framework.* Bioversity International and IFAD, Rome, 39 pp. https://cgspace.cgiar.org/handle/10568/102462

Pignataro, L. (2019) La rivincita della rucola, e quella della Piana del Sele conquista la igp. Retrieved on 20 May 2020. https://www.lucianopignataro.it/a/rucola-piana-del-sele-igp/162137/

Pingali, P.L. (2012) Green revolution: Impacts, limits, and the path ahead. *PNAS* 109(31): 12302–12308. https://doi.org/10.1073/pnas.0912953109

The Plant List. Retrieved on 8 May 2020. http://www.theplantlist.org/1/

Ravi, S.B., Hrideek, T.K., Kumar, A.T.K., Prabhakaran, T.R., Bhag, M. and Padulosi, S. (2010) Mobilizing neglected and underutilized crops to strengthen food security and alleviate poverty in India. *Indian Journal of Plant Genetic Resources* 23: 110–116.

RBG Kew (2016). The state of the world's plants report – 2016. Royal Botanic Gardens, Kew.

Schutter de, O. and Vanloqueren, G. (2011) The new green revolution: How twenty-first-century science can feed the world. *Solutions* 2: 33–44.

Singh, B. and Rana, J.C. (2020) Reviving the spiritual roots of agriculture for sustainability in farming and food systems: Lessons learned from peasant farming of Uttarakhand Hills in North-western India Ishwari. *American Journal of Food and Nutrition* 8(1): 12–15.

Stevenson, J.R., Villoria, N., Byerlee, D., Kelley, T. and Maredia, M. (2013). Green Revolution research saved an estimated 18 to 27 million hectares from being brought into agricultural production. *PNAS* 110(21): 8363–8368. https://doi.org/10.1073/pnas.1208065110

Swaminathan, M.S. (1996) *Sustainable agriculture: Towards an evergreen revolution.* Delhi: Konark Publ.

Swaminathan, M.S. (2020) A call to remember the forgotten crops. Retrieved on 10 May 2020. https://www.mssrf.org/content/call-remember-forgotten-crops-ms-swaminathan

Taranto, S. and Padulosi, S. (2009) Testing the results of a joint effort. *LEISA Magazine* 25(2): 32–33.

Thrupp, L.A. (2000) Linking agricultural biodiversity and food security: The valuable role of agrobiodiversity for sustainable agriculture. *International Affairs* 76(2): 265–281.

Tribe, D. (1994) *Feeding and greening the world: The role of international agricultural research.* Wallingford: CAB International.

Ulian, T., Diazgranados, M., Pironon, S., Padulosi, S., Liu, U., Davies, L., Howes, M.-J.R., Borrell, J., Ondo, I., Pérez-Escobar, O.A., Sharrock, S., Ryan, P., Hunter, D., Lee, M.A., Barstow, C., Łuczaj, Ł., Pieroni, A., Cámara-Leret, R., Noorani, A., Mba, C., Womdim, R.N., Muminjanov, H., Antonelli, A., Pritchard, H.W. and Mattana, E. (2020) Unlocking plant resources to support food security and promote sustainable agriculture. *People Plants Planet* 2(5): 421–445. ISSN: 2572-2611

UNSCN (2018) Non-communicable diseases, diets and nutrition. www.unscn.org/uploads/web/news/document/NCDs-brief-EN-WEB.pdf

Van Huis, J.A., Van Itterbeeck, H.K., Mertens, E., Halloran, A., Muir, G. and Vantomme, P. (2013) *Edible insects future prospects for food and feed security.* FAO Forestry Paper 171. Rome: FAO.

Vanhaute, E. (2011) From famine to food crisis: What history can teach us about local and global subsistence crises. *The Journal of Peasant Studies* 38(1): 47–65.

Willett, W., Rockström, J., Loken, B., et al. (2019) Food in the Anthropocene: The EAT–Lancet commission on healthy diets from sustainable food systems. *Lancet.* doi:10.1016/S0140-6736 (18)31788-4

Willis, K.J. (ed.) (2017). State of the world's plants 2017. Report. Royal Botanic Gardens, Kew.Wolf, M. (2018) *Plant guide for tepary bean (Phaseolus acutifolius).* Tucson: USDA-Natural Resources Conservation Service, Tucson Plant Materials Center.

2

LANDMARK NUS EVENTS AND KEY PUBLICATIONS

Stefano Padulosi and Danny Hunter

Introduction

Looking back at the last 50 years, we can comfortably say that both public and private institutions have been slow in recognizing, appreciating, and supporting research on neglected and underutilized species (NUS). Many are the reasons behind such a tepid position regarding the promotion of these species, but above all stands the strong influence exerted by the Green Revolution's legacy in setting research priorities in agriculture at both the national and international levels. The focus on a few staple crops to feed the world has been inhibiting the development of research for non-commodity/non-staple crops around the world. Notoriously, the Consultative Group on International Agricultural Research (CGIAR), the largest organization engaged in agricultural research spends almost its entire budget on a few strategic staple crops, viz. lentils, cowpeas, groundnuts, millets, wheat, coconuts, maize, pigeon peas, potatoes, rice, soybeans, sweet potatoes, and yams.[1] Most of these crops have been receiving dedicated funding since the early 1960s, whereas very limited resources have been directed towards developing NUS. Low attention to NUS is also registered among National Agricultural Research Systems (NARS) as reported in national country reports dedicated to the conservation and use of Plant Genetic Resources for Food and Agriculture (PGRFA) (FAO, 1996, 2011). Although a number of global institutions and NARS have been playing over the years a commendable role in support of NUS in terms of implementing supportive policies, programs and projects for their sustainable conservation and use, we would like to emphasize that it is Indigenous Peoples and other ethnic groups who have been so far the true custodians of these local resources and the promoters of their continued traditional use (Padulosi et al., 2019).

DOI: 10.4324/9781003044802-3

Period: 1970–1980

This period is characterized by a predominant R&D focus on major commodities (staple crops and industrial crops). Though the work of agricultural research institutes (including the CGIAR) have yielded important impacts in terms of hunger and poverty reduction, it is also causing the marginalization of hundreds of nutritious crops (especially minor cereals, pulses, vegetables, and fruits) that are no longer competitive with the high yielding varieties massively promoted by governments. Deployment of a broader basket of species to mitigate the impact of crop failures and, hence, to fight periodic food insecurity is far from becoming a key issue of this decade. A milestone in the advocacy for opening the door to research on NUS is the USA National Academy of Sciences' (NAS) document drawing attention to underutilized species (NAS 1975); the paper was the outcome of a NAS-commissioned "extensive survey of underexploited tropical plants" as possible crops for the future. Among the 36 species addressed by this report, the winged bean (*Psophocarpus tetragonolobus*) was particularly recommended to the research community for further attention due to its "exceptional merits".

Period: 1981–1990

This is when attention on NUS starts building up. This is due to increased recognition of the importance that crops' wild relatives gained within the CGIAR as well as within FAO (Hunter and Heywood, 2011). An example of this is the International Institute of Tropical Agriculture (IITA), which started a programme supported by the Italian government in 1990 to survey and collect wild *Vigna* species (Padulosi et al., 1991). A similar programme supported by the German Agency for Technical Cooperation GTZ (today GIZ) was also launched at IITA to survey, collect and study Bambara groundnuts (*Vigna subterranea*) in Africa (Begemann, 1988). In 1987, the International Conference on "New Crops for Food and Industry" was held at the University of Southampton and as a result of one of its recommendations, the International Center for Underutilized Crops (ICUC) was established the following year. In 1988, the University of Purdue (USA) organized the first of a series of International Symposia on New Crops, mainly looking for alternatives to major crops, targeting US farmers (Janick and Simon, 1990).

Also relevant is the CSC/ICAR Delhi International Workshop, held in 1987, on the maintenance and evaluation of life-support species in Asia and the Pacific region (Paroda, 1988). The publication in 1989 of *Lost Crops of the Incas* (NRC, 1989) was the first of several books from a highly prestigious source that focuses on NUS in Latin America (subsequent volumes of the book would be dedicated to NUS in Africa, viz. NRC, 1996, 2006, 2008). India recognized the importance of underutilized species as a means to attain sustainable agricultural production, improve the nutritional value of food for large sections of the population and reduce the country's dependence on food imports. Thus, in 1982, the All-India Coordinated Research Project on Underutilized Plants was launched, focusing on a list of priority species (Tyagi et al., 2017). International collaboration on underutilized species was also promoted through newly funded biodiversity projects such as that of the Department for International Development (DFID), meant to support ICUC research (focusing on underutilized tropical fruit trees[2]). Another important publication is *New Crops for Food and Industry* (Wickens et al., 1989).

The drawbacks of the Green Revolution were beginning to be acknowledged in literature at this time as well (Smale, 1997). The concept of sustainable agriculture made its first appearance in scientific papers (Dimitri and Richman, 2000), as did innovative approaches based on the deployment of greater diversity in farmers' fields using species thus far considered underutilized (see f.e. the case of alley-cropping introduced in sub-Saharan Africa by IITA, revolving around the use of leguminous crops such as *Leucaena spp.* to maintain or restore fertility in farmers' fields [FAO, 1995]).

Period: 1991–2000

In 1992, the Convention of Biological Diversity (CBD) stressed the concept of sustainability,[3] rooted in agricultural and cultural diversity supportive of nutritional needs, incomes and greater protection from biotic and abiotic stresses. It also drew attention to the significant contribution of Indigenous Peoples to the conservation and sustainable use of biodiversity (Article 8J). It is important also to recall the establishment of the Work on Agricultural Biodiversity proposed by the Conference of the Parties to the Convention, which decided to establish a multi-year programme of work on agricultural biological diversity (Decision III/11),[4] aimed – *inter alia* – aimed at "promoting the conservation and sustainable use of genetic resources of actual and potential value for food and agriculture" (Box 2.1).

The CBD had a tremendous impact in raising the awareness in people at the highest levels of the value of biodiversity, including underutilized species: it helped introduce new values such as the environmental services provided by biodiversity, which would have a profound impact in the years to come in influencing countries' strategies in agricultural activities, to make them more conducive than in the past to safeguarding less commercialized crops and species.

In 1992, the Asian Vegetable Research and Development Center (AVRDC -The World Vegetable Center), launched a number of projects focusing specifically on traditional African vegetables (Dinssa et al., 2016). The CGIAR revised its mission statement, limiting it no more to food security, but broadened it by including poverty reduction and protection of the environment; opportunities to work on species not necessarily used in food production were highlighted. An IDRC (International Development Research Centre)-supported study recommended IPGRI's greater involvement in medicinal plants in view of the fact that many of these species were underutilized and neglected by R&D despite their high-income generation potential (Leaman et al., 1999).

Box 2.1

The four key elements of the CBD programme of work on agricultural biodiversity:

1 Assessments: to provide an overview of the status and trends of the world's agricultural biodiversity, their underlying causes and knowledge of management practices
2 Adaptive Management: to identify adaptive management practices, technologies and policies that promote the positive effects and mitigate the negative impacts of agriculture on biodiversity, and enhance productivity and the capacity to sustain livelihoods, by expanding knowledge, understanding and awareness of the multiple goods and services provided by the different levels and functions of agricultural biodiversity
3 Capacity Building: to strengthen the capacities of farmers, Indigenous and local communities, and their organizations and other stakeholders, to manage agricultural biodiversity sustainably so as to increase their benefits, and to promote awareness and responsible action
4 Mainstreaming: to support the development of national plans and strategies for the conservation and sustainable use of agricultural biodiversity and to promote their mainstreaming and integration in both sectoral and cross-sectoral plans and programmes.

The decade was marked by a remarkable increase in support from Overseas Development Agencies (ODA) for NUS. Italy, the IDRC, the Asian Development Bank, the European Commission, the Netherlands and other donors joined Germany and the UK in financing ad hoc projects and networks dealing with NUS (e.g., MEDUSA ["Network on the Identification, Conservation and Use of Wild Plants in the Mediterranean Region"],

BAMNET [International Bambara Groundnut Network], UTFANET [Underutilized Tropical Fruit in Asia Network], SEANUC [Southern and East Africa Network on Underutilized Crops], PROSEA) [Plant Resources of South East Asia] (Williams and Haq, 2000).

In 1994, IPGRI launched "The Underutilized Mediterranean Species"(UMS) project, which established four NUS-focused networks, guided by the outcome of a Mediterranean-wide survey of 400 researchers and aimed at advancing knowledge on challenges, needs and opportunities in the promotion of NUS as well as promoting close cooperation for NUS use enhancement across regional research institutions (Padulosi et al., 1994; Padulosi, 1998).

Nevertheless, funding still remained limited against the background of large R&D gaps to fill. Therefore, international cooperation was advocated in several meetings as the only way to achieve a visible impact in this domain.

Other Symposia on New Crops were organized by the University of Purdue (Janick and Simon, 1993; Janick 1996; Janick, 1999; Janick and Whipkey, 2002); such meetings provided an important platform to the scientific community for sharing experiences and lessons directly related to NUS and their development of new crops.

The FAO process of the International Conference and Program for Plant Genetic Resources, leading to the 1996 Leipzig FAO IV Technical Conference on Plant Genetic Resources for Food and Agriculture (PGRFA), raised the visibility of NUS at the UN level. Preparatory national and regional meetings to the conference (especially the European meeting held in Nitra, Slovakia on 15 October 1995) contributed through country-driven, bottom-up approaches to the development of the Global Plan of Action (GPA) for PGRFA (FAO, 1996). The GPA, listing 20 activities across an array of themes from conservation to the sustainable use of PGRFA, provided unprecedented visibility to NUS, while an activity specifically dedicated to them (Activity 12: cfr. "Promoting development and commercialization of underutilized crops and species"[5]) was also included.

In 1994, IPGRI launched a series of projects focusing on African leafy vegetables that would end in 2004; these efforts were an important contribution to improving the methodological approach in NUS promotion especially in Kenya, contributing to changing "the food of the poor" stigma so often associated with these crops and helping draw policymakers' attention to their use enhancement (Gotor and Irungu, 2010).

The 1996, the FAO State of the World Report outlined a worrying situation with regard to the conservation of NUS: less than 22% of the estimated six million germplasm samples held in gene banks around the world were non-commodity crops, and of this portfolio (inclusive of NUS), species were scarcely represented in terms of intra-specific diversity. More than 80% of these, on average, were made up of less than ten accessions (Padulosi et al., 2002).

In 1998, in the framework of Italy's campaign in support of the development of the FAO International Treaty on Plant Genetic Resources, a panel of experts gathered in Florence, Italy to discuss the development of an alternative list of species to Annex I of the International Treaty. At this meeting, specific discussions were held on the possibility of including NUS in the treaty (Padulosi, 2000). It is interesting to note that Annex I of the approved treaty contains today a total of 80 genera, of which only 15 include some underutilized species. The discussion on the inclusion of NUS in Annex I was re-opened in 2013, but the debate is still ongoing today.

Also, an important endorsement of the value of NUS (particularly fruit trees, vegetables and medicinal and aromatic plants) was recorded at the World Conference on Horticultural Research held in Rome in 1998 (Morico et al., 1999).

In 1999, the IFAD-supported workshop organized in Chennai, India, by the CGIAR PGR Policy Committee, chaired by Prof. M.S. Swaminathan, covered NUS specifically (GRPC, 1999). The meeting yielded large support from attending CGIAR centres and donor representatives. It is interesting to note that this meeting

24 Stefano Padulosi and Danny Hunter

represented the first time ever that the CGIAR discussed NUS in a formal way. Important publications focusing in large part on Asia and the Pacific were also published (Bhag Mal 1994; de Groot and Haq, 1995; Smartt and Haq, 1997).

How underutilized species also play an important role in enriching the landscape and local cultures was covered at an international workshop focusing on the Mediterranean region, organized by the Italian National Research Council in Naples, Italy in 1997 (Padulosi, 1997).

Period: 2001–2020

The Global Forum on Agricultural Research (GFAR) discussed NUS at a meeting in Dresden, Germany (GFAR, 2001). As a follow-up to that meeting, a small group comprising IPGRI, ICUC, IFAD and the German Ministry of Economic Cooperation and Development (BMZ) recommended the establishment of a Global Facilitating Unit (GFU),

> to pursue the goal of drawing attention to the potential contribution hitherto underutilized species could make to food security and livelihood of marginalized and poor communities so that an increasing number of research institutions, extension services, policy makers and donors include the development of underutilized species in their programmes and plans.[6]

The unit was established in 2002 and was housed at IPGRI in Rome. The GFU was a major effort that aimed at increasing the contribution of underutilized species to food security and poverty alleviation of the rural and urban poor through facilitating access to information on underutilized species, performing policy analysis and providing advice to policymakers on how to create an enabling policy environment for underutilized species and enhancing public awareness on these species.[7]

In addition, Germany also funded a GTZ multi-regional project "People and Biodiversity in Rural Areas" that supported national partners in improving existing value-chains of underutilized crops and breeds, and that analyzed the economic potential of other underutilized species and breeds in selected regions. Several publications on the topic were published and workshops organized, such as the excellent publication by GTZ on *Promising and underutilized crops and breeds* (Thies, 2000).

In 2003, the PROTA (Plant Resources of Tropical Africa) Network was launched to provide access to information on 7,000 tropical African plants, most of them little known or underutilized (Schmelzer and Omino, 2003).

AVRDC launched its strategy for 2001–2010, one objective of which was to increase the diversity of indigenous and underutilized vegetables for better nutrition, health and income (AVRDC, 2001).

The first global project on NUS (Grant 533) was also implemented from 2001 to 2005 by IPGRI, thanks to the support of IFAD (focusing on Ecuador, Bolivia,

Peru, Yemen, Egypt, Nepal and India). This project would provide a unique opportunity to test the hypothesis that NUS are strategic crops in supporting poverty reduction and the empowerment of the poor. The results of the first phase of the project showed extremely encouraging results in the areas of income generation and empowerment of local communities, particularly in India and Latin America (Padulosi et al., 2003). Follow-up phases of this first UN-funded programme on NUS would be implemented until 2020.[8]

IPGRI published its strategy on NUS (Eyzaguirre et al., 1999), recommending interventions in eight main strategic areas, namely, (1) gathering and sharing information, (2) priority setting, (3) promoting production and use, (4) maintaining diversity, (5) marketing, (6) strengthening partnerships and capacities, (7) developing effective policies and (8) improving public awareness, which would then be translated into five languages (IPGRI, 2002). Interesting to point out that this is the first time the acronym "NUS" was used to refer to neglected and underutilized species (wild and cultivated) (see also Chapter 1). A joint ICUC-IPGRI analysis of the status of underutilized species was also published in 2002 (Williams and Haq, 2000).

A BMZ-funded workshop on underutilized species was organized by GFU, GTZ and InWEnt in Leipzig, Germany (Gundel et al., 2003) during which the need for mainstreaming NUS in the R&D agenda was stressed as necessary action to successfully exploit the potential of these species.

An issue of the LEISA magazine entirely dedicated to underutilized species was published in 2004 (LEISA, 2004), earning a largely positive consensus among stakeholders (see Padulosi and Hoeschle-Zeledon, 2004).

At the 7th Meeting of the Conference of the Parties to the CBD in 2004, a relevant pro-NUS recommendation of the Subsidiary Body on Scientific, Technical and Technological Advice (SBSTTA) was endorsed. This recommendation suggested activities that contributed to improved food security and human nutrition through the enhanced use of crop and livestock diversity, and the conservation and sustainable use of NUS. The SBSTTA underlined that the identification of constraints and success factors in marketing underutilized species is a key aspect for their promotion and that capacity building at different levels was highly needed.

For the first time, in 2004, a major donor (EU) made a specific call within its 6th Framework for "*[r]esearch to increase the sustainable use and productivity of annual and perennial under-utilized tropical and sub-tropical crops and species important for the livelihoods of local populations*". The EU recognized that these crops have potential for wider use and could significantly contribute to food security, agricultural diversification and income generation.

In 2005, the International Horticultural Assessment commissioned by US-AID was published. This work (which engaged 750 participants, 60 countries and three regional workshops, and which involved a major survey) was a strong endorsement of the value of underutilized crops to revitalizing an agricultural sector in crisis. More than a third of promising horticultural species were underutilized according to the report: of the 226 listed fruits, vegetable crops, herbs, spices and ornamentals, 79 belonged to the NUS category. Excerpts of the report related to horticulture in sub-Saharan Africa and Latin America were also supportive of underutilized species. For instance, in sub-Saharan Africa, the report stressed that despite diverse biophysical constraints such as droughts and low soil fertility, the region called for expanded cultivations of its underutilized and indigenous crops adapted to harsh conditions (such as the leafy vegetables *Cleome gynandra*, *Solanum aethiopicum*, *Solanum macrocarpon*, *Moringa oleifera* and *Hibiscus sabdariffa* and fruits like *Ziziphus mauritania*). In Latin America and the Caribbean, underutilized fruit trees represented opportunities to generate new markets. One recommendation suggested compiling regional knowledge about the cultivation and traditional uses of these species at a regional and national level.

In 2005, at the end of a broadly participative process, the CGIAR published its research priorities for the period 2005–2015.[9] In this document, NUS were given high visibility under System Priorities 1b ("Promotion, conservation and characterization of underutilized plant genetic resources to increase the income of the poor") and 3a ("Increasing income from fruit and vegetables" in consideration that many of the latter are considered underutilized species). In addition, underutilized species were also considered indirectly through Priorities 3d ("Sustainable income generation from forests and trees") and 4d ("Sustainable agro-ecological intensification in low- and high-potential environments"). The emergence of niche and high-value markets for underutilized crops in developed countries provides a potential pathway out of poverty for farmers in developing countries, hence, underutilized plant genetic resources (UPGR) were relevant also to Priority 5b ("Making international and domestic and domestic markets work for the poor").

In April 2005, 100 R&D experts and policy makers with varied backgrounds from 25 countries took part in a consultation at the M.S. Swaminathan Research Foundation in Chennai, India. This meeting represented a major milestone in support of agricultural biodiversity, including NUS. The consultation, jointly organized by IPGRI, GFU and The M.S. Swaminathan Research Foundation (MSSRF), was called to discuss how biodiversity could help the world achieve the UN's Millennium Development Goals – in particular, the goal of freedom from hunger and poverty. The "Chennai Platform for Action" that resulted from this consultation, in its ten recommendations, emphasized the importance of underutilized species and called upon policymakers to promote specific interventions in support of these species.[10]

During this period, increased visibility on underutilized species was provided by dedicated websites developed by both international and national agencies.

In 2006, a ICUC-IPGRI-GFU electronic consultation to design a strategic framework for R&D on underutilized species was carried out; this was followed by two regional strategy workshops held in Colombo, Sri Lanka (March 16–17), and Nairobi, Kenya (May 24–25) (Jaenicke and Höschle-Zeledon, 2006).

An International Conference on Indigenous Vegetables was held in Hyderabad, India, on December 2006 (Chadha and Lumpkin 2007). The International Society for Horticultural Science (ISHS) launched a Working Group on Underutilized Species within its Commission on Plant Genetic Resources, which was jointly chaired by GFU and ICUC. In 2004, the ACP-EU project on NUS value-chains for Africa was launched (Rudebjer, 2014).

In 2009, a new organization, Crops For the Future (CFF) was established through the merger of the International Centre for Underutilized Crops (ICUC) in Sri Lanka and the Global Facilitation Unit for Underutilized Species (GFU) in Rome (Gregory et al., 2019). Based in Kuala Lumpur, Malaysia, the mission of this new agency consisted of four major objectives, viz. (1) increase the knowledge-base for neglected crops, (2) advocate policies that do not discriminate against crop diversity; (3) increase awareness of the relevance of neglected crops for rural livelihoods and (4) strengthen capacities in relevant sectors. In 2001, the CFF Research Centre was created in Malaysia to provide research support to the global CFF organization.[11]

In 2010, the FAO Second State of the World Report on PGRFA (FAO 2010) reported on still poor conservation statuses for NUS, calling for urgent actions to be taken to safeguard these resources, important allies in the fight against food insecurity around the world. In 2011, the second ISHS International Symposium on "Underutilized Plants: Crops for the Future, Beyond Food Security" was held in Kuala Lumpur, Malaysia (Jaenicke, 2013).

On December 2012, an international conference on the "Lost Crops of XXI Century" was held in Cordoba, Spain (Padulosi et al., 2013). The conference, attended by the Spanish Minister of Agriculture, the Director General of FAO, the Director General of Bioversity and the Chairman of the Slow Food Movement, was held to debate how to fight hunger and rural poverty through the greater deployment of NUS (called here "promising crops").[12] The conference also celebrated the launch of the 2013 UN International Year of Quinoa. An important collective document, the Cordoba Declaration[13] was released by the participants, and called for a number of important actions in support of the use enhancement of NUS, viz:

- Raising awareness of these crops and their strategic roles
- Conserving genetic and cultural diversity
- Promoting their use in small-scale family farming to improve rural livelihoods
- Developing value-chains from production-to-consumption and to gastronomy
- Changing incorrect perceptions and developing an evidence base
- Enhancing research and capacities for promotion
- Building inter-sectoral and interdisciplinary collaboration
- Creating a conducive policy environment
- Establishing an ombudsman to represent the rights of future generations in national and international decision-making

28 Stefano Padulosi and Danny Hunter

Among these, the last was a particularly innovative idea, and helped underscore the fact that the conservation of agricultural biodiversity and other relevant natural resources are relevant both to present and future generations. A highly strategic policy recommendation also proposed in the document was the launching of a new international dialogue on PGRFA to explore ways in which the International Treaty on PGRFA (IT-PGRFA) could further support the conservation, exchange and sustainable use of NUS.

The following year, a major pan-African conference on NUS was organized by Bioversity in Accra, Ghana, through the support of the European Union (ACP-EU project) and the IFAD-NUS Project. The "Accra Statement for a food-secure Africa" that emerged from the conference set out nine action points to promote greater use of NUS (Bioversity International 2014), viz:

1   Include NUS in national and international strategies and frameworks that address global issues
2   Establish a list of priority NUS on which to focus R&D
3   Support research on NUS and their agronomic, environmental, nutritional and socioeconomic contributions to resilient production systems
4   Support the development of value-chains and small agri-businesses for NUS
5   Strengthen collaboration and information-sharing between researchers, extension specialists, the private sector, farmers and their organizations
6   Promote the cultivation of NUS through campaigns to raise awareness of the commercial opportunities they offer and their agronomic and nutritional benefits
7   Increase support for conservation of NUS *in situ*, on-farm and *ex situ*, and strengthen seed systems
8   Empower custodian farmers and support farmers' rights to share the benefits from NUS
9   Strengthen the capacity of individuals and organizations in R&D and education of NUS.

An important paper on agrobiodiversity highlighting the role of NUS in strengthening food security, health and income generation appeared in *Agronomy for Sustainable Development* (Kahane et al., 2013).

Bioversity's revised strategy for NUS was published in 2013 (Padulosi et al., 2013) offering an analysis of the NUS status of conservation and use, based on country reports produced for the FAO 2010 SWR on PGRFA. This document stressed eight key areas of action to realize the effective promotion of NUS (Box 2.2).

India approved, in 2013, its Food Security Bill, which included "coarse grains" (millets, sorghum and maize) in the Public Distribution System (PDS). This move, which represented an unprecedented recognition of NUS, never made by any previous government, was done in appreciation of the following aspects: (1) the importance of finger millets and others coarse cereals of high

Box 2.2 What needs to be done: Strategic actions needed to bring NUS back to the table (from Padulosi et al., 2013)

Type of action	What needs to be done
1 Change perceptions	The imbalance in agricultural policies and practices, currently skewed towards the major commodity crops, needs to change. It is important to make smallholder farmers and consumers in both urban and rural areas of poor countries aware of the benefits of conserving and using NUS and to make scientists and policymakers aware of the need to both protect and promote NUS.
2 Develop capacity	Organizations involved in research, education, development, policy departments, the private sector and farming communities need to develop skills and capacity in conserving and using NUS. Young scientists need to acquire the skills to holistically address the inter-disciplinary research challenges related to NUS and integrate nutrition into agricultural development. Policymakers and institutional leaders need to understand the role and benefits of NUS so as to integrate NUS in research and development (R&D) strategies and programmes, including adaptation to climate change.
3 Enhance research	It is extremely important to invest in research on NUS and on improving them. Data on NUS use, propagation and growth characteristics, resistance traits and intra-specific variation must be systematically collected in databases and disseminated and shared, and methodology strengthened. Scientific knowledge and indigenous knowledge need to be integrated. Instruments and processes for sharing lessons learned at the national and international levels need to be strengthened. Conservation must be linked to use.
4 Improve conservation	Species that are not suited to conservation in genebanks need to be conserved on farm. In parallel, greater ex situ conservation of NUS in genebanks, when practical, is needed. Enhancing use of NUS needs to be in harmony with traditional rights, cultural identities, ecosystem integrity and the principles of gender equity and benefit both the rural poor and urban consumers. A global on-farm conservation programme for NUS needs to be established.
5 Involve stakeholders	Communities and farmer organizations need to be consulted to ensure that research programmes are relevant and appropriate. Frameworks for involving communities in addressing challenges, needs and opportunities need to be set up. Farmers need access to seed, as well as training in maintaining and exchanging quality seed and planting material according to phytosanitary regulations.

6 Add value and upgrade market chains	Sustainable markets for NUS need to be developed and strengthened at the local, national and international levels while ensuring that benefits are shared fairly. Support to develop effective value-chains for NUS (through enhanced vertical and horizontal integration) is also needed. Domestic demand for NUS and value-added NUS products needs to be expanded and trade barriers for NUS products in developed countries need to be removed. The links between farmers, researchers and consumers need to be strengthened, and programmes need to highlight the growing importance of NUS in gastronomy.
7 Create a supportive policy environment	Legal frameworks are needed to protect NUS (wild or cultivated). National governments also need to put in place policies to effectively conserve and use NUS. There need to be incentives for managing NUS on-farm. Policies, guided by principles of equity and fairness, need to safeguard germplasm for crop improvement and sharing, and provide better access to international markets.
8 Increase cooperation	NUS are local and traditional, but are globally significant and thus require scientific and political attention beyond the local and national levels. More needs to be done to strengthen cooperation among stakeholder groups and create national, regional and international synergies.

nutritional value for the diversification of the food basket; (2) the inclusion of millets would expand the quantum of food that can be procured at the same time and (3) promote climate resilient farming, which would more appropriately cater to the food habits of different regions (Padulosi et al., 2015).[14] Notably, these crops were set to be used also to reinforce several nutrition-related schemes and activities, including the Integrated Child Development Services, mid-day meals and community canteens.

The Second International Conference on Nutrition Framework for Action called, in its recommendation (no. 10), for the "the diversification of crops including underutilized traditional crops". Two seminal papers on the holistic value-chain approach developed by IFAD-NUS were published (Padulosi et al., 2014, 2015) along with a strategic document from Bioversity on the way forward for the promotion of NUS (Padulosi et al., 2013).

In 2016, the Royal Botanical Gardens (RBG) at Kew developed its internal strategy on useful plants, which focuses on NUS, while use enhancement along the lines of Bioversity's work on holistic value-chains is being developed (Ulian et al., 2015).

In July 2016, the FAO International Treaty for PGRFA organized a meeting at Volterra, Italy, to address the development of a "toolbox" for the use enhancement of agricultural biodiversity, including NUS.

In November 2016, Bioversity and its partners from Africa, organized in Benin an expert meeting on NUS value-chains in sub-Saharan Africa for strengthening agricultural diversification, the UN Agenda 2030 and climate-change responses; among the recommendations that emerged from this meeting the following needs were seen as the most critical to advancing the NUS agenda in the region: (1) strengthening capacities of young scientists working on NUS; (2) mainstreaming NUS in school and university curricula; (3) greater PPP to boost local NUS entrepreneurs; (4) continued advocacy and lobbying with decision-makers at national and international levels for developing supportive policies for NUS and (5) facilitating access to NUS knowledge and material among workers.

Over the course of 2015 and 2016, three expert meetings were organized under the leadership of CFF in Paris, Kuala Lumpur and Rome, to discuss the implementation of a newly developed Global Action Plan for Agricultural Diversification (GAPAD), designed to meet the needs of a heating world. The alliance formed around GAPAD aimed at promoting agricultural diversification and NUS crops, in particular, which were recognized as a strategic means to contribute to the UN's Sustainable Development Goals (SGDs), with special reference to SDGs 2, 7, 12, 13, 15 and 17.

In 2016, an international consultation was jointly organized by FAO and the Australian Centre for International Agricultural Research under FAO RAP's Regional Initiative on Zero Hunger Challenge (RI-ZHC), in Bangkok, Thailand. Its purpose was to identify promising NUS crops that are nutritionally dense, climate resilient, economically viable and locally available or adaptable, and to provide strategic advice to decision-makers. These promising NUS were to be referred to as "Future Smart Food" (FSF). The consultation, attended by representatives from eight countries and several international agencies, aimed at four specific objectives: (i) validating the preliminary scoping report on crop-related NUS in the selected countries; (ii) ranking and prioritizing high-potential NUS based on established priority criteria, (iii) identifying five to six crop-related NUS per country and (iv) strategizing to enhance production and the utilization of selected crops in local diets (Li et al., 2018). The meeting identified ten recommendations for policymakers (Box 2.3) A similar meeting was also organized in Bangkok in 2017 by APAARI and was attended by representatives of 16 countries from the Asia-Pacific region (Tyagi et al., 2018a,b).

Two more projects worth mentioning have also sprung up over the last few years, viz. the GFAR's Collective Action on "Harnessing forgotten foods for improved livelihoods" (GFAR, 2017) and the campaign of the USA-based Lexicon on "Rediscovered Food", which includes *inter alia* some well-packaged introduction on 25 forgotten nutritious NUS and success stories of farmers, scientists and chefs from 14 countries.[15]

In September 2020, a major publication by the RBG (Ulian et al., 2020) was released in conjunction with the State of the World's Plants and Fungi 2020 report published by the same organization (Antonelli et al., 2020). Drawing on the results of the report, the paper reiterates that narrow reliance of humankind on a small handful of food crops, in spite of 7,039 documented edible species. In fact, the majority of these edible plants belong to NUS and should be better promoted in recognition of their role in improving the quality, resilience and self-sufficiency of food production. In order to unleash such potential, the authors call for renovated efforts especially with regard to: (1) filling knowledge gaps on the biology and ecology of NUS; (2) gaining a better understanding on the impacts of climate change on NUS to allow their better deployment in future climate-change coping strategies; (3) promoting the better integration of methods and tools developed by farmers and researchers for the cultivation of major crops, with the traditional knowledge of uses and practices regarding NUS; (4) providing further aid to strengthen research programs focusing on NUS, including supporting information exchange; (5) promoting participatory decision-making processes in NUS enhancement and (6) fostering the development and implementation of legal and policy frameworks accompanied by economic incentives and subsidies in support of NUS.

Box 2.3 Recommendations to policymakers emerging from the FAO regional meeting for Asia and the Pacific on smart food foods/NUS held in Bangkok in 2016

1 Urgent call for decision-makers to raise awareness of the nutrition-sensitive and climate-resilient benefits of NUS to address hunger, malnutrition and climate change.
2 Recognize, identify and promote complementarities of NUS with existing staple crops for nutrition enhancement, climate-change resilience and diversification of cropping systems, and reliable NUS as FSF to popularize these species.
3 Establish a National Coordinating Committee on FSF involving concerned ministries and appoint a Strategic Coordinator at the inter-ministerial level.
4 Create an enabling environment by strengthening national institutional support for mainstreaming FSF into national policies and programmes, using appropriate incentives, procurement of FSF for food programmes (e.g., mid-day meal/school-meal schemes) to enhance national consumption, local production and facilitate marketing.
5 Establish nationally coordinated research for development programmes targeting FSF with high potential, and expand coverage of national

agriculture statistics and national food composition data on FSF for evidence-based decision-making.

6 Document and validate best-bet FSF case studies, compile Indigenous knowledge related to FSF, undertake clinical and field studies to demonstrate the health benefits and climate resilience of FSF, and assemble quantitative data for public dissemination.

7 Enhance public awareness of the importance of FSF by developing nutrition and climate-change education materials and curricula on the importance of FSF for consumers, traders, producers, health professionals, researchers, teachers (e.g., school curricula), farmers, women and youth.

8 Identify key entry points in the value-chain and encourage value-chain development for specific FSF, including innovative and targeted interventions for promotion (e.g., ready-to-use food products) and increase funds for research, development and extension capacities on FSF production and processing technologies.

9 Strengthen multidisciplinary and multi-sectoral collaboration through existing coordination mechanisms, and build partnerships at national and regional levels, including academia, civil society and the private sector, to enhance research and consumption and to attract the private sector to boost production, processing, value addition, product development and marketing of FSF.

10 Establish a regionally coordinated network on FSF to facilitate the exchange of information, policy, technologies and genetic resources, as well as FSF promotion, in target countries.

Conclusions

Although we have come a long way from those isolated pro-NUS activities of the early 1980s, all the way to the international conferences and projects dedicated to NUS that have been launched lately, we are still far from having reached the "graduation" of NUS (with the exception of a few crops like quinoa, which has become truly a commodity crop). While we can affirm that methodologies for the use enhancement of NUS have been successfully developed, what is needed these days, rather, is to scale the many successful approaches recorded around the world in order to have a wider impact. Mainstreaming NUS into government actions is today the key pursuit for workers engaged in the movement and, to achieve that, enabling policies covering a range of priority areas (from conservation to resilient food systems, sustainable cultivations, equitable marketing and nutrition security) are urgently needed (the inclusion of minor millets into the PDS system in India is, to that regard, a most illuminating example). These policies will require lobbying from all stakeholders who should be working closely together to achieve the shared vision. One of the propositions that has been

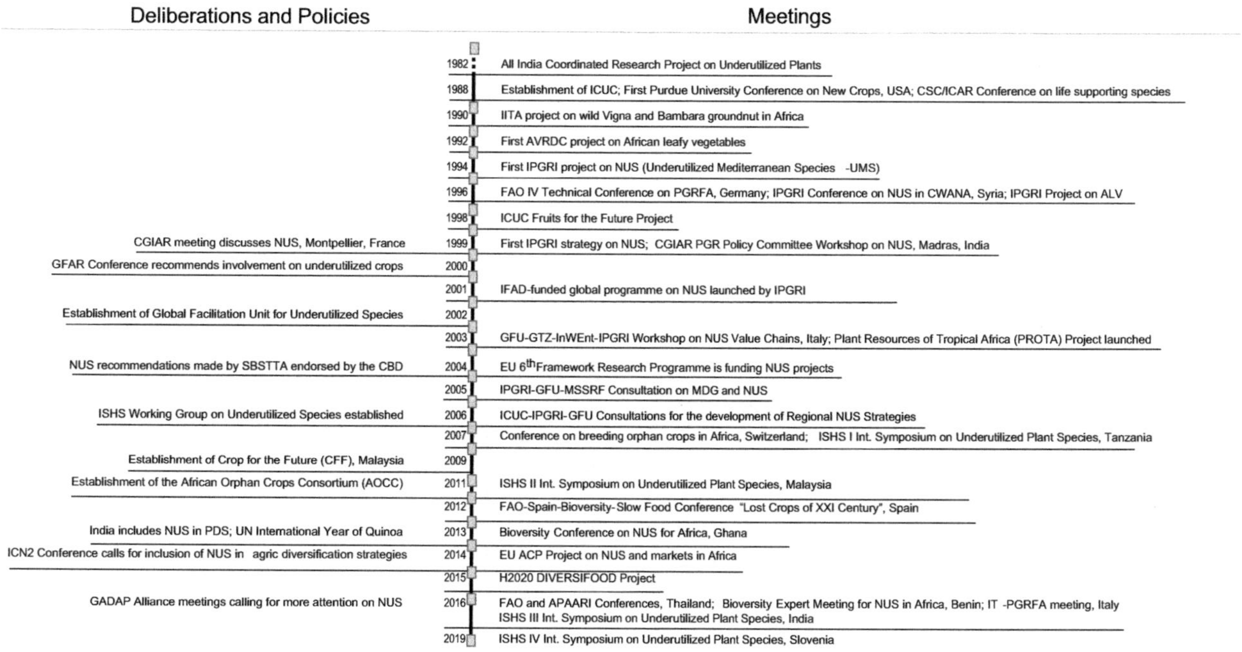

Deliberations and Policies — **Meetings**

Year	Meetings
1982	All India Coordinated Research Project on Underutilized Plants
1988	Establishment of ICUC; First Purdue University Conference on New Crops, USA; CSC/ICAR Conference on life supporting species
1990	IITA project on wild Vigna and Bambara groundnut in Africa
1992	First AVRDC project on African leafy vegetables
1994	First IPGRI project on NUS (Underutilized Mediterranean Species -UMS)
1996	FAO IV Technical Conference on PGRFA, Germany; IPGRI Conference on NUS in CWANA, Syria; IPGRI Project on ALV
1998	ICUC Fruits for the Future Project
1999	First IPGRI strategy on NUS; CGIAR PGR Policy Committee Workshop on NUS, Madras, India
2001	IFAD-funded global programme on NUS launched by IPGRI
2003	GFU–GTZ-InWEnt–IPGRI Workshop on NUS Value Chains, Italy; Plant Resources of Tropical Africa (PROTA) Project launched
2004	EU 6th Framework Research Programme is funding NUS projects
2005	IPGRI–GFU–MSSRF Consultation on MDG and NUS
2006	ICUC-IPGRI-GFU Consultations for the development of Regional NUS Strategies
2007	Conference on breeding orphan crops in Africa, Switzerland; ISHS I Int. Symposium on Underutilized Plant Species, Tanzania
2011	ISHS II Int. Symposium on Underutilized Plant Species, Malaysia
2012	FAO-Spain-Bioversity-Slow Food Conference "Lost Crops of XXI Century", Spain
2013	Bioversity Conference on NUS for Africa, Ghana
2014	EU ACP Project on NUS and markets in Africa
2015	H2020 DIVERSIFOOD Project
2016	FAO and APAARI Conferences, Thailand; Bioversity Expert Meeting for NUS in Africa, Benin; IT -PGRFA meeting, Italy; ISHS III Int. Symposium on Underutilized Plant Species, India
2019	ISHS IV Int. Symposium on Underutilized Plant Species, Slovenia

Deliberations and Policies:

- CGIAR meeting discusses NUS, Montpellier, France
- GFAR Conference recommends involvement on underutilized crops
- Establishment of Global Facilitation Unit for Underutilized Species
- NUS recommendations made by SBSTTA endorsed by the CBD
- ISHS Working Group on Underutilized Species established
- Establishment of Crop for the Future (CFF), Malaysia
- Establishment of the African Orphan Crops Consortium (AOCC)
- India includes NUS in PDS; UN International Year of Quinoa
- ICN2 Conference calls for inclusion of NUS in agric diversification strategies
- GADAP Alliance meetings calling for more attention on NUS

FIGURE 2.1 Key milestones (technical meetings, key deliberations and policies) in the NUS promotion pathway over the last 40 years.

advocated[16] for raising awareness about the importance of NUS among decision-makers, is the launching of an "International Year for NUS" by the United Nations. For this recommendation to translate into reality, care should be given to ensuring that the global promotion of NUS is done with the full participation of local communities, in respect of local needs and guided by sustainability and equity principles.

Notes

1 http://www.cgiar.org/our-strategy/crop-factsheets/
2 http://www.frp.uk.com/project_dissemination_details.cfm/projectID/5562/projectCode/R7187/disID/1965
3 http://www.biodiv.org/convention/articles.asp
4 https://www.cbd.int/agro/pow.shtml
5 http://www.fao.org/ag/agp/agps/GpaEN/gpaact12.htm
6 http://bit.ly/2vVYS8R
7 http://www.fao.org/docs/eims/upload/210543/doneuus02report.pdf
8 Viz: Grant 899 (2007-2010, focusing on Bolivia, Peru, Yemen and India); Grant 1241 (2011-2015, focusing on Bolivia. Nepal and India, supported also by Grant 1434); Grant 2000000-526 (2014-2015, focusing on Guatemala, Mali and India) and Grant 2000000-978 (2016-2020, focusing on Guatemala, Mali and India) and co-funded also by the EU.
9 ftp://ftp.fao.org/docrep/nonfao/ah893e/ah893e00.pdf
10 http://www.underutilized-species.org/documents/PUBLICATIONS/chennai_declaration_en.pdf
11 https://www.nottingham.edu.my/CFFRC/index.aspx
12 https://tkbulletin.wordpress.com/2012/12/12/this-week-in-review-cordoba-seminar-hears-from-fao-director-general-on-value-of-underutilized-crops-sustainable-diets/
13 http://www.nuscommunity.org/fileadmin/templates/nuscommunity.org/upload/documents/Publications/2012_Cordoba-Declaration-on-Promising-Crops-for-the-XXI-Century.pdf
14 http://bit.ly/2wKepWM
15 https://www.thelexicon.org/
16 https://www.bioversityinternational.org/news/detail/spotlight-on-forgotten-crops/#:~:text=Annual%20Report%20story.,This%20year%20neglected%20and%20underutilized%20species%20found%20the%20spotlight:%202013,the%20world's%20%E2%80%9Cforgotten%20crops%E2%80%9D.

Bibliography

(cited in text plus selected papers published over the 50 years deemed particularly relevant for NUS; entries marked with ⋆ include priority setting and/or lists of priority species).

Anonymous (2012) Cordoba Declaration on Promising Crops for the XXI Century. Retrieved on 10 May 2020. http://www.fao.org/fileadmin/templates/food_composition/documents/Cordoba_NUS_Declaration_2012_FINAL.pdf

Anthony, K., Meadley, J. and Röbbelen, G. (1993a) *New Crops for Temperate Regions*. Chapman & Hall, London.

Anthony, K., de Groot, P. and Haq, N. eds. (1993b) *Underutilized Fruits and Nuts in Asia*. CSC, London.

36 Stefano Padulosi and Danny Hunter

Anthony, K., Haq, N. and Cilliers, B. eds. (1995) *Genetic Resources of Underutilized Crops of Southern and Eastern Africa*. FAO/ICUC/CSC, Nelspruit, S. Africa.

Antonelli, A., Fry, C., Smith, R.J., Simmonds, M.S.J. and Kersey, P.J. (2020) *State of the World's Plants and Fungi 2020*. Royal Botanic Gardens, Kew. https://doi.org/10.34885/172

Arora, R.K. (1985) *Genetic Resources of Less Known Cultivated Food Plants*. NBPGR, New Delhi, India.

Arora, R.K. (2014) *Diversity in Underutilized Plant Species: An Asia-Pacific Perspective*. Bioversity International, New Delhi.

AVRDC (2001) *The AVRDC Strategy for 2001–2010*. AVRDC-The World Vegetable Center, Taiwan.

Bala, R., Hoeschle-Zeledon, I., Swaminathan, M.S. and Frison, E. (2006) *Hunger and Poverty: The Role of Biodiversity*. Report of an International Consultation on the Role of Biodiversity in Achieving the UN Millennium Development Goal of Freedom from Hunger and Poverty, Chennai, India, 18–19 April 2005.

Baldermann, S., Blagojević, L., Frede, K., Klopsch, R., Neugart, S., Neumann, A., Ngwene, B., Norkeweit, J., Schröter, D., Schröter, A., Schweigert, F.J., Wiesner, M. and Schreiner, M. (2016) Are Neglected Plants the Food for the Future? *Critical Reviews in Plant Sciences*. doi: 10.1080/07352689.2016.1201399

Begemann, F. (1988) Ecogeographic differentiation of bambara groundnut (*Vigna subterranea*) in the collection of the International Institute of Tropical Agriculture (IITA) Technical University of Munich, Germany.Béné, C., Fanzo, J., Haddad, L. et al. (2020) Five Priorities to Operationalize the EAT–Lancet Commission Report. *Nature Food* 1, 457–459. https://doi.org/10.1038/s43016-020-0136-4

Bioversity International (2014) *Accra Statement for a Food-Secure Africa*. Bioversity International, Rome, Italy, 6 pp.

CGIAR (1999) *Background Papers. Consultative Workshop on Enlarging the Basis of Food Security: Role of Underutilized Species*. Genetic Resources Policy Committee of CGIAR, M.S.S.R.F., Chennai, India.

CGIAR (2001). Mid Term Meeting, Dresden, Germany, May 21–26, 2000: Summary of Proceedings and Decisions. 125 pp.

Chada, M.L. and Lumpkin, T. (2007) Proceedings of the Ist International Conference on Indigenous Vegetables and Legumes: Prospectus for Fighting Poverty, Hunger and Malnutrition. Hyderabad, India, December 12–15, 2006. Acta Hortic, 752.

Chishakwe, N.E. (2008a) *The Role of Policy in the Conservation and Extended Use of Underutilized Plant Species: A Cross-National Policy Analysis*. GFU, Rome, Italy, 31 pp.

Chishakwe, N.E. (2008b) *An Overview of the International Regulatory Frameworks that Influence the Conservation and Use of Underutilized Plant Species*. GFU, Rome, Italy, 31 pp.

Chomchalow, N., Gowda, C.L.L. and Laosuwan, P. eds. (1993) Proceedings of the FAO/UNDP Project RAS/89/040 Workshop on Underexploited and Potential Food Legumes in Asia, 31 October–3 November 1990, Chiang Mai, Thailand. FAO, Bangkok, Thailand.

Chweya, J.A. and Eyzaguirre, P.B. eds. (1999) *The Biodiversity of Traditional Leafy Vegetables*. IPGRI, Rome, Italy.

Dansi, A., Vodouhè, R., Azokpota, P., Yedomonhan, H., Assogba, P., Adjatin, A., Loko, Y.L., Dossou-Aminon, I. and Akpagana, K. (2012) Diversity of the Neglected and Underutilized Crop Species of Importance in Benin. *The Scientific World Journal*, 2012, 1–19. https://doi.org/10.1100/2012/932947

Dawson, I.K. and Jaenicke, H. (2006) *Underutilized Plant Species: The Role of Biotechnology*. Position Paper No. 1. ICUC, Colombo, Sri Lanka, 27 pp.

Dawson, I.K., Guarino, L. and Jaenicke, H. (2007) *Underutilized Plant Species: Impacts of Promotion on Biodiversity*. Position Paper No. 2. ICUC, Colombo, Sri Lanka, 23 pp.

de Groot, P. and Haq, N. eds. (1995) *Promotion of Traditional and Underutilized Crops*. CSC, London.

Dimitri, C. and Richman, N.J. (2000) Organic Foods: Niche Marketers Venture into the Mainstream. *Agricultural Outlook* (June–July), 11–14.

Dinssa, F.F., Hanson, P., Dubois, T., Tenkouano, A., Stoilova, T., Hughes, J. d'A. and Keatinge, J.D.H. (2016) AVRDC – The World Vegetable Center's Women-Oriented Improvement and Development Strategy for Traditional African Vegetables in Sub SaharanAfrica. *European Journal of Horticultural Science* 81(2), 91–105. doi: 10.17660/eJHS.2016/81.2.3.

Durst, P. and Bayasgalanbat, N. eds. (2014) *RAP Publication 2014/07 Promotion of Underutilized Indigenous Food Resources for Food Security and Nutrition in Asia and the Pacific*. Food and Agricultural Organization of the United Nations, Regional Office for Asia and the Pacific, Bangkok, Thailand.

Eyzaguirre, P., Padulosi, S. and Hodgkin, T. (1999) IPGRI's strategy for neglected and underutilized species and the human dimension of agrobiodiversity. In Padulosi, S. (ed.) *Priority Setting for Underutilized and Neglected Plant Species of the Mediterranean Region*. Report of the *IPGRI Conference*, 9–11 February 1998, ICARDA, Aleppo. Syria. International Plant Genetic Resources Institute, Rome, Italy.

FAO (1988) *Traditional Food Plants*. FAO Food and Nutrition Paper 62. FAO, Rome, Italy.

FAO (1995) Leucaena Psyllid: A Threat to Agroforestry in Africa. *Proceedings of Workshop Held in Dar-es-Salaam*, United Republic of Tanzania, 10–24 October 1994. FAO, Rome, Italy.

FAO (1996) *Global Plan of Action for the Conservation and Sustainable Use of Plant Genetic Resources for Food and Agriculture*. FAO, Rome, Italy.

FAO (1998) The State of the World's Plant Genetic Resources for Food and Agriculture (and documents from regional consultations and country reports).

FAO (2010) *The Second Report on the State of the World's Plant Genetic Resources for Food and Agriculture*. Commission on Genetic Resources for Food and Agriculture. Food and Agriculture Organization of the United Nations, Rome, Italy. Retrieved on 15 June 2017. http://www.fao.org/docrep/013/i1500e/i1500e00.pdf

FAO (2011) *Second Global Plan of Action for Plant Genetic Resources for Food and Agriculture* adopted by the FAO Council, Rome, Italy, 29 November 2011. Commission on Genetic Resources for Food and Agriculture.

FAO (2016) *Voluntary Guidelines for Mainstreaming Biodiversity into Policies, Programmes and National and Regional Plans of Action on Nutrition*. FAO, Rome, Italy.

FAO (2019) The State of the World's Biodiversity for Food and Agriculture. In Bélanger, J. and Pilling, D. (eds.) *FAO Commission on Genetic Resources for Food and Agriculture Assessments*. FAO, Rome, Italy, 572 pp.

GFAR 2001. Proceedings of Meeting on Underutilized and Orphan Commodities. May 2000, Dresden, Germany. GFAR, Rome, Italy.

GFAR (2017) Retrieved on 13 March 2020. https://www.gfar.net/sites/default/files/files/Forgotten%20foods%20concept_Sept2017(1).pdf

Giuliani, A. (2007) *Analysis of the Research and Development Activities Involving Underutilized Plant Species Carried Out by the CGIAR Centres*. November 2007. SGRP.

Global Panel on Agriculture and Food Systems for Nutrition (2016) *Food Systems and Diets: Facing the Challenges of the 21st Century*. London. http://glopan.org/sites/default/files/ForesightReport.pdf

Gotor, E. and Irungu, C. (2010) The Impact of Bioversity International's African Leafy Vegetables Programme in Kenya. *Impact Assessment and Project Appraisal* 28(1), 41–55. doi: 10.3152/146155110X488817

Gregory, P., Mayes, S., Chai, H.H., Ebrahim, J., Julkifle, A., Kuppusamy, G., Ho, K., Lin, T., Massawe, F., Syaheerah, T. and Azam-Ali, A. (2019) Crops for the Future

(CFF): An Overview of Research Efforts in the Adoption of Underutilized Species. *Planta* 250. doi: 10.1007/s00425-019-03179-2

Gruère, G., Giuliani, A. and Smale, M. (2006) Marketing underutilized plant species for the benefit of the poor: a conceptual framework. EPTD discussion papers 154, International Food Policy Research Institute (IFPRI).

Gruere, G., Nagarajan, P.L. and King, E.D.I. (2007) Collective Action and Marketing of Underutilized Plant Species: The Case of Minor Millets in Kolli Hills, Tamil Nadu, India. CAPRi Working Papers 69, International Food Policy Research Institute (IFPRI).

Gündel, S., Höschle-Zeledon, I., Krause, B. and Probst, K. (2003) Underutilized Plant Species and Poverty Alleviation. International Workshop on Underutilized Plant Species, Leipzig, Germany, 6 - 8 May, 2003. InWEnt, Zschortau, Germany, and GFU, Rome, Italy

Hernandez-Bermejo, J. and León, J.E. (1992) *Cultivos marginados, otra perspectiva de 1492.* FAO, Rome, Italy

Hunter, D. and Heywood, V. eds. (2011) *Crop Wild Relatives: A Manual of in Situ Conservation.* Earthscan, London, Washington, DC.

Hunter, D., Borelli, T., Beltrame, D., Oliveira, C., Coradin, L., Wasike, V., Wasilwa, L., Mwai, J., Manjella, A., Samarasinghe, G., Madhujith, T., Nadeeshani, H., Tan, A., Ay, S., Güzelsoy, N., Lauridsen, N., Gee, E. and Tartanac, F. (2019) The Potential of Neglected and Underutilized Species for Improving Diets and Nutrition. *Planta.* doi: 10.1007/s00425-019-03169-4

Imrie, B.C., Bray, R.A., Wood, I.M. and Fletcher, R.J. eds. (1997) New Crops, New Products, New Opportunities for Australian Agriculture. Volume 1: Principles and Case Studies. *Proceedings of the First Australian New Crops Conference, 8–11 July 1996,* University of Queensland, Australia. RIRDC, Barton, Australia/97/21.

IPGRI. (1998–2000) *Monographs on Underutilized Crops.* IPGRI, Rome, Italy.

IPGRI (2002) *Neglected and Underutilized Plant Species: Strategic Action Plan of the International Plant Genetic Resources Institute,* Rome, Italy.

Jaenicke, H. and Höschle-Zeledon, I. eds (2006) *Strategic Framework for Underutilized Plant Species Research and Development, with Special Reference to Asia and the Pacific, and to Sub-Saharan Africa.* International Centre for Underutilized Crops, Colombo, Sri Lanka and Global Facilitation Unit for Underutilized Species, Rome, Italy, 33 pp.

Jaenicke, H. (2013) Research and Development of Underutilized Plant Species: Crops for the Future – Beyond Food Security. *Acta Hortic.* 979, 33–44. doi: 10.17660/ActaHortic.2013.979.1

Janick, J. and Simon, J.E. eds. (1990) Advances in New Crops. *Proceedings of the First National Symposium New Crops – Research, Development, Economics,* 23–26 October 1988. Timber Press, Portland, OR and Indianapolis.

Janick, J. and Simon, J.E. eds. (1993) New Crops. *Proceedings of the Second National Symposium New Crops – Exploration, Research, and Commercialisation,* 6–9 October 1991. Wiley, Chichester and Indianapolis.

Janick, J. ed. (1996) Progress in New Crops. *Proceedings of the Third National Symposium: New Crops, New Opportunities, New Technologies,* 22–25 October 1996. ASHS Press, Alexandria, VA and Indianapolis.

Janick, J. ed. (1999). *Proceedings of the Fourth National Symposium New Crops and New Uses: Biodiversity and Agricultural Sustainability.* ASHS Press, Alexandria, VA.

Janick, J. and Whipkey, A. eds. (2002). *Proceedings of the fifth National Symposium New Crops and New Uses: Strength in Diversity.* ASHS Press, Alexandria, VA.

Kahane, R., Hodgkin, T., Jaenicke, H., Hoogendoorn, C., Hermann, M., Keatinge, J.D.H. (Dyno), d'Arros Hughes, J., Padulosi, S. and Looney, N. (2013) Agrobiodiversity for Food Security, Health and Income. *Agronomy for Sustainable Development* 147. doi: 10.1007/s13593-013-0147-8. http://link.springer.com/article/10.1007/s13593-013-0147-8

Kasolo, W., Chemining'wa, G. and Temu, A. (2018) *Neglected and Underutilized Species (NUS) for Improved Food Security and Resilience to Climate Change: A Contextualized Learning Manual for African Colleges and Universities.* ANAFE.

Kour, S., Bakshi, P., Sharma, A., Wali, V., Jasrotia, A. and Kumari, S. (2018) Strategies on Conservation, Improvement and Utilization of Underutilized Fruit Crops. *International Journal of Current Microbiology and Applied Sciences* 7, 638–650. https://doi.org/10.20546/ijcmas.2018.703.075

Leaman, D.J, Fassil, H. and Thormann, I. (1999) *Conserving Medicinal and Aromatic Plant Species: Identifying the Contribution of the International Plant Genetic Resources Institute. Study Commissioned by the International Development Research Centre (IDRC).* IPGRI, Rome, Italy.

Li, X., Siddique, K., Akinnifesi, F., Callens, K., Broca, S., Noorani, A. and Solh, M. (2018) *Future Smart Food-Rediscovering Hidden Treasures of Neglected and Underutilized Species for Zero Hunger in Asia.* Food and Agriculture Organisation of the United Nations, Rome, Itlay (*).

Łuczaj, Ł., Pieroni, A., Tardío, J., Pardo-de-Santayana, M., Sõukand, R., Svanberg, I. and Kalle, R. (2012) Wild Food Plant Use in 21st Century Europe, the Disappearance of Old Traditions and the Search for New Cuisines Involving Wild Edibles. *Acta Societatis Botanicorum Poloniae* 81(4), 359–370. https://doi.org/10.5586/asbp.2012.031

Mal, Bhag (1994) *Underutilized Grain Legumes and Pseudocereals – Their Potentials in Asia.* RAPA/FAO, Bangkok, Thailand.

Maundu, P.M. (1997) The Status of Traditional Vegetable Utilization in Kenya. In: Guarino, L. (ed.) *Traditional African Vegetables. Promoting the Conservation and Use of Underutilized and Neglected Crops.* Institute of Plant Genetics and Crop Plant Research, Gatersleben, Germany/International Plant Genetic Resources Institute, Rome, Italy.

Mbosso, C., Boulay, B., Padulosi, S., Meldrum, G., Mohamadou, Y., Berthe Niang, A., Coulibaly, H., Koreissi, Y. and Sidibé, A. (2020) Fonio and Bambara Groundnut Value Chains in Mali: Issues, Needs, and Opportunities for Their Sustainable Promotion. *Sustainability* 12, 4766.

Meldrum, G., Padulosi, S., Lochetti, G., Robitaille, R. and Diulgheroff, S. (2018) Issues and Prospects for the Sustainable Use and Conservation of Cultivated Vegetable Diversity for More Nutrition-Sensitive Agriculture. *Agriculture* 8, 112.

Morico, G., Grassi, F. and Fideghelli, C. (1999) Horticultural Genetic Diversity: Conservation and Sustainable Utilization and related International Agreements. *Acta Horticulture* 495, 233–244. doi: 10.17660/ActaHortic.1999.495.9

NAS (1975) *Underexploited Tropical Plants with Promising Economic Value.* National Academy of Sciences, Washington, DC

National Research Council (1989) *Lost Crops of the Incas: Little-Known Plants of the Andes with Promise for Worldwide Cultivation.* The National Academies Press, Washington, DC. https://doi.org/10.17226/1398

National Research Council (1996) *Lost Crops of Africa: Volume I: Grains.* The National Academies Press. https://doi.org/10.17226/2305

National Research Council (2006) *Lost Crops of Africa: Volume II: Vegetables.* The National Academies Press. https://doi.org/10.17226/11763

National Research Council (2008) *Lost Crops of Africa: Volume III: Fruits*. The National Academies Press. https://doi.org/10.17226/11879

ODI (1997) Neglected Species, Livelihoods and Biodiversity in Difficult Areas: How Should the Public Sector Respond? Natural Resources Perspective Paper 23, ODI, London, UK.

Padulosi S.; Laghetti, G., Pienaar, B., Ng, N.Q. and Perrino, P. (1991). Survey of wild *Vigna* in Southern Africa. *FAO/IBPGR Plant Genetic Resources Newsletter*, 83(84): 5–8.

Padulosi, S., Ager, H. and Frison, E.A. (1994) Report of the IPGRI Workshop on Conservation and Use of Underutilized Mediterranean Species. Valenzano, Bari, 26–30 March 1994. International Plant Genetic Resources Institute, Rome, Italy.

Padulosi, S. (1997) The Neglected Wild and Cultivated Plant Richness of the Mediterranean. In Monti, L. (ed.) *Proceedings of the CNR International Workshop on "Neglected Plant Genetic Resources with a Landscape and Cultural Importance for the Mediterranean Region"*, 7–9 November 1996, Naples (Italy).

Padulosi, S. (1998) The Underutilized Mediterranean Species Project (UMS): An Example of IPGRI's Involvement in the Area of Underutilized and Neglected Species. Third Regional Workshop of MEDUSA, Coimbra, Portugal 27–28 April 1998. CHIEAM, Crete, Greece.

Padulosi, S. ed. (1999a) *Priority Setting for Underutilized and Neglected Plant Species of the Mediterranean Region*. Report of the *IPGRI Conference*, 9–11 February 1998, ICARDA, Aleppo. Syria. International Plant Genetic Resources Institute, Rome, Italy (*).

Padulosi, S. (1999b) A Comprehensive vs. Limited List of Crops: The Role of Underutilized Crops and Opportunities for International Centres, Donor Communities and Recipient Countries. In Broggio, M. (ed.) *Exploring Options for the List Approach - Proceeding International Workshop: Inter-dependence and Food Security Which List of PGRFA for the Future Multilateral System?* Istituto Agronomico per l'Oltremare, 2 October 1998, Firenze, Italy. (*).

Padulosi, S. (1999c) Criteria for Priority Setting in Initiatives Dealing with Underutilized Crops in Europe. In Gass, T., Frese, F. Begemann, F. and Lipman, E. (compilers) *Implementation of the Global Plan of Action in Europe – Conservation and Sustainable Utilization of Plant Genetic Resources for Food and Agriculture. Proceedings of the European Symposium*, 30 June–3 July 1998, Braunschweig, Germany. International Plant Genetic Resources Institute, Rome.

Padulosi, S. (2000) A Comprehensive vs. Limited List of Crops: The Role of Underutilized Crops and Opportunities for International Centres, Donor Communities and Recipient Countries. Paper presented at the *IAO Workshop on "Interdependence and Food Security: Which List of PGRFA for the Future Multilateral System?"* Istituto Agronomico per l'Oltremare, 1–2 October, Firenze, Italy.

Padulosi, S., Hodgkin, T., Williams, J.T. and Haq, N. (2002) Underutilized Crops: Trends, Challenges and Opportunities in the 21st Century. In Engels, J.M.M., et al. (eds.) *Managing Plant Genetic Resources*. CABI-IPGRI, pp. 323–338. http://eprints.soton.ac.uk/53786/

Padulosi, S. and Hoeschle-Zeledon, I. (2004) Underutilized Plant Species: What Are They? *LEISA* 20(1), 5–6.

Padulosi S., Noun, J.R., Giuliani A., Shuman F., Rojas, W. and Ravi, B. (2003) Realizing the benefits in neglected and underutilized plant species through technology transfer and Human Resources Development. In P.J. Schei, Odd T. Sandlund and R. Strand (eds.) *Proceedings of the Norway/UN Conference on Technology Transfer and Capacity Building*, June 23–27, 2003, Trondheim, Norway, pp. 117–127.

Padulosi, S., Bergamini, N. and Lawrence, T. eds. (2011) On-farm Conservation of Neglected and Underutilized Species: Status, Trends and Novel Approaches to Cope with Climate Change. *Proceedings of an International Conference*, 14–16 June 2011, Friedrichsdorf, Frankfurt. Bioversity International, Rome, Italy.

Padulosi, S., Heywood, V., Hunter, D. and Jarvis, A. (2011) Underutilized Species and Climate Change: Current Status and Outlook. *Crop Adaptation to Climate Change*, 507–521. https://doi.org/10.1002/9780470960 929.ch35

Padulosi, S., Galluzzi, G. and Bordoni, P. (2013) Una Agenda para las species olividadas e infrautilizadas. AMBIENTA no 102 (March 2013) Ministero de Agricultura, Alimentacion e Medio Ambiente, Spain, pp. 26–37. https://cgspace.cgiar.org/handle/10568/35559

Padulosi, S., Amaya, K., Jäger, M., Gotor, E., Rojas, W. and Valdivia, R. (2014) Holistic Approach to Enhance the Use of Neglected and Underutilized Species: The Case of Andean Grains in Bolivia and Peru. *Sustainability* 2014(6), 1283–1312. doi: 10.3390/su6031283. ISSN 2071-1050

Padulosi, S., Bhag Mal, King, O.I. and Gotor, E. (2015) Minor Millets as a Central Element for Sustainably Enhanced Incomes, Empowerment, and Nutrition in Rural India. *Sustainability* 7(7), 8904–8933. doi: 10.3390/su7078904. http://www.mdpi.com/2071-1050/7/7/8904

Padulosi, S., Cawthorn, D.-M., Meldrum, G., Flore, R., Halloran, A. and Mattei, F. (2019) *Leveraging Neglected and Underutilized Plant, Fungi, and Animal Species for More Nutrition Sensitive and Sustainable Food Systems*. In Ferranto, P., Berry, M.E. and Anderson, J.R. Eds. 'Encyclopedia of Food Security and Sustainability'- General and Global Situation. Vol III, 361-370 pp. Elsevier.

Padulosi, S., Roy, P. and Rosado-May, F.J. (2019) *Supporting Nutrition Sensitive Agriculture through Neglected and Underutilized Species – Operational Framework*. Bioversity International and IFAD, Rome, Italy, 39 pp. https://cgspace.cgiar.org/handle/10568/102462

Paroda, R.S. (1988) Life Support Plant Species: Diversity and Conservation. *Proceedings of CSC/ICAR International Workshop on Maintenance and Evaluation of Life Support Species in Asia and the Pacific Region*. NBPGR, New Delhi, India, 4–7 April 1987.

Polar, V. and Flores, P. (2008) Priority Setting for Research on Neglected and Underutilized Species. September 2008. SGRP (★).

Putter, A. ed. (1994) *Safeguarding the Genetic Basis of Africa's Traditional Crops*. CTA/IPGRI.

Quah, S.C., Kiew, R., Bujang, I., Kusnan, M., Haq, N. and de Groot, P. eds. (1996) *Underutilized Tropical Plant Genetic Resources, Conservation and Utilization*. Universiti Pertanian Malaysia Press, Kuala Lumpur.

Raneri, J.E., Padulosi, S., Meldrum, G. and King, O.I. (2019) Promoting Neglected and Underutilized Species to Boost Nutrition in LMICs. In UNSCN Nutrition 44 – Food Environments: Where People Meet the Food System, pp. 10–25. Retrieved on 16 July 2020. https://www.unscn.org/uploads/web/news/UNSCN-Nutrition44-WEB-version.pdf

Rudebjer, P. (2014) Strengthening Capacities and Informing Policies for Developing Value Chains of Neglected and Underutilized Crops in Africa. EU ACP Grant FED/2013/330-241. Bioversity International. Retrieved on 16 July 2020. http://www.nuscommunity.org/fileadmin/templates/nuscommunity.org/upload/documents/Capacity_and_value_chains_project/ACP_CapDev_factsheet_web.pdf

Schippers, R. (2000) *African Indigenous Vegetables – An Overview of the Cultivated Species*. Chatham: Natural Resources Institute/ACP-EU Technical Centre for Agricultural and Rural Cooperation.

Schmelzer, G.H. and Omino, E.A. (2003) Plant Resources of Tropical Africa. *Proceedings of the First PROTA International Workshop*. PROTA Foundation.

Sidibé, A., Meldrum, G., Coulibaly, H., Padulosi, S., Traore, I., Diawara, G., Sangaré, A.R. and Mbosso, C. (2020) Revitalizing Cultivation and Strengthening the Seed

Systems of Fonio and Bambara Groundnut in Mali through a Community Biodiversity Management Approach. *Plant Genetic Resources: Characterization and Utilization* 1–18. doi: 10.1017/S1479262120000076

Smale, M. (1997) The Green Revolution and Wheat Genetic Diversity: Some Unfounded Assumptions. *World Development* 25(8), 1257–1269.

Smartt, J. and Haq, N. eds. (1997). *Domestication, Production and Utilization of New Crops.* International Centre for Underutilized Crops, Southampton.

Tapia, M.E. (1997) *Cultivos Andinos Subexploitades y su aparte a la Alimentacion.* FAO, Santiago, Chile.

Temu, A., Rudebjer, P., Yaye, A.D. and Ochola, A.O. (2016) Curriculum Guide on Neglected and Underutilized Species: Combating Hunger and Malnutrition with Novel Species. *African Network for Agriculture, Agroforestry and Natural Resources Education (ANAFE),* Nairobi, Kenya and Bioversity International, Rome, Italy.

Thies, E. (2000) Promising and Underutilized Species. *Crops and Breeds Eschborn,* July 2000. GTZ.

Tyagi, R., Pandey, A., Agrawal, A., Varaprasad, K., Paroda, R., & Khetarpal, R. (2017). *Regional expert consultation on underutilized crops for food and nutritional security in Asia and the Pacificthematic, strategic papers and country status reports.* Bangkok, Thailand: Asia-Pacific Association for Agricultural Research Institutions (APAARI).Tyagi, R.K., Agrawal, A., Pandey, A., Varaprasad, K.S., Paroda, R.S. and Khetarpal, R.K. (2018a) *Proceedings and Recommendations of 'Regional Expert Consultation on Underutilized Crops for Food and Nutritional Security in Asia and the Pacific'.* Asia-Pacific Association for Agricultural Research Institutions (APAARI), Bangkok, Thailand, 13–15 November 2017, ix + 56p.

Tyagi, R.K., Agrawal, A., Pandey, A., Varaprasad, K.S., Paroda, R.S. and Khetarpal, R.K. (2018b) Regional Expert Consultation on Underutilized Crops for Food and Nutritional Security in Asia and the Pacific – Thematic, Strategic Papers and Country Status Reports. Asia-Pacific Association for Agricultural Research Institutions (APAARI), Bangkok, Thailand, 13–15 November 2017, x+349 p.

Ulian, T., Hudson, A., Mattana, E. and Hechenleitner, P. compilers (2015). Project MGU – The Useful Plants Project Review Workshop Report, 22–24 July 2014. Royal Botanic Gardens, Kew.

Ulian, T., Diazgranados, M., Pironon, S., Padulosi, S., Liu, U., Davies, L., Howes, M.-J.R., Borrell, J., Ondo, I., Pérez-Escobar, O.A., Sharrock, S., Ryan, P., Hunter, D., Lee, M.A., Barstow, C., Łuczaj, Ł., Pieroni, A., Cámara-Leret, R., Noorani, A., Mba, C., Womdim, R.N., Muminjanov, H., Antonelli, A., Pritchard, H.W. and Mattana, E. (2020) Unlocking Plant Resources to Support Food Security and Promote Sustainable Agriculture. *People Plants Planet* 2(5), 421–445. ISSN: 2572-2611.

Vietmeyer, N. (2008) Underexploited Tropical Plants with Promising Economic Value: The Last 30 Years. *Tree for Life Journal* 3, 1.

Wickens, G.E., Haq, N. and Day, P. (1989) *New Crops for Food and Industry.* Chapman and Hall, London.

Will, M. (2008) *Promoting Value Chains of Neglected and Underutilized Species for Pro-Poor Growth and Biodiversity Conservation.* GFU, Rome, Italy, 109 pp.

Williams, J.T., ed. (1993) *Underutilized Crops: Pulses and Vegetables.* Chapman & Hall, London.

Williams, J.T., ed. (1995) *Underutilized Crops: Cereals and Pseudocereals.* Chapman & Hall, London.

Williams, J.T. and Haq, N. (2000) *Global Research on Underutilized Crops. An Assessment of Current Activities and Proposals for Enhanced Cooperation.* ICUC, Southampton.

3

MAINSTREAMING NUS FOR NUTRITION-SENSITIVE AGRICULTURE

A holistic approach

Stefano Padulosi, Gennifer Meldrum,
Shantanu Mathur and Danny Hunter

Introduction

Neglected and underutilized species (NUS) can be strategic allies in developing nutrition-sensitive agricultural food systems and value chains. This statement is supported by the numerous benefits linked to these resources, as highlighted in Chapter 1 (Fanzo et al., 2013; Baldermann et al., 2016; Bioversity International, 2017; Fanzo, 2019; Hunter et al., 2020). The central message reiterated throughout this book is that the diversification of food and production systems is urgently needed to adapt to climate change and fight malnutrition in all its forms, and to that regard, NUS offer a tremendous opportunity for impact. How to translate this vision into reality is not an easy task as many barriers need to be addressed, including the lack of or very poor germplasm conserved in *ex situ* gene banks, lack of improved varieties, poor seed systems, low competitiveness in production systems and markets and little awareness from consumers of their livelihood benefits. Tools to assist practitioners, and more generally all actors, involved in the use-enhancement and promotion of NUS are needed. As an answer to this call, in 2019, Bioversity and IFAD published an operational framework to assist experts of IFAD and other agencies at the national and international levels, in promoting the wider use of biodiversity, especially NUS, to improve the lives of people, particularly the more vulnerable members of society (Padulosi et al., 2019). This document complements other valuable strategic papers recently published (e.g., Hunter et al., 2015; Kennedy et al., 2017; Romanelli et al., 2020), briefly presented later in the chapter.

The scope of this chapter is to present this framework and enrich it with reflections prompted by the 2020 global COVID-19 pandemic, for which we believe NUS can prove to be a valuable instrument to strengthen the resilience of local communities, their weakened economies and food systems.

DOI: 10.4324/9781003044802-4

44 Stefano Padulosi et al.

Key messages we would like to convey through this chapter include:

- The deployment of local crops in food and production systems is a valid contribution in addressing malnutrition and making food systems more resilient and adapted to changes such as those brought about by the current pandemic.
- NUS are highly adapted to agro-ecological niches and marginal areas and are able to withstand climatic stresses. These traits are of great value to feed the world's growing population and, as such, should be better recognized and leveraged in support of agricultural strategies and development plans.
- The enhanced use of NUS offers important opportunities for income generation and empowerment of both rural and urban communities, benefiting also vulnerable members of society.
- Bringing NUS out of their condition of marginalization requires a viable market that creates incentives for local value-chain actors, to secure their continued cultivation, processing and marketing. Upgrading NUS value-chains must, therefore, be a central element of their overall promotion strategy.
- A holistic value-chain approach is needed for the promotion of NUS. 'Siloed' formulas focusing on specific aspects (e.g., only on the conservation, cultivation or marketing), as typically practiced for commodity crops, cannot work for NUS. These resources have been for too long at the margins of research and development, and multiple technological gaps need to be filled. The holistic approach we are advocating for NUS should be interdisciplinary, multi-sectorial, highly participatory and gender sensitive.
- Nutrition education and behavioral change communication are key elements in the design of projects addressing NUS.
- Bringing back NUS to the table cannot be achieved without consistent investment in human resources of National Agricultural Research Systems (NARS). Capacity building and training are urgently needed if we are to seize the multiple opportunities linked to NUS and build more resilient food systems.

Nutrition-sensitive agriculture and value chains

According to FAO, nutrition-sensitive agriculture is "a food-based approach to agricultural development that puts nutritionally rich foods, dietary diversity, and food fortification at the heart of overcoming malnutrition and micronutrient deficiencies" and nutrition-sensitive value chains are those that

> leverage opportunities to enhance supply and/or demand for nutritious food, as well as opportunities to add nutritional value (and/or minimize food and nutrient loss) at each step of the chain, thereby improving the availability, affordability, quality and acceptability of nutritious food. For lasting impacts on nutrition, this approach must be placed in a sustainability context as well.
>
> *(FAO, 2014a,b)*

Numerous publications have addressed issues and the strategies needed to make agriculture and value chains more nutrition sensitive (e.g., Jaenicke and Virchow, 2013; Gelli et al., 2015; FAO, 2017; De la Peña and Garrett, 2018). Among these, worth mentioning is the outcome of the Second International Conference on Nutrition (ICN2), which specifically calls upon governments to "[P]romote the diversification of crops including underutilized traditional crops, more production of fruits and vegetables, and appropriate production of animal-source products as needed, applying sustainable food production and natural resource management practices"[1] (FAO and WHO, 2014).

While referring to basic concepts contained therein, through this chapter we will focus exclusively on interventions to leverage the value of NUS for better attaining these goals.

In the NUS operational framework published by Bioversity and IFAD, we indicate five basic steps for making a project focusing on NUS more nutrition-sensitive, viz:

- Step 1. Explicitly incorporate improved nutrition into the project's outcomes and integrate nutrition-relevant indicators into the project logical framework
- Step 2. Include a situation analysis on the nutrition context, addressing nutrient gaps of the targeted beneficiaries
- Step 3. Include nutrient-dense NUS in agricultural development interventions to complement the nutritional role played by staple crops, guided by a diversity-based sustainable diet approach
- Step 4. Trace the impact pathway – i.e., the steps from breeding and production to seed systems and consumption needed for the intervention – to improve nutrition along the NUS value chain. Design and implement project actions that will affect that pathway in a systematic way. For example, determine if a change in dietary habits is needed to encourage the consumption of some wild edibles and, if so, implement actions to promote such changes. Allocate dedicated financial resources to implementing nutrition-sensitive activities
- Step 5. Through policy engagement and partnerships, address opportunities and constraints that affect the pathway and the effectiveness of the intervention, such as the institutional environment, gender and/or environmental sustainability, and define implementing arrangements for the delivery of nutrition-sensitive activities.

The central element of efforts needed to mobilize NUS in support of nutrition-sensitive agriculture is the holistic value-chain, as represented in Figure 3.1. This approach consists of highly interlinked interventions in each of the segments of a typical value chain for agricultural products, namely input supply, production, harvest, processing, marketing and final consumption. For those products realized by farmers exclusively for home consumption, marketing interventions would not be necessary. The holistic value-chain approach has

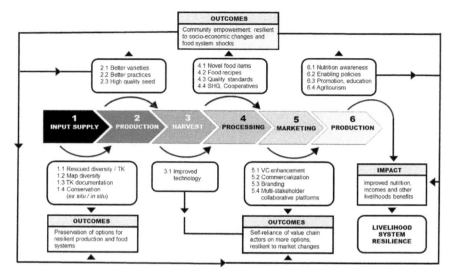

FIGURE 3.1 Holistic value-chain approach (from Padulosi et al., 2019).

been designed to tackle simultaneously three key challenges: climate change, nutrition and the economic insecurity of local communities. As shown in Figure 3.1, the synergistic interaction among the outcomes resulting from these multiple interventions is expected to ultimately strengthen the resilience of livelihoods as a whole, which should be the ultimate goal of our efforts. A description of the key entry points along the value chains in order to realize such an impact with regard to NUS is provided in the following section.

Entry points to strengthen nutritional outcomes in NUS value chains

Area 1: input supply

This is the first entry point that is often overlooked by agricultural development projects, interested more in the income-generation dimension with little or no attention paid to the inter- and intra-specific diversity needed to strengthen nutrition. Our experience is that too many rural development projects focus minimally (or not at all) on plant genetic resources. Most often, attention is dedicated to only a handful of commodity crops (and few varieties therein) with the intent of boosting agricultural food production and fighting poor nutrition, neglecting the wealth of wild and cultivated species that can be mobilized for climate-smart and nutrition-sensitive agriculture. In such a segment of the value chain, the following interventions are, thus, highly recommended:

- Include NUS in the design of projects (especially species that might identify key nutrient gaps that cause local malnutrition problems) and ensure that their diversity is well represented at both inter- and intra-specific level. Provision of germplasm should be accompanied also by enhanced cultivating practices, where both scientific and traditional knowledge should be leveraged. This is very relevant for NUS, whose knowledge is highly vulnerable due to the abandonment of cultivation. The selection of species to focus upon should be carried out in a highly participatory fashion, involving all stakeholders and giving ample space to women, who are the strategic nexus between biodiversity and nutrition in households (King and Padulosi, 2017).

- An assessment of the nutrition needs and seasonal dynamics of hunger in target communities should be also carried out in conjunction with the previous activity so as to develop an inventory of local food species (wild and cultivated) to allow all players to gain a good understanding of spectrum of species that can be potentially targeted by the project. The result of these surveys will be then used to develop seasonal calendars that can guide local populations in making best use of existing overlooked nutritious food species and to strengthen their capacity in fighting nutrition insecurity, especially during the 'hungry season' (see also Chapter 14).

- Support an *ex situ/in situ* integrated conservation of the genetic resources of NUS and their associated traditional knowledge. This is a fundamental intervention that must be factored into projects if we are to secure continued access to the genetic diversity of NUS by user communities and safeguard, at the same time, associated relevant practices (agronomic, value addition, food preparation, ceremonies, rituals, etc.) that allow the cultivation, use and appreciation of these traditional resources across generations (Padulosi et al., 2012).

- Encourage visibility of NUS in local and regional media such as newspapers, radio and TV programs, as well as social media. Great opportunities exist in this regard by involving renowned chefs who can help remove the 'food of the poor' stigma associated with many NUS (Amaya et al., 2019).

- Promote the use of NUS in schools and universities through curriculum development —describing the multiple values of NUS — and associated activities such diverse school gardens (for more in this regard, see Rudebjer and Padulosi, 2018; Hunter et al., 2020). The agronomists of the future need to be informed on the value of crop diversification for supporting nutrition security and, at the same time, as future consumers, need to be aware of the multiple values of NUS, too often disregarded by younger generations as the food of the poor. Campaigns on the importance of NUS should be done in an enticing fashion, that is, by linking NUS to nutrition and health benefits as well as to other important benefits, such as the sustainability of our planet, keeping options for the future, building resilient food systems in difficult times (including those experienced during the COVID-19 pandemic) and maintaining cultural and territorial identity. Lastly, consumers should all be made aware that by bringing NUS back to the table, there will be more options for tastier, more exciting, more attractive and more enjoyable foods!

48 Stefano Padulosi et al.

- By nature of their definition, NUS are largely undervalued by the scientific community and, as such, there is often limited data available on their potential, such as micronutrient content or climate-change resilience. In cases where micronutrient content is unknown, it can be estimated with a similar food or categorized according to the common nutritional traits of its food group characteristics (e.g., dark green leafy vegetables can contain iron and vitamin A). Local and scientific knowledge associated with prioritized species is essential, as some may have toxic or anti-nutrient properties, and must be consumed in small amounts or be processed in special ways to reduce toxin content.

Area 2: production

Once NUS species have been selected, the second entry point for nutrition along the value chain is related to production activities, such as selecting better varieties, developing better practices and improving seed. In order to support these activities, decentralized seed-selection and production systems should be pursued (de Boef et al., 2013; Vernooy et al., 2015). Increased use of NUS should be promoted to help farmers take maximum advantage of the hardiness of these species, especially in vulnerable, drought-prone areas where such crops are able to exploit residual soil moisture and scarce rainfall regimes.

NUS should be included by projects and research managers in crop selection programs/participatory variety selection, and national and international players should cooperate to build up a knowledge base on the performance of these species, which is currently lacking. More studies should be carried out on their agronomic performance, including climate-change adaptation, which will allow for comparisons with staple crops. Such a process should involve all value-chain actors and encourage the active participation of women, Indigenous Peoples and youth. Evaluation activities should go hand-in-hand with the characterization of germplasm for nutritional content to support the selection of resilient, nutritious and economically interesting species (Shanthakumar et al., 2010).

More efforts should be deployed in improving cultivation practices for NUS, aiming at the elimination of drudgery in the field and the introduction of innovative cultivation approaches based on the blending of traditional knowledge and scientific findings (Padulosi et al., 2015).

Area 3: harvest

Improved harvesting methods and storage are extremely relevant for NUS, especially for perishable, nutritious-dense fruits and vegetables. In view of the great progress achieved in these disciplines for commodity crops, there is ample scope for leveraging such knowledge for the benefit of NUS. Our experience has often demonstrated that modest investments in support of innovation to improve

NUS for nutrition-sensitive agriculture **49**

harvest operations can be highly valuable, as in the case of amaranth, where a simple modification in the sieve of the harvesting machine used by farmers in Peru has resulted in a significant reduction of grain loss during harvest operations (Padulosi et al., 2014).

Area 4: processing

Improved processing can play a key role in making nutritious NUS foods more convenient, increasing their shelf life and facilitating transportation and storage. However, processing enhancement of NUS should be done in ways that respect their nutritional value and do not prioritize convenience or aesthetic values only. For instance, in the case of fonio and minor millets, polishing operations have been observed to eliminate important micro-nutrients. Grains will look more appealing to consumers, but will be of reduced nutritional value. Improved processing can be achieved by building on Indigenous technology and local wisdom and/or by blending it with scientific findings. Our work on minor millets in India has also taught us that simple, inexpensive processing technology mobilized by local communities has played a strategic role in boosting local consumption of these crops at the household level, while also supporting women-led cottage industries and local entrepreneurship, and has contributed to building greater self-esteem and empowerment among vulnerable women groups (King and Padulosi, 2017).

Area 5: marketing

Marketing aspects for NUS are crucial to bringing about sustainability in their use enhancement. Nutritious local crops should be effectively marketed if we are to strengthen nutrition-sensitive agriculture and value-chains. A highly strategic component of this domain of action (but equally relevant also for the other areas) has to do with making food more diverse and healthier and, at the same time, delicious and appealing to the consumer, but doing so in ways that involve and incentivize all actors in the food systems, taking into account multiple agendas and values (Béné et al., 2020). Some of the actions related to marketing NUS more effectively include:

- Supporting the identification and promotion of nutrition-friendly packaging solutions, including approaches accessible by local vendors and small markets. Strengthen collectives (e.g., cooperatives, farmer-producer companies, confederations of producers/MSMEs) that are required to build reliable distribution systems for NUS that are typically produced locally and by farmers not well-linked to other value-chain actors.
- Dissemination of nutrition information as well as any other information on useful traits present in NUS (including resilience) for promoting greater adoption of these species by value-chain actors.

50 Stefano Padulosi et al.

Area 6: consumption

This is the last segment of the value chain and the one most strategic in influencing the wider utilization of NUS purchased in markets or self-produced/bartered and consumed in farmers' households. Key actions at this stage include:

- Building greater awareness among consumers on the importance of NUS for crafting more nutritious and healthy diets.
- Developing products, and marketing healthy and sustainable options, as appealing and delicious, rather than focusing on messages of health, sustainability or abstemiousness (Sunstein, 2015; Vermeulen et al., 2020).
- Supporting the development of more attractive food recipes and food products that have NUS as ingredients, by organizing cooking sessions with community members and forming alliances between growers, value-chain actors, chefs and food movements to promote local foods that are also culturally appealing; and challenging niche-identity associations of these products through marketing and branding.
- Raising awareness of decision-makers to get their buy-in regarding NUS in government actions and seeking their commitment to developing enabling policies that mainstream NUS into agricultural development programs. Policy-makers should be engaged from the start of NUS projects in order to solicit a sense of ownership in the promotion of NUS. Among the policy interventions for NUS, the most strategic are those favoring their inclusion in procurement programs (e.g., targeting school feeding programs) (Hunter et al., 2019), or those regarding the inclusion of NUS in school/university curricula, which can have important multiplier effects (Rudebjer and Padulosi, 2018; Amaya et al., 2019; Hunter et al., 2020).
- Making healthy and sustainable options highly visible and increasing the quantity of them available in menus, supermarkets, at catered events, schools, canteens or other public places (Vermeulen et al., 2020).
- Supporting food festivals and using these occasions to recognize the invaluable work done by custodian farmers, Indigenous Peoples, women and youth organizations for safeguarding the wealth of NUS and traditional knowledge, and, at the same time, support a better understanding of the Indigenous approach to marketing.
- Introducing marketing intelligence systems that allow the sharing of information on offer and demanding estimates via mobile phones across value-chain actors, as successfully practiced in Kenya and other African countries. Such an approach should be complemented by the development of mobile apps for local communities that can provide information about where to find NUS in the community, how to use them and how to buy and sell them. These apps could be particularly helpful in situations when people's movements are limited, such as conditions experienced during the COVID-19 pandemic. Promoting local foods through convenient mobile apps can the winning

NUS for nutrition-sensitive agriculture **51**

card for bringing NUS to the masses; to that end, it could be useful to replicate (or join) currently available systems that link consumers with farmers markets (see f.e. apps like Farmstand and Locavore).[2]

Cross-cutting issues

In addition to the entry points mentioned above, there are interventions that can be considered that cut across the entire value chain. The NUS operational framework (Padulosi et al., 2019) stresses, in particular, those related to capacity building and empowering vulnerable groups.

Capacity building

Scaling up best practices to more effectively use NUS will need a robust investment in capacity building that benefits different stakeholders and covers highly interlinked themes, including adaptation to climate change of agricultural production, food and nutrition security, value chains and marketing issues, raising public awareness and policy advocacy. Strengthening the governance and technical capacity of farmer associations, especially Indigenous food community associations, will need to be pursued as part of a broader public–private partnership initiative and investment goal in the agricultural marketing of NUS.

Dealing with NUS calls for a broad interdisciplinary understanding of their roles, which is too often not registered among R&D experts and rural advisory services in most NARS (Raneri et al., 2019). During project design we should, thus, create opportunities for an interdisciplinary dialogue among stakeholders where knowledge about NUS is provided and shared. These interactions will allow all stakeholders to become acquainted with the roles of NUS, especially regarding their resilience, nutritional values and income-generation opportunities and understand the importance of building synergy across such disciplines for sustained impact (Ulian et al., 2020). These participatory reviews can be supported through *ex ante* analysis to capture existing values of NUS in the socioeconomic context of the target areas through available data and food system modelling.

Strengthening the capacities of community members, including women, youth and Indigenous Peoples, is needed in order toto boost production efficiency, improve post-production, technologies, business and entrepreneurship skills as well as enhance markets and market information, sustainable investments in physical infrastructure and access to produce.

With regard to access to markets, the networks of local, weekly markets present in most countries can be leveraged to sell NUS. This will be sustainable only if NUS marketing bottlenecks are effectively addressed, such as the possible isolation of production areas from marketplaces and the existence of poor infrastructure for storage, processing and packaging. Specialized technical

52 Stefano Padulosi et al.

assistance for NUS products and market development, along with agri-business capacity-building and mentoring of producer groups, should be also pursued. Innovative technologies for NUS processing and their commercial viability need to be tested in a commercial context to promote scalability.

The promotion of wild species should be done with caution so as to avoid overharvesting; domestication can be an effective intervention to mitigate that risk (Padulosi et al., 2002). In view of the considerable efforts needed in the domestication process (long-term field and laboratory experiments, marketing research, etc.), it should receive the support of government research institutions, universities and the private sector and could also benefit from being included in agricultural and rural development projects (Heywood, 1999).

Locally available biodiversity may not be able to address all the issues related to poor diet quality and malnutrition. A combination of that, along with other approaches, including the introduction of nutritious species from elsewhere, may be required to fill possible gaps.

Cross-collaboration between agricultural extension and rural health services can be effective for promoting resilient and nutritious NUS in diets and production systems. Training extension agents on the agronomy, marketing and nutrition aspects of NUS is a strategic way to leverage their role for the promotion of NUS (Raneri et al., 2019). Strengthening their familiarity with the wild and semi-domesticated plants in the local environment will support their role in assisting communities in the identification of priority species. Understanding local diets and nutrition issues in target populations supports the prioritization process. Sensitization on local consumption preferences and perceptions, and how to engage with farmers to collect this type of information, is also important for understanding demand-side issues. Building the capacity to recognize important actors in the value-chain and establishing a proactive attitude to reach out and engage these actors to overcome bottlenecks will help advance the use-enhancement of these nutritious species.

The dearth of data on crop adaptation can be addressed through the use of participatory varietal selection done in farmers' fields. Crowdsourcing can be a highly effective way to select the best varieties whose performance is assessed in the context of where the need is, and not at the research stations (van Etten et al., 2017).

The use-enhancement of NUS can be a cost-effective and culturally appropriate way to improve the resilience of local food systems, as well as farming household incomes and nutrition. NUS can be integrated into existing extension programs that focus on commodity species. For example, NUS of fruits and vegetables should be considered in any project working to increase the availability and marketing of major staples (Padulosi et al., 2018).

Empowerment of vulnerable groups

The use-enhancement of NUS is a robust way to empower women, Indigenous Peoples, rural youth and other vulnerable groups who depend on these species for their livelihoods. Economic empowerment should be at the core of these actions.

This includes designing interventions targeted at strengthening community institutions and farmer groups as a tool for empowering marginalized communities. Groups should be supported in the areas of adaptive extension, leveraging local agrobiodiversity, market access, agro-processing and value addition, paying particular attention to business skills development and market orientation. These actions should be designed through a culturally sensitive lens. Use-enhancement of NUS along the value chain should be driven by equity considerations and the inclusion of vulnerable groups (Rosado-May et al., 2018).

Additional frameworks and policies that enhance the mainstreaming of NUS value-chains for improved nutrition

While Indigenous Peoples and local communities have been nurturing and using NUS for many generations, it is only in the more recent post-Green Revolution period that NUS have become a focus of global policy in general (see Chapter 1) and, only even more recently, that they have been considered a portfolio of foods that can substantially contribute to more sustainable solutions to global malnutrition. The operational framework supporting nutrition-sensitive agriculture through the greater use of NUS described in this chapter is an additional resource to the growing basket of tools that aims to mainstream NUS into more projects, programs and investments for improved nutrition, and which are important instruments for strengthening collaboration between the agriculture, nutrition and health and environmental sectors.

A major milestone in the recognition of the important role NUS, and biodiversity more broadly, can play in improving nutrition and human health was the establishment of the Framework for a Cross-Cutting Initiative on Biodiversity for Food and Nutrition in 2006, within the Convention on Biological Diversity (CBD). In 2004, the CBD's Conference of the Parties (COP) formally recognized the linkages between biodiversity, food and nutrition, and the need to enhance sustainable use of biodiversity to combat hunger and malnutrition. The COP requested the CBD's Executive Secretary, in collaboration with FAO and the former International Plant Genetic Resources Institute – now Bioversity International – to undertake a cross-cutting initiative on biodiversity for food and nutrition, which was adopted by the COP in 2006 (WHO/CBD, 2015). The recently completed GEF-funded Biodiversity for Food and Nutrition Project has been a major vehicle for the implementation of this cross-cutting initiative and has demonstrated, at a country level, how to prioritize NUS for mainstreaming into relevant policies, programs and markets so as to improve nutrition. It has also developed a range of methods, tools and other resources to support mainstreaming (Hunter et al., 2020; see Chapter 13).

Around the same time, the Commission on Genetic Resources for Food and Agriculture (CGRFA) also requested that FAO evaluate the relationship between biodiversity and nutrition. In 2005, via the Intergovernmental Technical Working Group on Plant Genetic Resources for Food and Agriculture, eight high-priority actions and six lower-priority actions were identified. Subsequently, the CGRFA, at its 14th session in 2013, formally recognized nutrients and diets, as well as food, as ecosystem services, in order to further increase the awareness of human nutrition as a concern for the environmental sector, and the awareness among human nutritionists of the importance of biodiversity. It also requested the preparation of guidelines for mainstreaming biodiversity into all aspects of nutrition, including education, interventions, policies and programs (WHO/CBD, 2015). The *Voluntary Guidelines for Mainstreaming Biodiversity into Policies, Programmes and National and Regional Plans of Action on Nutrition* (FAO, 2015) was adopted at the 15th Session of the CGRFA in 2015 and provides a framework to mobilize NUS for improving nutrition.

The Second International Conference on Nutrition (ICN2), jointly convened by the FAO and the WHO in 2014, focused on policies aimed at eradicating malnutrition in all its forms and transforming food systems to make nutritious diets available to all. Participants at ICN2 endorsed the Rome Declaration on Nutrition and the Framework for Action, with Recommendation 10 being particularly for the NUS community (ibid. p. 45). "[P]romote the diversification of crops including underutilized traditional crops, more production of fruits and vegetables, and appropriate production of animal source products as needed, applying sustainable food production and natural resource management practices".

In 2015, the WHO and CBD jointly published a "state of knowledge" review, *Connecting Global Priorities: Biodiversity and Human Health* (WHO/CBD, 2015), which contained a chapter dedicated to NUS, biodiversity and nutrition (Hunter et al., 2015). This was followed, in 2017, by the publication by Bioversity International of the *Mainstreaming Agrobiodiversity in Sustainable Food Systems* report (Bioversity International, 2017), which included a chapter on the importance of NUS and food biodiversity for healthy, diverse diets (Kennedy et al., 2017). Both publications highlight approaches and opportunities to better mainstream NUS in order to improve nutrition and diets. Most recently, the WHO has published further *Guidance on Mainstreaming Biodiversity for Nutrition and Health*, which focuses on the importance of NUS and their value chains (Romanelli et al., 2020).

Collectively, this presents an ever-improving, enabling policy environment for the promotion of NUS, especially in the context of addressing malnutrition challenges. However, as Chapter 1 and other chapters of this book highlight, there are still many constraints and barriers on both the supply and demand side that need to be addressed. One such need that is urgent is better collaboration across the NUS community, not just in the agriculture sector but with others who champion NUS in sectors such as health, environment and education. Greater coherence between global and national policies is also needed to ensure that the NUS community collaborates more effectively in developing and implementing projects and investments, to ensure the most efficient use of limited resources.

Conclusions

The importance of promoting NUS for making agriculture and value chains more effective in supporting nutrition has been particularly appreciated over the last decade, and the recent COVID-19 pandemic has made even more apparent the resilience role that NUS can play in buffering communities against both supply and demand shocks. The world needs NUS and we need societies to steer food systems in the right direction, a pathway that is sustainable for our own lives and for the planet.

Notes

1 Recommendation 10 of ICN2 2014/3 Corr.1
2 More example at https://foodtank.com/news/2015/01/harvesting-the-best-10-apps-for-healthy-organic-shopping/

References

Amaya, N., Padulosi, S. and Meldrum, G. (2019) Value Chain Analysis of Chaya (Mayan Spinach) in Guatemala. *Economic Botany* 1–15. doi: 10.1007/s12231-019-09483-y
Baldermann, S., Blagojević, L., Frede, K., Klopsch, R., Neugart, S., Neumann, A., Ngwene, B., Norkeweit, J., Schröter, D., Schröter, A., Schweigert, F.J., Wiesner, M. and Schreiner, M. (2016) Are Neglected Plants the Food for the Future? *Critical Reviews in Plant Sciences.* doi: 10.1080/07352689.2016.1201399
Béné, C., Fanzo, J., Haddad, L. et al. (2020) Five priorities to operationalize the EAT–Lancet Commission Report. *Nature Food* 1, 457–459. doi: 10.1038/s43016-020-0136-4
Bioversity International (2017) *Mainstreaming Agrobiodiversity in Sustainable Food Systems: Scientific Foundations for an Agrobiodiversity Index.* Bioversity International, Rome, Italy, 157 pp.
de Boef, W.S., Subedi, A., Peroni, N., Thijssen, M.H. and O'Keeffe, E. (eds.) (2013) *Community Biodiversity Management: Promoting Resilience and the Conservation of Plant Genetic Resources.* Issues in agricultural biodiversity, Earthscan, New York. doi:10.4324/9780203130599.
De la Peña, I. and Garrett, J. (2018) *Nutrition-Sensitive Value Chains, A Guide for Project Design* (Vol I and Vol II). IFAD. https://bit.ly/2PWtTzV and https://bit.ly/2D8qoBf
Fanzo, J. (2019) Biodiversity: An Essential Natural Resource for Improving Diets and Nutrition. In *Agriculture for Improved Nutrition: Seizing the Momentum.* Chapter 4. Fan, Shenggen, Yosef, Sivan and Pandya-Lorch, Rajul (Eds.). International Food Policy Research Institute (IFPRI) and CABI, Wallingford.
Fanzo, J., Hunter, D., Borelli, T. and Mattei, F. (2013) *Diversifying Food and Diets: Using Agricultural Biodiversity to Improve Nutrition and Health.* Issues in Agricultural Biodiversity, Earthscan/Routledge, London.
FAO (2014a) http://www.fao.org/3/a-as601e.pdf. Retrieved on 11 July 2020.
FAO (2014b) *Developing Sustainable Food Value Chains – Guiding Principles.* Rome, Italy.
FAO (2015) *Voluntary Guidelines for Mainstreaming Biodiversity Into Policies, Programmes and National and Regional Plans of Action on Nutrition.* FAO, Rome, Italy.
FAO (2017) *Nutrition-Sensitive Agriculture and Food Systems in Practice. Options for Intervention.* Food and Agricultural Organization of the United Nations, Rome, Italy.

FAO and WHO (2014) *Outcome of the Second International Conference on Nutrition (ICN2). Framework for Action.* Food and Agricultural Organization of the United Nations, Rome, Italy, 19–21 November 2014.

Gelli, A., Hawkes, C., Donovan, J., Harris, J., Allen, S.L., De Brauw, A., Henson, S., Johnson, N., Garrett, J. and Ryckembusch, D. (2015) *Value Chains and Nutrition: A Framework to Support the Identification, Design, and Evaluation of Interventions.* Documento de debate 01413 del IFPRI. IFPRI, Washington, DC.

Heywood, V.H. (1999) *Use and Potential of Wild Plants in Farm Households.* Food and Agriculture Organization of the United Nations, Rome, Italy. 113 pp.

Hunter, D., Borelli, T., Beltrame, D., Oliveira, C., Coradin, L., Wasike, V., Mwai, J., Manjella, A., Samarasinghe, G., Madhujith, T., Nadeeshani, H., Tan, A., Tuğrul Ay, S., Güzelsoy, N., Lauridsen, N., Gee, E. and Tartanac, F. (2019) The Potential of Neglected and Underutilized Species for Improving Diets and Nutrition. *Planta* 250(3), 709–729. doi: 10.1007/s00425-019-03169-4

Hunter, D., Borelli, T. and Gee, E. (2020) *Biodiversity, Food and Nutrition: A New Agenda for Sustainable Food Systems.* Issues in Agricultural Biodiversity, Earthscan from Routledge London and New York.

Hunter, D., Burlingame, B. and Remans, R. et al. (2015) Biodiversity and nutrition. In *Connecting Global Priorities: Biodiversity and Human Health, a State of Knowledge Review.* (World Health Organization and Secretariat of the Convention on Biological Diversity, 2015).

Jaenicke, H. and Virchow, D. (2013) Entry Points Into a Nutrition-Sensitive Agriculture. *Food Security* 5, 679–692. doi: 10.1007/s12571-013-0293-5

Kennedy, G., Stoian, D., Hunter, D., Kikulwe, E., Termote, C., Alders, R., Burlingame, B., Jamnadass, R., McMullin, S. and Thilsted, S. (2017) Food Biodiversity for Healthy, Diverse Diets. In de Boef, W., Haga, M., Sibanda, L., Prof. M.S. Swaminathan, Winters, P. (eds.) *Mainstreaming Agrobiodiversity in Sustainable Food Systems: Scientific Foundations for an Agrobiodiversity Index.* Bioversity International, Rome, Italy, pp. 23–52

King, O.I. and Padulosi, S. (2017) Agricultural Biodiversity and Women's Empowerment: A Successful Story from Kolli Hills, India. In Del Castello, A. and Bailey, A. (eds.) *Creating Mutual Benefits: Examples of Gender and Biodiversity Outcomes from Bioversity International's Research.* Bioversity International, Rome, Italy, 4 p. ISBN: 978-92-9255-063-9

Padulosi, S., Amaya, K., Jäger, M., Gotor, E., Rojas, W. and Valdivia, R. (2014) Holistic Approach to Enhance the Use of Neglected and Underutilized Species: The Case of Andean Grains in Bolivia and Peru. *Sustainability* 2014(6), 1283–1312. doi: 10.3390/su6031283. ISSN 2071-1050.

Padulosi, S., Begriming, N. and Lawrence, T. Eds. (2012) On Farm Conservation of Neglected and Underutilized Species: Trends and Novel Approaches to Cope with Climate Change. *Proceedings of an International Conference*, Frankfurt, 14–16 June 2011. Bioversity International, Rome, Italy.

Padulosi, S., Mal, B., King, O.I. and Gotor, E. (2015) Minor Millets as a Central Element for Sustainably Enhanced Incomes, Empowerment, and Nutrition in Rural India. *Sustainability* 7(7), 8904–8933. doi: 10.3390/su7078904

Padulosi, S., Leaman, D. and Quack, P. (2002) Challenges and Opportunities in Enhancing the Conservation and Use of Medicinal and Aromatic Plants. *Journal of Herbs, Spices & Medicinal Plants* 9(2/3 and 4), 243–268.

Padulosi, S., Phrang, Roy and Rosado-May, F.J. (2019) *Supporting Nutrition Sensitive Agriculture through Neglected and Underutilized Species - Operational Framework.* Rome: Bioversity International and IFAD. 39 pp. https://cgspace.cgiar.org/handle/10568/102462

Padulosi, S., Sthapit, B., Lamers, H., Kennedy, G. and Hunter, D. (2018) Horticultural Biodiversity to Attain Sustainable Food and Nutrition Security. *Acta Horticulturae* 1205, 21–34. doi: 10.17660/ActaHortic.2018.1205.3

Raneri, J.E., Padulosi, S., Meldrum, J. and King, O.I. (2019) Promoting neglected and underutilized species to boost nutrition in LMICs. In *UNSCN Nutrition 44—Food Environments: Where People Meet the Food System*; UNSCN: Rome, Italy. pp. 10–25.

Romanelli, C., Miaero, M., Mahy, L., Haddad, J., Lundberg, K., Hunter, D., Savage, A.J., Egorova, A. and Cooper, D. (2020) *Guidance on Mainstreaming Biodiversity for Nutrition and Health*. WHO, Geneva.

Rosado-May, F., Cuevas-albarrán, V., Moo Xix, F., Chan, J.H. and Arroyo, J. (2018) Intercultural Business: A Culturally Sensitive Path to Achieve Sustainable Development in Indigenous Maya Communities. doi: 10.1007/978-3-319-71312-0_32.

Rudebjer, P. and Padulosi, S. (2018) Introduction to Neglected and Underutilized Species (NUS). In *Neglected and Underutilized Species (NUS), for Improved Food Security and Resilience to Climate Change: A Contextual Learning Manual for African Colleges and Universities*, Kasolo, W., Chemining'wa, and Temu, A. (Eds.). ANAFE, Nairobi, 151 pp.

Shanthakumar, G., Mal, B., Padulosi, S. and Bala Ravi, S. (2010) Participatory Varietal Selection: A Case Study on Small Millets in Karnataka. *Indian Journal of Plant Genetic. Resources* 23(1), 117–121.

Sunstein, C.R. (2015) *The Ethics of Influence: Government in the Age of Behavioral Science*. Cambridge University Press, Cambridge.

Ulian, T., Diazgranados, M., Pironon, S., Padulosi, S., Liu, U., Davies, L., Howes, M.-J.R., Borrell, J., Ondo, I., Pérez-Escobar, O.A., Sharrock, S., Ryan, P., Hunter, D., Lee, M.A., Barstow, C., Łuczaj, Ł., Pieroni, A., Cámara-Leret, R., Noorani, A., Mba, C., Womdim, R.N., Muminjanov, H., Antonelli, A., Pritchard, H.W. and Mattana, E. (2020) Unlocking Plant Resources to Support Food Security and Promote Sustainable Agriculture. *People Plants Planet* 2(5), 421–445. ISSN: 2572-2611

Van Etten, J., van de Gevel, J. and Mercado, L. (2017) First Experiences with a Novel Farmer Citizen Science Approach: Crowdsourcing Participatory Variety Selection through On-Farm Triadic Comparisons of Technologies (Tricot). *Experimental Agriculture*. doi: 10.1017/S0014479716000739

Vermeulen, S.J., Park, T., Khoury, C.K. and Béné, C. (2020) Changing Diets and the Transformation of the Global Food System. *Annals of the New York Academy of Science*. doi: 10.1111/nyas.14446

Vernooy, R., Sthapit, P. and Sthapit, B. Eds. (2015) *Community Seed Banks Origins, Evolution and Prospects*. Issues in Agricultural Biodiversity. Earthscan from Routledge, London.

WHO/CBD (2015) *Connecting Global Priorities: Biodiversity and Human Health, a State of Knowledge Review*. CBD, Geneva. 344 pp.

4
NUS AND PEOPLES SENSE OF DIETARY DIVERSITY AND WELL-BEING

Gigi Manicad

Introduction

Agrobiodiversity provides diverse options for people's strategies in cultivating, collecting and processing a diversity of plants for their food, nutrition, livelihood security and resilience. Only in the last two decades has the attention of agricultural research started to slowly broadening from staple crops and vegetables to "neglected and underutilized species" (NUS) or "orphan crops".

This chapter is largely based on three baseline studies in Myanmar, Peru and Zimbabwe. The objective is to understand people's perspectives of NUS in relation to their concept of nutrition and their food security strategies. Particular attention is paid to women's access and use of NUS and the biodiversity source of household diets. A combination of methods were used:

- Household surveys of socio-demographics and agricultural crops, and a modified version of Household Dietary Diversity Score (HDDS) (FAO, 2010).
- A gender-disaggregated community resource flow map (Manicad, 2002, 2004), which consisted of area mapping farms and the landscape, and the collection and use of plants. A focus group discussion identified people's folk taxonomy, uses, locations, seasonality, collection and propagation of NUS (Oxfam Novib, 2018).
- Participatory Rural Appraisal, including a seasonal calendar developed to identify the NUS availability and locations for the entire year (ANDES, 2016).

The methods were applied twice, once each during the plenty and lean seasons of food availability. The studies involved Indigenous and local peoples, who are mostly engaged in subsistence farming and employ largely rain-fed agricultural

DOI: 10.4324/9781003044802-5

FIGURE 4.1 Lares Valley women's focus group discussion on NUS. Photo Credit: ANDES.

practices. The three countries represented diverse crop systems in distinct agro-ecologies, from the delta in Myanmar to the high altitudes of Peru and the semi-arid regions of Zimbabwe (see Table 4.1).

Context and conceptual challenges

One of the most widely accepted characteristic of NUS is the neglect of science and markets. During the surveys, the term "neglect" and "underutilisation" consistently created confusion amongst local people. In Peru, NUS have "strong sacred value as they are conceived as fruits of the Pachamama (Mother Earth), and are crops sent by the Apus (natural divinities), or are blood of the Apus and species used to purify the body" (ibid.).

However, while NUS may be sacred to some, they also carry the stigma of being "poor people's food". In Zimbabwe, Tsine (or "Black Jack") is perceived as a food consumed by those suffering from hunger and as curative item for people who are HIV positive (CTDT, 2015).

A key defining feature of NUS is that these are highly localised plants involving specialised local knowledge of mostly women, who are often marginalised and impoverished. Through lessons learnt, the confusion over NUS terminology was better managed through a combination of the following techniques: (i) prior to the

FIGURE 4.2 Lares Valley women's focus group discussion on NUS. Photo Credit: ANDES.

survey, we harmonised the community custodians' local concepts of NUS with a set of working definitions of it. (ii) We properly sequenced the material resource-flow map to first enumerate all the plants collected. This was followed by key informant interviews to classify which of the collected plants were NUS; and then we characterised each NUS according to the local knowledge of folk taxonomy, ethnobotany, seasonal availability, locations, harvesting, propagation and processing methods. This was followed by (iii) the collection of specimens, (iv) verification with the wider community and (v) discussions with research organisations about, for instance, the NUS' scientific names and its nutritional composition.

NUS consumption during lean periods

A "lean" or "hunger" period refers to recurrent food shortages amongst the poorest households. Reflecting the seasonality of agricultural production, the lean period often occurs just before the upcoming harvest when food supply is at its lowest and when it coincides with the low demand for agricultural labour. In the coastal Irrawaddy region of Myanmar, the lean period also coincides with the cyclone season. Hence, the lean season often combines the stresses of lack of food, lack of wages to buy food and, in some cases, environmental calamities. The lean period is a predictable, seasonal "sink or swim" time when households either survive from one season to the next, or they lose assets, accumulate debt and sink further into poverty.

NUS, dietary diversity and well-being **61**

The survey on the lean period, referred to as "hunger" or "scarcity" periods, proved highly sensitive. In Peru for example, "scarcity" ("*escasez*" in Spanish) does not have an equivalent in Quechua and is often associated with poverty. Asking someone about their food scarcity or hunger were considered taboo in all three countries.

These sensitivities were managed by keeping the survey private within households, and ensuring that trusted local people were part of the study team. Table 4.1 shows that a substantial percentage of the sample populations confirmed that they regularly experienced hunger. However, the percentages for each country cannot be compared due to the difficulties in covering the topic and the likely different interpretations across cultures of the hunger period.

Most respondents referred to the lean period as the time when they had low access to food supply and as a period of physical stress and mental anxieties. In

TABLE 4.1 Survey information of the area and the people engaged in subsistence farming

	Myanmar	*Peru*	*Zimbabwe*
Area and main food crops	*Ayeyarwady region*: lowland delta and rice *Southern Shan State*: Upland and rice, maize, pulses	*Puna*: 3,500 m above sea level and roots, tubers, beans, grains *Keshua*: 2,300–3,500 m above sea level and maize, rice, barley *Yunga*: below 2,300 m above sea level and citrus, banana, squash	*Chiredzi*: semi-arid, drought prone *Goromonzi*: relatively high rainfall, moderate temperature *Tsholotsho*: low rainfall, height temperature and pearl millet *UMP*: moderate rainfall and temperature All areas except Tsholotsho grow sorghum and maize in addition to pearl millet
Indigenous/ethnic origins	Mamar, Rakhinde, Kayin, Pa O, Shan, Le Su, Lar Hu	Quechua	Kalanga, Ndebele, San, Shanganes, Shona
Population in specific project areas and % sample households and % of female respondents	7,702 people, 10% of households, with unknown percentage of female respondents	5,027 people, about 14%–15% of households, with 64% female respondents	728,476 people, about 0.4% households, with 78% female respondents
Respondents who regularly experienced lean or hunger periods	77%	87%	40%
HDDS during lean and sufficient periods	Lean: 9 Sufficient: 6.4	Lean: 7.7 Sufficient: 7.4	Lean: 2.5 Sufficient: 3.2
Total number of NUS initially identified	66	78	46

Source: Oxfam Novib (2018).

Myanmar, respondents associated hunger with a shortage of rice. This could be attributed to the importance of rice in their culture, diet and income. It is also likely that the lack of carbohydrates directly manifest in the sensation of hunger.

The three countries have common some coping strategies. This includes consumption of food preserved or stored during the season of plenty; aspects of reciprocity, such as borrowing or receiving food from relatives and friends; and buying food on credit or shifting to cheaper and lower quality and taste. More vulnerable households turn to harmful strategies such as reducing their portion sizes or skipping meals. In some households, adults may reduce or skip meals in favour of children; in other households, working members are prioritised. In more desperate circumstances, some households in Myanmar and Zimbabwe resort to eating premature crops. This is an example of households losing assets during lean periods, such as their next harvest and seeds for the subsequent growing season. For example, the over-harvesting of perennial crops such as taro is likely to reduce yields, and the consumption of cereal seeds is a major threat to seed security. In Zimbabwe, many respondents also rely on food aid.

All respondents referred to the collection and consumption of NUS as an important part of their coping strategies. Most households collect semi-wild and wild plants to supplement their diets and for market sales. NUS are cultivated and collected from home gardens, on farms and roadsides and in forests, wetlands and pasture areas. They also referred to the traditional knowledge of the elderly for understanding the availability of different NUS during specific times and at specific locations. The responsibilities for executing the coping strategies, including the collection of NUS, are shared within the household, with women having a prominent role.

People associate nutrition with well-being

In all three countries, the local people stated that they have limited knowledge of nutrition since they lack access to their governments' nutrition education programme. In Myanmar, the respondents said that they are less aware of the nutritional impacts of their food choices. The communities showed a very limited understanding of micronutrients such as vitamins. However, in all three countries, there were good indications that local people traditionally have a functional knowledge of the association between dietary diversity and well-being.

NUS are particularly important for the people's sense of well-being and are appreciated for both the eating satisfaction, as well as their health and healing values. These knew cures for gastroenteritis, anaemia, colds and flu, prostrate and liver inflammation, muscle cramps and various pains. A number of NUS consumed as food were also valued for their contraceptive uses, or for easing menstruations (ANDES, 2016). In Myanmar, NUS were consumed because of their associated good health benefits. The uses of NUS as medicines are most helpful during hunger periods when money is tight. In general, however, people increasingly prefer to purchase medicines from pharmacies (Metta and Searice, 2015). In Zimbabwe, the most important NUS were believed to have curative properties against malaria, dehydration, pain and HIV/AIDS (CTDT, 2015). These findings concur with that of other studies (e.g. Pieroni and Price, 2006) on the lack of distinction between food and medicine in traditional food systems. The surveys indicated that inter-generational transfer of knowledge might need to be better understood and further enhanced.

In the Lares Valley, the equivalent concept of well-being is "*Sumaq kausay*", which means "good living". They refer to *Sumaq kausay* as the balance and reciprocal relation between the individual and collective, and nature and ecosystem services. Aspects of cultural identity, leisure, celebrations and knowledge heritage are also integrated in their concept of well-being (ANDES, 2016).

Diverse land use, biodiversity and dietary diversity

In all three countries, most of the respondents' sense of well-being was connected to healthy ecosystems with diverse land use, including farming and gardening. People derived their dietary diversity from the biodiversity on farms and within landscapes. This was articulated in the NUS flow maps in Myanmar, Zimbabwe and Peru, which illustrated where people gather domesticated, semi- and non-domesticated plants. Many were NUS, which vary across time and space. In Peru, a high dietary diversity resulted from a combination of more intensive food production, gathering of plants within diverse ecologies and the purchase and barter of a wider variety of food. In the Lares Valley, women's networks operate five traditional barter markets involving farms within the three levels of agro-ecologies from the top, middle and lower zones of the region (see Table 4.1). This enables the exchange of a variety of vegetables, fruits, spices, condiments and beverages. The barter markets enhance local resilience based on diverse land use combined with the traditional social relationships of reciprocity and redistribution (ANDES, 2016).

Similar links between dietary diversity and agrobiodiversity were established in a study involving Indigenous and local peoples in the East Usambara Mountains in Tanzania (Powell et al., 2017). In this study, local farmers indicated that maintaining multiple land uses increases crop diversity, which helps ensure their food security and dietary diversity. In another study in Mount Malindang in the Philippines, the Subanens also displayed a diverse collection of biodiversity on farms and in the wild from diverse ecologies within their landscape (Manicad, 2004). Similar to Peru, the study further alluded to the people's integrated landscapes perspectives with the management of biodiversity on farms and in the wild. In both the Subanens in the Philippines and the

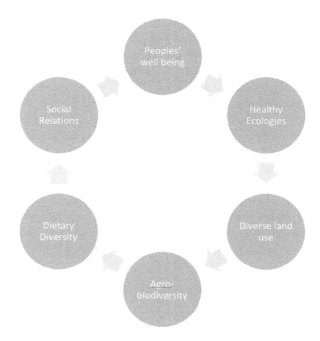

FIGURE 4.3 People's interconnected sense of well-being, biodiversity and dietary diversity.

Quechua in Peru, traditional beliefs and social relations also help regulate the sustainable and equitable harvesting and distribution of biodiversity in the wild (ibid and ANDES, 2016).

The studies in Myanmar, Peru and Zimbabwe indicated that land degradation, farming, environmental pollution, extensive mining and deforestation threaten their well-being and food sources. In Peru and Zimbabwe, malnutrition was associated with the lack of consumption of local wild foods and the occurrence of diseases in cultivars. In Myanmar, the respondents directly linked the inaccessibility of diverse diets as a main cause of malnutrition in their communities.

The respondents in the three countries also referred to changing climates as a threat to the diversity of their food sources. They pointed to increased pest and disease infestations, and the worsening scarcity of wild plants. In Zimbabwe, they pointed to climate change and the total drying up of wetlands in Uzumba, Maramba and Pfungwe (UMP), which in turn resulted in the disappearance of indigenous vegetables that used to grow there during the lean seasons (CTDT, 2015). This indicates that not only does climate change threaten livelihoods and food security, it also threatens people's coping strategies.

FIGURE 4.4 Zimbabwe seed fair with NUS. Photo credit: CTDT.

Conclusion

The baseline studies in Myanmar, Peru and Zimbabwe contribute to the understanding of NUS as integrated in local peoples' land use and agrobiodiversity management. Local people associate agrobiodiversity, including NUS, with dietary diversity, which they closely link with their sense of well-being. These baseline studies indicate that NUS are part of people's strategies for food and nutrition security during both the lean and the plenty seasons. For the further enhancement of NUS for people's well-being, a few courses of action are suggested:

1 Research and development interventions on NUS need to be guided by the principle of equity. For effective interventions on NUS, it is important to understand and address the underlying gender and social relations within households and between generations, including the youth.
2 Considering the rich cultural context and the sensitivities within which NUS are managed, training and engaging local communities and researchers can contribute to a more robust and user-oriented results. This also creates capacities to implement and follow through on plans for intervention.

Joint planning and periodic reviews of methods, measurements and analysis can mitigate the trade-off between diverse local research and global comparisons.

3   This chapter concurs with a study in Tanzania (Powell et al., 2017), wherein local people tend to see nutrients as abstract concepts. People tend to relate more to the functionality of nutrition. While laboratory analysis of the nutritional composition and stability during food preparation of NUS is important, nutrition education programmes should seek to complement and build from the local people's concepts and associations of nutrition and NUS. Building from existing knowledge systems and enhancing cultural culinary practices are likely to be more effective.

4   An integrated approach to people's biodiversity management and food management strategies should also maintain specialised sets of objectives and activities for NUS. The NUS priority needs for the most vulnerable groups require particular emphasis. It is likely that local people who are interested in staple crops might not be the same ones specialised in NUS. Moreover, the seasonality and locations of NUS are likely to differ from those of main crops. Local level interventions on NUS should not mimic the marginalisation of NUS by conventional research and markets. At an operations level, the risk of NUS interventions being side-lined by the emphasis on main crops should be avoided.

5   The conservation and sustainable use of NUS should be integrated into national, regional and global agendas related to the nexus of climate change, food and nutrition security, and agrobiodiversity management. NUS are potentially important for climate resilience. The conservation and sustainable use of NUS requires interdisciplinary studies and multi-stakeholder dialogue on planning and integrating complementary approaches between *in situ* and *ex situ* management.

Acknowledgments

The author is deeply grateful to the *Association ANDES* of Peru, the Community Technology Development Trust (CTDT) of Zimbabwe, the Metta Development Foundation of Myanmar and the Southeast Asia Regional Initiative for Community Empowerment (SEARICE) for generously sharing their respective country's studies and insights. The author is also grateful to the Sowing Diversity = Harvest Security (SD=HS) programme of Oxfam.

References

ANDES (2016) *Nutrition, Coping Strategies, Knowledge and Use of Neglected and Underutilized Species*. Cusco: Andes.

CTDT (2015) *Zimbabwe Baseline Report*. Harare: CTDT.

FAO (2010) *Guidelines for Measuring Household and Individual Dietary Diversity*. Rome: FAO.

Manicad, G. (2002) "PRA in Agricultural Research" in *ISNAR Engendering Participatory Research. A Learning Module in Facilitating the Agricultural Innovation Process*. The Hague: ISNAR.

Manicad G. (2005) "People, Landscapes and Worldviews: A Pilot Study on Analysing Non-Domesticated and Semi-Domesticated Biodiversity for Local Use". A Study Commissioned by the Centre for Genetic Resources. Wageningen University and Research Centre, the Netherlands.

Metta and SEARICE (2015) In: Metta Development Foundation (ed.) *Myanmar Baseline Report*. Quezon City: Philippines.

Oxfam Novib (2018) *Women, Seeds and Nutrition. Technical Report of the Baseline Survey of Myanmar, Peru, Vietnam and Zimbabwe*. The Hague: Oxfam Novib.

Pieroni, A. and Price, L. (eds) (2006) *Eating and Healing. Traditional Food and Medicine*. New York: Haworth Press.

Powell, B., Kerr, R.B., Young, S.L. and Johns, T. (2017) The Determinants of Dietary Diversity and Nutrition: Ethnonutrition Knowledge of Local People in the East Usambara Mountains, Tanzania. *Journal of Ethnobiology and Ethnomedicine* 13(1): 23.

5

DIVERSITY AND SMALL MILLETS IN INDIA

Annadana Seetharam and Prabhakar Bhat

Introduction

Millets refer to a group of cereal grasses grown largely as grain and feed crops. Millets require less water than any other grain crop and provide assured harvests in arid, semi-arid and mountainous regions in the tropics and sub-tropics, where monsoon failure and droughts are frequent, soil fertility is poor and land terrain is difficult. Millet grains are a main staple for farm households in several old world countries and among the poorest people. Millet straw is a valuable livestock feed, besides having other uses as building material, fuel, etc., in those farming systems. India has the largest diversity of millet crops in the world, composed of major millets (viz. sorghum and pearl millet) and several small grain millets (viz. finger millet [*ragi*], foxtail millet [*kangni*], kodo millet [*kodo*], proso millet [*cheena*], barnyard millet [*sawan*] and little millet [*kutki*]). There are varieties especially in little millet and proso millet, which mature in 60–70 days, yet provide reasonable and assured harvests even under most adverse conditions. India, being the primary or secondary centre of origin and domestication of small millets, is a storehouse of highly valuable genetic variability. Asia and Africa are the major producers of these species at the global level, and India is possibly the largest producer of millets, with an estimated production of around 16 million tonnes annually.

Exports and imports of millets are negligible, indicating low demand and/or unreliable availability of marketable surpluses in local/world markets. Urbanites by-and-large have forgotten the amazing foods and dishes made out of millets. Being eco-friendly crops, they are suitable for fragile and vulnerable environments and for sustainable and green agriculture. The promotion of these crops can lead to more efficient natural resource management and a holistic approach to sustaining precious agrobiodiversity.

DOI: 10.4324/9781003044802-6

Diversity and small millets in India **69**

TABLE 5.1 Nutritional composition of millet grains (NIN, 2007)

Crop	Protein (g)	Carbohydrates (g)	Fat (g)	Fibre (g)	Minerals (g)	Calcium (mg)	Phosphorus (mg)
Wheat	11.8	71.2	1.5	12.9	1.5	41	306
Rice	6.8	78.2	0.5	5.2	0.6	10	160
Sorghum	10.4	72.2	1.9	12.0	1.6	25	222
Bajra	11.6	67.5	5.0	16.0	2.3	42	296
Finger millet	7.3	72.0	1.3	18.8	2.7	344	283
Proso millet	12.5	70.4	1.1	14.2	1.9	14	206
Foxtail millet	12.3	60.9	4.3	14.0	3.3	31	290
Kodo millet	8.3	65.9	1.4	15.0	2.6	27	188
Little millet	8.7	75.7	5.3	12.0	1.7	17	220
Barnyard millet	11.6	74.3	5.8	13.5	4.7	14	121

Nutritional relevance

Millet grains are very nutritious as they contain good quality protein and are rich in minerals, dietary fibre, phytochemicals and vitamins. A cursory look at the proximate composition of various food grains (Table 5.1) would reveal the distinct nutritional superiority of millets over major food grains such as wheat and rice. This needs to be exploited in terms of its health and nutritional benefits for alleviating nutritional deficiency, the result of changed food habits and a shrunken food basket. Millet, with its unique grain properties, exhibit considerable opportunities for diversification of its food use through processing and value addition.

Fall in millet area and production

Securing the existing cultivation area for millets is a matter of concern. As shown in Table 5.2, in the past six decades, the area used for growing millets has gone down by 62.57%, dropping from 36.34 million ha (1955–1956) to around 13.84 million ha (2017–2018). Yet, the production targets of all millet crops were maintained as a result of an increase in productivity per hectare – over four times in pearl millet and more than two times in sorghum, finger millet and small millet crops – over the decades.

Even though small millets are cultivated in almost every Indian state, the distribution of individual millets is not uniform. The major finger millet growing states are Karnataka, Tamil Nadu, Andhra Pradesh, Orissa, Bihar, Maharashtra and Uttar Pradesh. Karnataka has the largest area under finger millet, accounting for nearly 60% of the total area. Kodo, little millet and foxtail millet are grown widely in Karnataka, Tamil Nadu, Andhra Pradesh, Orissa, Bihar, Madhya Pradesh and Maharashtra. In Madhya Pradesh, both kodo and little millet are

TABLE 5.2 Area (million ha), production (million tonnes) and productivity (kg/ha) of millet crops in India (DMD, GOI, 2018)

Crop/year	Category	1955–1956	1965–1966	1975–1976	1985–1986	1995–1996	2005–2006	2017–2018
Sorghum	Area	17.36	17.68	16.09	16.10	11.33	8.68	4.97
	Production	6.73	7.58	9.50	10.20	9.33	7.63	4.66
	Productivity	387	429	591	633	823	880	938
Pearl Millet	Area	11.34	11.97	11.57	10.65	9.32	9.58	7.21
	Production	3.43	3.75	5.74	3.66	5.38	7.68	9.26
	Productivity	302	314	496	344	577	802	1,284
Finger Millet	Area	2.30	2.70	2.63	2.41	1.77	1.53	1.15
	Production	1.85	1.33	2.80	2.52	2.50	2.35	1.96
	Productivity	800	492	1,064	1,049	1,410	1,534	1,703
Small Millets	Area	5.34	4.56	4.67	3.16	1.66	1.06	0.51
	Production	2.07	1.56	1.92	1.22	0.78	0.47	0.42
	Productivity	388	341	412	386	469	443	828
All Millets Crops*	Area	36.34	36.91	34.92	32.32	24.08	20.85	13.84
	Production	14.08	14.22	19.96	17.60	17.99	18.13	16.30
	Productivity	469	394	641	603	820	915	1,188

predominant, whereas foxtail millet is important in Andhra Pradesh, Karnataka and Telangana. Barnyard millet and proso millet are grown largely in hills of Uttar Pradesh, Uttarakhand, the north-eastern regions and plains of north Bihar and western Uttar Pradesh and Maharashtra.

Millets and food security

The bulk of millets produced in the country are consumed at the farm level and very little finds its way to organised markets. Millets provide much-needed food and feed security, especially to vulnerable groups. However, it is important to note that food security at the national level will only be effective when regionally important crops are allowed to play their due role in meeting the food and fodder needs of the region. Such a vulnerable situation calls for an urgent intervention from policy makers; an ideal correction would be to produce the required food within the state/region itself, by promoting and developing locally and/or regionally important rainfed crops. The promotion of millets can lead to much more efficient natural resource management and, ultimately, to a more holistic approach in sustaining precious agrobiodiversity.

Small millets diversity in India

Small millets are grown throughout India and have thousands of years of cultivation history in diverse ecologies. During this process, a vast amount of diversity has formed within each species, resulting in numerous local varieties. Finger millet or ragi - *Eleusine coracana* (L.) Gaertn. subsp. *coracana*, is native to the highlands of eastern Africa; its closest wild relative is E. *coracana* subsp. *africana* (Kenn.-O'Byrne). Its introduction in India took place at least 3,000 years ago; its long history of cultivation and selection has made Indian finger millet highly variable, resulting in many unique local varieties (Seetharam, 2015). In view of this, India is recognised as a secondary centre of diversity for this crop. The highly variability found in finger millet is divided into five cultivated races viz. coracana, vulgaris, elongata, plana and compacta; they are distinct essentially on the basis of inflorescence morphology.

Foxtail millet, *Setaria italica* (L.) P. Beauv., and its closest wild relative, *Setaria italica* subsp. *viridis* (L.) Thell., originated in the highlands of China some 5,000 thousand years ago and later spread to India as a cereal crop (Prasad Rao et al., 1987). Proso millet (*Panicum miliaceum* L.) is native to Manchuria (China) and moved to Europe some 3,000 years ago. Little millet (*Panicum sumatrense* Roth) is native to India and is widely cultivated across the country in many states. Indian barnyard millet (*Echinochloa frumentacea* Link), domesticated and cultivated in India, is highly variable. Kodo millet (*Paspalum scrobiculatum* L.), domesticated in India some 3,000 years ago, is widely distributed in wet damp habitats, grown in several states from Tamil Nadu in the south to Uttar Pradesh in the north. By and large, small millets in India are important to hill and tribal agriculture.

Collection and conservation of germplasm

In the past, small millets breeders had limited access to germplasm and worked with very limited accessions that lacked diversity, which blunted opportunities for yield improvement. This situation was, to some extent, rectified in the 1960s, when first attempts were made by Indian Council of Agricultural Research (ICAR) to pool the collections under the "PL 480 Project". Conservation activities gained further momentum with the National Bureau of Plant Genetic Resources (NBPGR) in New Delhi, which has been playing a key role in augmenting the initial small millets collection. Recognising the importance of conservation and greater access to germplasm, the All-India Coordinated Millets Improvement Project (AICMIP) established a germplasm unit in Bangalore in 1979 to cater to these species. This unit, since then, has been making efforts to collect pool germplasm from various sources and make it available to breeders. The unit is also recognised as one of the NBPGR's National Active Germplasm Sites (NAGS) and has the mandate for the collection, conservation, evaluation and documentation of small millets germplasm within the country. Presently, the Bangalore unit maintains one of the largest collections, with more than 15, 000 accessions of six small millet species, incl. 7,122 accessions of finger millet, 2,821 of foxtail millet, 1,537 of kodo millet, 939 of proso millet, 1,657 of little millet and 988 of barnyard millet (AICSMIP, 2012).

Utilisation of germplasm

Full utilisation of germplasm depends on two factors: (a) evaluation and characterisation and (b) identification of useful gene sources. These two areas have received attention during the last 25 years and the majority of millet accessions have been screened for desirable agronomic and grain quality traits. A good data base is available for most accessions conserved.

In order to improve the efficiency of germplasm use, core subsets of the collection have been formed and made available to breeders. The identification of several sources of stable resistance to blast disease in finger millet and their deployment in breeding research has been highly rewarding with regard to the development of high-yielding blast-resistant varieties of finger millet (Ravikumar et al., 1990, 1991; Byre Gowda et al., 1999). The secondary gene pool of subsp. *africana* has been of interest from the breeding point of view as a source of useful genes for improving tillering abilities, fodder yield and quality, drought tolerance, finger number and length. Pre-breeding is required for the introgression of characters from *E. africana* to *E. coracana*, so as to derive lines useful in breeding programs.

Exotic collections of finger millet, especially from Africa, have been used in recombination breeding, resulting in the release of many superior, high-yielding varieties (HYVs) in many states. The African germplasm has thick stem, dark leaves, robust growth, large ears and high grain density and is a source of resistance to blast disease (Naik et al., 1993). Hybridisation between African and Indian elite varieties has been highly rewarding and has resulted in the release of many HYVs.

Several useful genetic stocks have also been identified in the germplasm of small millets with regard to higher protein content, desirable agronomic attributes with high carbon dioxide fixation and low leaf area, suitable for rain fed situations and germination under conditions of limited moisture and a hard soil crust (Sashidhar et al., 1983; Seetharam et al., 1984). Accessions capable of producing higher biomass, dual purpose types with superior stover quality are available for improving the grain and stover yield of cultivars (Schiere et al., 2004). There are finger millet accessions that have significantly higher grain calcium and protein content. In foxtail millet, new sources of dwarfing genes controlled by oligo genes have been also identified. These accessions are useful in breeding dwarf foxtail millet. The variability available for protein content (7.176–15.73 g) and seed fat content (4.0–7.1 g) in foxtail millet is enormous and can be exploited in breeding (Seetharam, 2015). Optimising the use of germplasm in small millets is a priority for crop improvement, in order to make these species competitive vis-à-vis other crops.

Progress in crop improvement

Small millets improvement efforts have been in progress since the beginning of the twentieth century (Seetharam, 2015). The launching of coordinated crop improvement programs during the late 1950s has contributed significantly to the development of new superior varieties and enhanced production and protection technologies in small millets. The release of improved varieties and production packages has contributed to a threefold increase in grain productivity in the country. Small millets have been the last crops prioritised in the agriculture developmental agenda in the country, with finger millet receiving a little more attention than other species. An attempt has been made here to trace the progress in crop improvement of small millets over the last nine decades.

Crop improvement efforts during the early years

In the 1950s and 1960s, crop improvement for small millets was confined to a few states such as Tamil Nadu, Andhra Pradesh, Karnataka and Uttar Pradesh. The emphasis was on varietal improvement through selecting better types from local varieties. In Tamil Nadu, a Millet Research Station was established in 1923 at Coimbatore, under the erstwhile Madras Presidency. Finger millet work in Karnataka dates back to 1900, initiated in Bangalore especially with finger millet, and in Uttar Pradesh at Kanpur and Gorakhpur in 1944.

The first finger millet variety released in the country was H22, as early as 1918 in Karnataka. Other finger millet varieties released were Co6 (1935); R0870, ES13, K1 and ES11 (1939); Hagari1 (1941); Co1, Co2, Co3 and Co4 (1942); VZM1 and VZM2 (1958) and T36 B (1949). Finger millet improvement received a boost in Karnataka during the 1950–1960 period, and several new varieties such as Aruna, Udaya, K1, Purna, ROH 2 and Cauvery were released. Similarly, many varieties were released for other small millets in other states as well (Seetharam, 2015), including little millet variety Co 1 (1954); foxtail millet varieties Co1, Co2, Co3 (1943), H1, H2 (1948), T 4 (1949); kodo millet varieties PLR 1 (1942), T2 (1949), Co1 (1953); proso millet variety Co1 (1954) and barnyard millet varieties T25 and T46 (1949).

During the 1950s, with food production remaining stagnant and with population growing, the importance of millet crops in Indian agriculture, as important resources for dry land agriculture, started gaining recognition. A project for the intensification of research on cotton, oil seeds and millets was launched during this period, with several centres working on millets alone. The importance of genetic resources as primary raw material for crop improvement was recognised prior to initiation of coordinated project. The first attempt to collect the germplasm of millets in the country was made in 1961 under the PL 480 Project "Storage, Maintenance and Distribution of Millets Germplasm", which resulted in the collection of nearly 3,000 genetic stocks of various species including 718 samples of finger millet, 584 of kodo millet, 431of little millet, 615 of foxtail millet, 250 of proso millet and 399 of barnyard millet.

Crop improvement efforts during the AICMIP Era

Millets, in general, started receiving attention with the launch of the AICMIP in 1969. Small millets also started receiving some attention in a few selected centres. Crop improvement received a major boost during 1978–1979 through an International Development Research Centre (IDRC)-funded project of five crop-specific lead research centres in the country (viz. Almora in Uttarakhand [barnyard millet], Dholi in Bihar [proso millet], Dindori in Madhya Pradesh [kodo millet], Semiliguda in Orissa [little millet] and Nandyal in Andhra Pradesh [foxtail millet]). The project continued until 1985 and was replaced subsequently by the "All-India Coordinated Small Millets Improvement Project" (AICSMIP) in 1986. The centres that were operating under the IDRC project became part of AICSMIP. Small millets research has focused all along on the development of varieties and agro-production and protection technologies suitable for different regions. Currently, there are 14 centres functioning under the AICSMIP, spread across the country. Their work is multidisciplinary and applied in nature.

Crop improvement is aimed at developing HYVs with resistance to blast disease quality fodder, early and medium maturity and white seed in finger millet, resistance to head smut in kodo millet and resistance to shoot fly in both proso and little millets. So far, a total of 293 varieties for the six small millet species have been released in the country from 1918 to 2020. Out of this, 86 were released before 1986 (pre coordinated-era) and 207 during 1986–2020 (post coordinated-project era) (Table 5.3).

Modern finger millet varieties have the genetic potential of producing five to six tonne/ha under optimum cultivation conditions. Recombination breeding has resulted in the release of many superior varieties in foxtail, proso and barnyard millets as well. Large-scale variety testing on farmers' fields in participatory mode (Gowda et al., 2000) and frontline demonstrations have helped in spreading HYVs among farmers.

TABLE 5.3 Improved varieties of small millets released in India from 1918 to 2020

Crop	No of varieties released		Total
	Pre-coordinated project era (1918–1985)	Post coordinated project era (1986–2020)	
Finger millet (1918–2020)	45	91	136
Foxtail millet (1942–2020)	12	26	38
Little millet (1954–2020)	6	26	32
Proso millet (1954–2020)	8	19	27
Barnyard millet (1949–2020)	4	18	22
Kodo millet (1942–2020)	11	27	38
Total	**86**	**207**	**293**

Seed production

Seed production and distribution, which is key to the successful adoption of HYVs is weak in many states. This has deprived farmers of many benefits arising from the cultivation of improved millets, in most parts of the country. Harnessing the yield advantages from these improved varieties is the need of the hour in order to make the cultivation of small millets truly competitive and economically viable.

Crop production and protection technologies

Packages of information about best practices such as the optimal time for sowing/planting, choice of varieties, time and method of application of fertilisers, etc. have been developed for different regions of the country. Management practices for aberrant weather conditions, for mitigating early, mid- and late season drought have been worked out. Remunerative cropping systems involving different pulse crops in millet for different regions have been created. Technology transfers attempted through frontline demonstrations on farmers' fields and large-scale station demonstrations have also helped narrow down the yield gap that exists between farmers' fields, demonstration plots and research station trials (Seetharam, 2015; Prabhakar, 2017).

Plant protection measures to control economically important diseases and pests have been enhanced for small millets (Seetharam, 2015). Several blast-resistant lines have been identified from the germplasm available at NAGS, and better crop protection practices for these diseases have been recommended.

Future prospects

After years of neglect, millets are finding a place in agricultural research and agendas of large private companies, and from there are reaching supermarkets everywhere. They are increasingly being recommended by doctors and nutritionists as important foods for health and wellness, and as being helpful in preventing many diseases related to modern lifestyles, including obesity and diabetes. Numerous elite food chains have begun selling millets and millet-based products as health foods. As it stand, today, the productivity of small millets can be increased by more than 50% by adopting improved production practices, allowing more people to take advantage of these nutritious and healthy crops. In recent years, milling technology has been considerably improved to enhance grain quality. Millet mills are available for cottage-level and large-scale processing. Millets can be further processed for making various food items such as flakes, quick food cereals, ready to eat snacks, supplementary foods, extrusion cooking items, malt-based products, weaning foods and, more importantly, health foods.

Farmers who had shifted from millets to other crops are keen now to go back to millets in view of the stable harvests ensured, the easy crop production, its drought resistance and eco-friendly production perspectives. The higher prices being offered in many parts of the country – especially for small millets, duly recognising their unique nutritional features – is making millets production a remunerative proposition vis-à-vis other crop options available in each region. The R&D efforts made in the area of grain processing and value addition through the development of novel diversified foods especially in sorghum, finger millet and other millets are opening new avenues for expanding the consumer base and enhanced the absorption of production for food use.

The immediate needs for supporting millet promotion are:

I Sustain and increase production of nutritionally rich millets and expand opportunities for value addition as incentives.
II Conserve the genetic resource base and provide superior germplasm possessing the required attributes for food, feed and other diversified use-needs.
III Overcome production constraints by providing technological inputs appropriate to each region.
IV Raise awareness among the general public, policy makers, traders, farming communities, consumers, entrepreneurs and donors on the role of nutrient-dense millets, thereby increasing demand and use.

Keeping all the above in view, research activities are to be restructured and should reflect closely the location-specific needs in each region. Millet R&D should not be viewed from crop and productivity angles only, but also should be more holistic, reflecting other benefits such as acquiring greater ecological balance and climate change adaptation of production systems, as well as strengthening the food, nutrition and health security of people at large.

Conclusion

After years of neglect, small millets suitably designated as climatic-resilient or climate-smart crops, or as nutri-cereals, are now receiving more attention in India's agricultural R&D agenda. Their rich crop diversity has made these plants well suited to contingency crop planning and also to address issues related to climate change. Highly versatile millets like foxtail, barnyard, proso and little millet would fit into any situation of climatic change and could save farmers from total crop failures. Future research priorities for small millets include the utilisation of trait-specific germplasm, basic and strategic research for resistance to biotic and abiotic stresses, varietal diversification, developing viable crop production and protection technologies and value addition.

Small millets are viewed as important crops for the health and wellness of people and can help in preventing many kinds of diseases related to modern lifestyles, including obesity and diabetes. Lately, plenty of elite food chains have begun selling millets and millets-based products as health food. Increased popularity of small millets as nutri-cereals paves away for further strengthening nutritional security in India.

References

All India Coordinated Small Millets Improvement project (AICSMIP) (2012) Annual Progress Report 2012–13. Project Coordinating Unit, AICSIMP, University of Agricultural Sciences, Bangalore.

Byre Gowda, M., Shankare Gowda, B.T. and Seetharam A. (1999) Selection for combining grain yield with high protein and blast resistance in finger millet (*Eleusine coracana* G.). *Indian. J. Genet.* 59(3): 345–349.

Directorate of Millets Development (2018) Govt of India, Agricultural Statistics at a glance.

Gowda, B.T.S., Halaswamy, B.H., Seetharam, A., Virk, D.S. and Witcombe, J.R. (2000) Participatory approach in varietal improvement: a case study in finger millet in India. *Curr. Sci.*, 79(3): 366–368.

Naik, B.J., Gowda B.T.S. and Seetharam, A. (1993) Pattern of variability in relation to domestication of finger millet in Africa and India, In *Recent Advances in Small Millets* (Eds., K.W. Riley, S.C. Gupta, A. Seetharam and J. Moshanga). *Proc. Second Intl. Small Millets Workshop.* Oxford: IBH Publishing Company, pp. 347–364.

National Institute of Nutrition (2007) Nutritive value of Indian foods, NIN, MILLET in your meals.

Prabhakar (2017) Small millets: Climate resilient crops for food and nutritional security. *Mysore J. Agric. Sci.* 51(1): 52–62.

Prasada Rao, K.E., deWet, J.M.J., Brink, D.E. and Mangesha, M.H. (1987) Intraspecific variation and systematics of cultivated *Setaria italica* (Graminae). *Eco. Bot.*, 41: 108–116.

Ravikumar, R.L., Dinesh Kumar, S.P. and Seetharam, A. (1991). Association of seed colour with seed protein, phenol, tannin and blast susceptibility in finger millet (*Eleusine coracana* Gaertn.). *SABRAO J.* 23: 91–94.

Ravikumar, R.L., Seetharam, A. and Gowda, B.T.S. (1990) Identification of sources of stable resistance to finger millet (*Eleusine coracana* Gaertn.) blast. *SABRAO J.* 22: 117–121.

Sashidhar, V.R., Prasad, T.G., Seetharam, A. and Udayakumar, M. (1983). Ear photosynthesis in long and normal glumed genotypes of finger millet (*Eleusine coracana* Gaertn.) *Indian J. Plant Physiol.* 4: 359–362.

Schiere, J.B., Joshi, A.L., Seetharam, A., Costing, S.J., Goodchild, A.V. and Deinum Band Van Keulen, H. (2004) Grain and straw for whole plant value: Implications for crop management and genetic improvement strategies. *Exp. Agric.* 40: 277–294.

Seetharam, A. (2015) Genetic improvement in small millets, In *Millets - Ensuring Climatic Resilience and Nutritional Security* (Eds., V.A. Tonapi and J.V. Patil). ICAR-IIMR, Hyderabad, pp. 233–260.

Seetharam, A., Aradhya, K.M., Sashidhar, V.R., Mahishi, D.M. and Gowda, B.T.S. (1984) Protein content in white and brown seeded finger millet genotypes. *SABRAO J.* 16: 65–67.

6

FROM DISCOVERY TO FOOD SYSTEM DIVERSIFICATION WITH AFRICAN NEGLECTED AND UNDERUTILIZED SPECIES

Céline Termote, Stepha McMullin, Hendre Prasad

Introduction

Africa's dominant food systems heavily rely on exotic species (National Research Council, 2008). Some of the main staples across the continent, such as maize (*Zea mays*) and cassava (*Manihot esculenta*), as well as a number of widely grown fruits and vegetables such as papaya (*Carica papaya*) and tomatoes (*Solanum lycopersicum*), were introduced from the 'New World'. Other species widely present on the continent such as mango (*Mangifera indica*) and irrigated rice (*Oryza sativa*) were imported from Asia.

Nonetheless, a number of species originating from the African continent have made it into global value-chains, including coffee (arabica [*Coffea arabica*] and robusta [*Coffea canephora*]), oil palm (*Elaeis guineensis*), and a cotton species (*Gossypium* spp.) (Van Damme and Termote, 2008). Africa is also a center of diversity for a number of commodities of importance, that are as yet under-researched and under-invested in. Staples such as sorghum (*Sorghum bicolor*), pearl millet (*Pennisetum glaucum*), yams (*Dioscorea* spp.), finger millet (*Eleusine coracana*), fonio (*Digitaria exilis*) and teff (*Eragrostis teff*); legumes such as cowpea (*Vigna unguiculata*) or lablab (*Lablab purpureus*); vegetables such as African egg-plant (*Solanum aethiopicum*) or okra (*Abelmoschus esculentus*); and fruits such as tamarind (*Tamarindus indica*) or baobab (*Adansonia digitata*) have entered national or regional value-chains, but are yet to be used to their full potential (Van Damme and Termote, 2008). At the same time, the continent harbors an enormous amount of locally important species – such as fumbwa (*Gnetum africanum*), spider plant (*Cleome gynandra*), weda (*Saba senegalensis*), safou (*Dacryodes edulis*) and African locust bean (*Parkia biglobosa*) to name but a few – that have potential to be further developed and to contribute to the livelihoods and nutrition security of smallholder family farmers.

DOI: 10.4324/9781003044802-7

However, scientists and policy makers often remain unaware of the great nutritional, socio-cultural, economic and environmental potential of these resources to African food systems (Akinola et al., 2020). Therefore, this chapter presents a number of examples on how to increase interest in these species through research, domestication, production, marketing, consumption and promotion of these valuable species.

Discovery

Ethnobotany is the science that documents human–plant interactions, or, in other words, how humans use the plant biodiversity surrounding them for food, feed, arts, medicines, etc., allowing new crops to be discovered (Van Damme and Termote, 2008).

Ethnobotanical surveys have been and are still being carried out all over the continent in order to document the rich knowledge on edible species held by the inhabitants of specific villages, tribes or regions (e.g., Termote et al., 2011 on wild edible plants in Tshopo District, DR Congo or Ojelel et al., 2019 for wild edible plants in Teso-karamoja, Uganda). Most of the studies only cover specific regions or tribes and it is much harder to find national overviews, such as the book by Maundu et al. (1999) *Traditional Food Plants of Kenya*. Other authors only deal with a specific food group such as fruits (e.g., Gueye et al., 2014, wild fruits from the Malinke in Senegal) or vegetables (e.g., Achigan-Dako et al., 2010, *Traditional Vegetables in Benin*).

While there is plenty of information available on African edible plants that merits inclusion in accessible national and regional databases, there are still many gaps in terms of coverage of ethnic groups, agro-ecological zones and/or regions, in order to unravel the continent's rich botanical heritage before it's lost (Akinola et al., 2020). In that regard, the open access PROTA book series,[1] presenting plant resources for tropical Africa in different volumes, is an interesting resource to start with.

Characterization

A number of these species have the potential to be further developed and promoted for niche markets (Van Damme and Termote, 2008). Such markets allow for price setting by the farmers and offer many more opportunities for small-scale, resource-poor farmers to earn decent livelihoods compared to global commodity markets for items such as coffee, tea or cocoa, where prices are set by international players at the expense of the farmer. As the resources for further morphological, genetic, nutritional, economic or socio-cultural characterizations as well as domestication trials and market development are limited, a prioritization of species for further development is required, a process that should include beneficiary communities.

TABLE 6.1 Species prioritization in Atacora, northern Benin, and percentage of participants who mentioned consuming specific plant parts during ethnobotanical surveys

Species name	Plant parts used	Use (%)
Adansonia digitata	Leaves	100
	Fruits	98.3
	Grains	89.2
Moringa oleifera	Leaves	95.4
Parkia biglobosa	Fruits	98
	Grains	93
Balanites aegyptiaca	Fruits	57
Detarium microcarpum	Fruits	88
Tamarindus indica	Fruits	99
Blighia sapida	Leaves	58.3
	Fruits	98.8
Hibiscus sabdariffa	Leaves	96.3
	Fruits/Flowers	90.4
	Grains	44.2
Cleome gynandra	Leaves	89.2
Ocimum gratissimum	Leaves	99.2
Vigna radiata	Leaves	26.3
	Grains	87.1

For example, in northern Benin, Bioversity International (BI) worked with local stakeholders to prioritize 11 multi-purpose species (Table 6.1), which are currently undergoing full characterization (nutritional composition, socio-cultural and economic value, and agronomic potential). Combined with studies on threats, conservation status and domestication potential, species knowledge products are in development and will support ongoing interventions using local biodiversity to improve diet quality and nutrition of vulnerable populations in northern Benin.

In Northern Kenya (Turkana county), BI researchers documented local knowledge on 64 wild edible plant species (WEPs) (unpublished data); however, almost none of these species were found in the diets of 240 mothers and small children surveyed in pastoralist and agro-pastoralist communities in Loima sub-county (Aluso, 2018). Nonetheless a number of them have the potential to improve dietary quality of women and small children in the area. Six of these wild foods (some naturalized) were selected based on availability of nutrient content data and their potential contribution to closing nutrient gaps (*Sterculia africana* [African-star chestnut], *Berchemia discolor* [wild almond], *Grewia tenax* [white crossberry], *Amaranthus hybridus* [amaranth], *Solanum americanum* [American nightshade] and *Celosia argentea* [celosia]), to model the lowest-cost, most-nutritious diets with and without these wild foods using the Cost of Diet linear programming tool developed by Save the Children, UK.[2] Results showed that the addition of the three wild vegetables resulted in cost reductions of 30%–71% compared to the models without wild foods, as well as making up for iron and zinc gaps in the models without wild foods (Sarfo et al., 2020).

Many NUS have higher nutrient values than their cultivated counterparts (Akinola et al., 2020); however, they do not always contribute as much as they could to closing nutrient gaps or improving livelihoods. In a study around the Lama Forest in southern Benin, 91% of households mentioned they sometimes collect WEPs, but only 8 of the 61 WEPs documented in the area were found in 37% of the 240 24h-dietary intake recalls carried out during the lean season. Fermented *Parkia biglobosa* kernels were consumed the most (Boedecker et al., 2014). Similar results were found in a study on WEP consumption in Tshopo District, DR Congo by Termote et al. (2012) with only 11 out of 77 known WEPs reported in 30% of the recalls carried out in Yaoseko village. Both studies found that people who consumed WEPs the previous day, compared to those that did not, had higher dietary diversity scores and/or higher nutrient intakes for that day. Other studies, such as Dovie et al. (2007) in South Africa, found much higher contributions of WEPs to micronutrient intakes.

Food system diversification with NUS

Increasing knowledge on the nutritional and livelihood benefits of NUS through nutrition education or counseling – whether or not in combination with trainings on how to grow, prepare, conserve and market them – has been successful in several projects throughout the continent. One of the earliest examples (Box 6.1), is the African Leafy Vegetables (ALV) project carried out in Kenya by BI and partners from 1996 to 2003, wherein the multifaceted approach of prioritizing, characterizing, improving production and simultaneously creating demand through nutrition education and media campaigns led to better livelihoods for hundreds of ALV farmers (Gotor and Irungu, 2010). Building on this early work, an integrated community-based approach for farm, market and diet diversity was developed in Vihiga county, Kenya.[3] A diagnostic survey documented dietary intake patterns and agricultural biodiversity. Subsequently, community groups were taken through a series of workshops wherein (a) the importance of nutrition and balanced diets was explained, (b) the diagnostic survey results were discussed and (c) community action plans (CAP) were drafted. All groups decided to grow ALVs and legumes in kitchen gardens and some even added poultry activities. During CAP implementation, the groups were guided through agricultural trainings and nutrition counseling. An end-line survey after one year showed that dietary diversity scores had significantly increased for women as well as for small children in the whole community. The percentage of children meeting minimum dietary diversity (consuming foods from a minimum of four out of seven food groups) also increased significantly (Boedecker et al., 2019). The approach is currently being further piloted in two new contexts in Kenya (in Turkana and Busia counties).

Box 6.1 African Leafy Vegetables in Kenya

Realizing that nutritious traditional leafy vegetables were rapidly disappearing from plates and fields, BI and its partners started implementing a project on ALV in the mid-1990s. Through ethnobotanical surveys, 210 different ALVs were documented, of which few were found to be regularly consumed by those surveyed. Germplasm was collected for conservation in the national gene-bank and characterized. Eight species, including Amaranthus (*Amaranth* spp.), spider plant (*Cleome gynandra*) and cowpeas (*Vigna unguiculata*), were prioritized for further research. Their nutritional composition was determined and the effects of storage and processing on their nutrient composition was studied.

Subsequently, the project worked with farmers and consumers to optimize agronomic protocols and increase knowledge on the nutritional benefits of ALVs. Four hundred and fifty farmers benefited from seed starter packages and training on growing ALVs. Farmers were linked with market outlets and public awareness strategies such as cooking contests; recipe booklets and radio station coverage were organized. As a result, ALVs started appearing in supermarkets in Nairobi (EIARD, 2013). Three years after the project ended, an impact study revealed that the production of ALVs in peri-urban Nairobi had increased more than ten-fold since 1997. Almost 23% more farmers grew a minimum of one ALV, and the average number of ALVs grown increased from 1.5 to 2.3 (Gotor and Irungu, 2010). Women played a key role in the production, marketing and consumption of ALVs, and incomes increased in regions where farmers were linked to markets. Increased awareness of the nutritional value of ALVs was identified as a key driver for the increased demand.

Other interesting models currently being pilot-tested in Kenya to promote production and consumption of NUS are f.e. (1) direct food procurement for school feeding programs, linking ALV farmers with schools in Busia county with the double goal of improving school meal quality, nutrition and performance of pupils as well as improving farmer incomes and livelihoods (Borelli et al., in press); and (2) multi-stakeholder platforms to connect ALV farmers with markets, facilitate business plan development and attract investments in Nakuru (SASS project[4] led by the European Centre for Development Policy Management) and Kisumu (Healthy Food Africa project[5] led by BI), with the aim to build inclusive value-chains that support food system resilience and diet quality.

To support this work, a range of tools to raise awareness on the nutritional and livelihood benefits of NUS have been developed; for example, in southern Benin, BI researchers developed a picture-based recipe book integrating NUS into traditional recipes to improve their nutritional quality (Bodjrènou et al., 2018). Another tool developed and refined within several BI projects is the seasonal food availability calendar to assist extension workers and consumers with year-round selection of foods from different food groups to meet minimum dietary diversity standards and, thus, increase chances of covering all individual nutrient needs (Lochetti et al., 2020).

Trees on farms – promoting diversity

The World Agroforestry (ICRAF) developed the *food tree and crop portfolio* approach to enhance the seasonal availability of nutritious foods in local food systems (McMullin et al., 2019). These nutritious food portfolios are location-specific recommendations for cultivating a greater diversity of indigenous and exotic food trees (those that provide fruits, leafy vegetables, nuts, seeds and oils) with complementary vegetable, pulse and staple crops that could address food harvest and micronutrient gaps in local households' diets. Additionally, the tool shows the nutritional value (vitamin A, vitamin C, iron and folate) of selected species using a scoring system to simplify nutrient content information, which is very useful in supporting decision-making for species selection. Indigenous and underutilized species are included in the portfolios as they are socio-culturally relevant in local food systems, and are adapted to individual landscapes, with a diversity of species enhancing resilience to more variable environmental conditions (Dawson et al., 2019). Tree-foods are often available when other crops fail, and also during the "lean" season. They complement and diversify the predominantly staple-based diets of rural households through the year (Kehlenbeck et al., 2013).

Another tool to support mainstreaming micronutrient rich tree-foods in local food systems is ICRAF's Priority Food Tree and Crop Food Composition Database,[6] which presents nutrient data of over 90 food-tree and crop species. The database presents the backbone of the portfolios but can also be used for dietary assessments, developing training materials, and selecting nutrient-rich species for domestication and breeding programs. However, for a range of indigenous species, nutrient content data is still missing due to a lack of research and private sector interest and investment in these species (Dawson et al., 2018).

One of the biggest challenges to growing a diversity of food-tree species, and NUS in general, is the lack of availability and access to quality planting material. Attention to delivery systems for planting material has been identified as an important success factor for mainstreaming orphan crops (McMullin et al., 2021). To effectively reach large numbers of smallholder farmers, well-functioning, decentralized systems for delivering seeds, seedlings and associated management information are required (Dawson et al., 2012).

Domestication and improvement efforts

ICRAF has been working on domesticating and improving its mandate trees, many of which fall into the category of NUS. The Genetic Resources Unit of ICRAF maintains more than 120 tree species either as seeds or *in-situ* collections. Most trees have a very long lifespan, are propagated using vegetative means and are difficult to breed using traditional methods. Traditionally, trees have been bred using simple methods like selecting good performing ones, known as "(plus)+" trees, which are then supplied as clonal improved material; but further steps are

84 Céline Termote et al.

needed for major increments in phenotypic traits and better performance. To enhance tree-breeding efforts, the African Orphan Crops Consortium (AOCC[7]) is developing genomics resources for 101 crop species, around half of which are trees while the remaining are annual species. The AOCC aims to improve the nutritional value, productivity and climate adaptability of these traditional food crops and to increase their use in African diets (Hendre et al., 2019; Jamnadass et al., 2020). A number of AOCC species have been considered by ICRAF for developing advanced breeding programs using modern genomics tools (Table 6.2).

TABLE 6.2 Mandate tree species list for domestication and improvement. Species marked as "$" have been whole genome sequenced by the Africa Orphan Crops Consortium.

	Species name	Common name
1	Adansonia digitata	Baobab, Mbuyu
2	Moringa oleifera[$]	Moringa, Drumstick tree, Mzunze
3	Sclerocarya birrea[$]	Marula, Mng'ongo
4	Uapaca kirkiana	Wild apple, Wild loquat, Mkusu
5	Allanblackia stulhmanii	Allanblackia, Kionzo, Bouandjo, veg tallow tree
6	Dacryodes edulis	African Plum, Safoutier
7	Vitellaria paradoxa	Shea Butter
8	Ziziphus mauritiana	Jujube, Mkunazi
9	Irvingia gabonensis	Sweet bush mango, Dika
10	Ricinodendron heudelotii	Ground Nut Tree, Muawa
11	Parkia biglobosa	Nere, African Locust
12	Vitex doniana	Chocolate berries, Mfudu
13	Tamarindus indica	Tamarind
14	Annona senegalensis	Wild Custard Apple, Mtokwe
15	Anacardium occidentale	Cashew, Korosho
16	Artocarpus altilis[$]	Breadfruit, Mshelisheli
17	Artocarpus heterophyllus[$]	Jack Tree, Fenesi
18	Dovyalis caffra	Kei Apple, Kei-appel
19	Faidherbia albida[$]	Acacia (Apple-ring)
20	Parinari curatellifolia	Mobola plum, Mbura
21	Saba senegalensis	Gumvines, Mbungu
22	Casimiroa edulis	White sapote
23	Detarium microcarpum	Sweet detar
24	Garcinia livingstonii	African Mangosteen, Mpekechu
25	Mangifera indica	Mango, Maembe
26	Morus alba	Mulberry
27	Psidium guajava	Guava, Mpera
28	Strychnos cocculoides	Kaffir Orange, African Orange, Kikwakwa
29	Strychnos spinosa	Kaffir Orange, African Orange, Kikwakwa
30	Syzygium guineense	Water berry, Mzuari

Conclusion

Africa is home to a lot of promising species with the potential to improve diets and livelihoods. Only a few African species have made it into global value-chains, but a large number of them have regional, national or local importance, albeit with their full potential still yet to be discovered. While knowledge gaps remain on the nutritional, economic, socio-cultural and ecological importance of these species, ongoing research into context-adapted models promoting NUS production and consumption are generating evidence on how these have the potential to enhance diet quality, nutrition and incomes. Given that NUS are also well adapted to their local environments and support farm resilience, they could play an important role in the transition towards healthy, sustainable and inclusive African food systems.

Notes

1 www.prota4u.org
2 https://www.heacod.org/en-gb/Pages/AboutCotD.aspx#:~:text=The%20Cost%20 of%20the%20Diet,of%20protein%2C%20fat%20and%20micronutrients.
3 https://cgspace.cgiar.org/bitstream/handle/10568/110074/community-led_ Nowicki_2020.pdf
4 www.ecdpm.org/sass
5 https://healthyfoodafrica.eu/
6 http://www.worldagroforestry.org/products/nutrition/index.php/home
7 http://africanorphancrops.org/

References

Achigan‑Dako, E.G., Pasquini, M.W., Assogba‑Komlan, F., N'danikou, S., Yédomonhan, H., Dansi, A. and Ambrose‑Oji, B. (2010) *Traditional vegetables in Benin*. Institut National des Recherches Agricoles du Bénin. Imprimeries du CENAP, Cotonou, Bénin.

Akinola, R., Pereira, L.M., Mabhaudhi, T., de Bruin, F.-M. and Rusch, L.A. (2020) Review of indigenous food crops in Africa and the implications for more sustainable and healthy food systems. *Sustainability*, vol. 12, 3493.

Aluso, L. (2018) *A dietary analysis of women in the pastoral and agropastoral livelihood zones in Turkana County*, Kenya. MSc Thesis, Ghent University at Ghent, Belgium.

Bodjrènou, F.S.U., Termote, C., Mitchodigni, I., Honnougan, F., Amoussa Hounkpatin, W., Ntandou Bouzitou, G., Kennedy, G. and Mutanen, M. (2018) *Livret de recettes des aliments de complément pour les enfants âgés de 6-23 mois*. Bioversity International, Bureau de l'Afrique de lOuest et du Centre, Cotonou, Benin.

Boedecker, J., Oduor, F., Lachat, C., Van Damme, P., Kennedy, G. and Termote, C. (2019) Participatory farm diversification and nutrition education increase dietary diversity in Western Kenya. *Maternal and Child Nutrition*, vol. 15, no. 3, e12803.

Boedecker, J., Termote, C., Assogbadjo, A.E., Van Damme, P. and Lachat, C. (2014) Dietary contribution of wild edible plants to women's diets in Benin – an underutilized potential. *Food Security*, vol. 6, 833–849.

Borelli, T., Wasike, V., Maniella, A., Hunter, D. and Wasilwa. L. (2021) "Linking farmers and schools to improve diets and nutrition in Busia County, Kenya". In Swensson, L., Tartanac, F., Hunter, D. and Schneider, S. (eds) Sustainable public food procurement for sustainable food systems and healthy diets.

86 Céline Termote et al.

Dawson, I.K., Harwood, C., Jamnadass, R. and Beniest, J. (2012) *Agroforestry tree domestication: A primer.* World Agroforestry Centre, Nairobi.

Dawson, I.K., Hendre, P., Powell, W., et al. (2018) Supporting human nutrition in Africa through the integration of new and orphan crops into food systems: Placing the work of the African Orphan Crops Consortium in context. ICRAF Working Paper, International Centre for Research in Agroforestry, Nairobi.

Dawson, I.K., McMullin, S., Kindt, R., Muchugi, A., Hendre, P., Lillesø, J.P. and Jamnadass, R. (2019) Integrating perennial new and orphan crops into climate-smart African agricultural systems to support nutrition. The CSA Papers.

Dovie, D.B., Shackleton, C.M. and Witkowski, E.T. (2007) Conceptualizing the human use of wild edible herbs for conservation in South African communal areas. *Journal of Environmental Management*, vol. 84, 146–156.

EIARD (2013) *African leafy vegetables come out of the shade.* Swiss Agency for Development and cooperation, Bern.

Gotor, E. and Irungu, C. (2010) The impact of Bioversity International's African leafy vegetables programme in Kenya. *Impact Assessment and Project Appraisal*, vol. 28, no. 1, 41–55.

Gueye, M., Ayessou, N.C., Koma, S., Diop, S., Akpo, L.E. and Samb, P.I. (2014) Wild fruits traditionally gathered by the Malinke ethnic group in the edge of Niokolo Koba park (Senegal). *American Journal of Plant Sciences*, vol. 5, 1306–1317.

Hendre, P.S., Muthemba, S., Kariba, R., Muchugi, A., Fu, Y., Chang, Y., Song, B., Liu, H., Liu, M., Liao, X., Sahu, X.K., Wang, S., Li, L., Lu, H., Peng, S., Cheng, S., Xu, X., Yang, H., Wang, J., Liu, X., Simons, A., Shapiro, H-Y., Mumm, R.H., Van Deynze, A. and Jamnadass, R. (2019) African Orphan Crops Consortium (AOCC): Status of developing genomic resources for African orphan crops. *Planta*, vol. 250, 989–1003.

Jamnadass, R., Mumm, R.H., Hale, I., Hendre, P., Muchugi, A., Dawson, I.K., Powell, W., Graudal, L., Yana-Shapiro, H., Simons, A.J. and Deynze, A.V. (2020) Enhancing African orphan crops with genomics. *Nature*, vol. 52, 356–360.

Kehlenbeck, K., Asaah, E. and Jamnadass, R. (2013) Diversity of indigenous fruit trees and their contribution to nutrition and livelihoods in sub-Saharan Africa: Examples from Kenya and Cameroon. In J. Fanzo, D. Hunter, T. Borelli and F. Mattei (eds.) *Diversifying food and diets: using agricultural biodiversity to improve nutrition and health.* London, Earthscan, 257–269.

Lochetti, G., Meldrum, G., Kennedy, G. and Termote, C. (2020) *Seasonal food availability calendar for improved diet quality and nutrition: Methodology guide.* Alliance of Bioversity International and CIAT, Rome.

Maundu, P.M., Ngugi, G.W. and Kabuye, C.H.S. (1999) *Traditional food plants of Kenya.* Kenya Resource Centre for Indigenous Knowledge, National Museums of Kenya, Nairobi.

McMullin, S., Njogu, A., Wekesa, B., Gachuiri, A., Ngethe, E., Stadlmayr, B., Jamnadass, R. and Kehlenbeck, K. (2019) Developing fruit tree portfolios that link agriculture more effectively with nutrition and health: A new approach for providing year-round micronutrients to smallholder farmers. *Food Security*, vol. 11, 1355–1372.

McMullin, S., Stadlmayr, B., Mausch, K., Revoredo-Giha, C., Burnett, F., Guarino, L., Brouwer, I.D., Jamnadass, R., Graudal, L., Powell, W. and Dawson, I.K. (2021) Determining appropriate interventions to mainstream nutritious orphan crops into African food systems. *Global Food Security*, vol. 28, 2021. doi:10.1016/j.gfs.2020.100465.

National Research Council (2008) *Lost crops of Africa: Volume III: Fruits.* The National Academies Press, Washington, DC.

Ojelel, S., Mucunguzi, P., Katuura, E., Kakudidi, E.K., Namaganda, M. and Kalema, J. (2019) Wild edible plants used by communities in and around selected forest reserves in and around Teso-karamoja region, Uganda. *Journal of Ethnobiology and Ethnomedicine*, vol. 15, no. 3. doi: 10.1186/s13002-018-0278-8

Sarfo, J., Keding, G., Boedecker, J., Pawelzick, E. and Termote, C. (2020) The impact of local agrobiodiversity and food interventions on cost, nutritional adequacy and affordability of women and children's diet in Northern Kenya: A modeling exercise. *Frontiers in Nutrition*, vol. 7, 129.

Termote, C., Bwama Meyi, M., Dhed'a Djailo, B., Huybregts, L., Lachat, C., Kolsteren, P. and Van Damme, P. (2012) A biodiverse rich environment does not contribute to better diets. A case study from DR Congo. *Plos One*, vol. 7, no. 1, e30533.

Termote, C., Van Damme, P. and Dhed'a Djailo, B. (2011) Eating from the wild: Turumbu, Mbole and Bali traditional knowledge on wild edible plants, district Tshopo, DR Congo. *Genetic Resources and Crop Evolution*, vol. 58, no. 4, 585–617.

Van Damme, P. and Termote, C. (2008) African Botanical heritage for new crop development. *Afrika Focus*, vol. 21, no. 1, 45–64

7

BARRIERS TO ADOPTING A DIVERSITY OF NUS FRUIT TREES IN LATIN AMERICAN FOOD SYSTEMS

Robin Van Loon, Elisabeth Lagneaux, Gabriela Wiederkehr Guerra, Fidel Chiriboga-Arroyo, Evert Thomas, Bruno Gamarra, Maarten van Zonneveld, Chris Kettle

Introduction

Latin America is home to an extraordinary diversity of nutritionally important fruit-tree species (Kermath et al., 2014) and, for millennia, local people have selected and domesticated useful species in their landscape (Levis et al., 2018). Despite the high diversity of nutritionally rich fruit trees, the homogenization and Westernization of consumption patterns in the region have driven the spread of poor diets, particularly in rural areas. Diets, especially of Indigenous communities are increasingly based on staples and processed food with lower nutritional values (Coimbra et al., 2013). Farming in Latin America is increasingly based on unsustainable and environmentally damaging practices, which drive tropical forest degradation and deforestation (Dobrovolski et al., 2011). Over 80% of Latin American farms are managed by smallholders (Leporati et al., 2014), who are incentivized by policy initiatives to prioritize fruit cash-crops at the expense of on-farm diversity (Sthapit et al., 2016). Although crops such as palm oil *Elaeis guineensis*), cacao (*Theobroma cacao*) and banana (*Musa* sp.) can play an important economic role in the region, monocultures make farmers increasingly vulnerable to socioeconomic and environmental shocks (Maas et al., 2020).

The advantages of adopting a greater diversity of fruit-tree species clearly extend beyond economic resilience. One premise of this chapter is that ecological benefits, including restoration of landscapes and delivery of ecosystem services – carbon sequestration, pollination, soil protection, and fauna habitat connectivity – will be generated through increasing the role of underutilized fruit species in local, national and international markets. This will also contribute to the conservation of important genetic resources (vanHove and VanDamme, 2013; Thomas et al., 2018; van Zonneveld et al., 2020). In the Americas alone, there are several thousand fruit-bearing tree species with the potential to generate

DOI: 10.4324/9781003044802-8

income and improve diets (Bioversity, 2004). Despite considerable evidence supporting their ecological, cultural, nutritional, and livelihood-related benefits (Santos, 2005; Jansen et al., 2020), many of these native fruit species remain neglected or underutilized (NUS). While it is clear that not all fruit species have the potential to become global market sensations, significant opportunities exist to increase the range of promising species at regional or national levels, improving community food security and local economies.

We reviewed a representative sample of 150 NUS across 27 families from various Latin American ecoregions. From this species list, we identify 25 high-potential species, considered to have the greatest scope expanding their commercial horizons (Table 7.1). We identify and discuss a range of barriers to adoption of native fruit-tree species, across farmer-facing obstacles, gaps in the value-chain and barriers in consumer demand (Table 7.2). We describe examples of how these specific barriers have been lifted in the past. Finally, we make suggestions on how these solutions can be extended to other contexts in the future, unlocking the potential of a broader range of Latin American biodiversity – thus meeting growing consumer demand for supply-chain transparency and benefiting farmers throughout the continent.

The barriers to adoption of NUS fruit-tree species in Latin American food systems

Multiple factors can inhibit the uptake of native fruit-tree species in diversified farmer production systems and consumers' diets (see Table 7.2). The barriers are compiled into three main categories: (i) farmer-facing constraints at the production stage; (ii) value-chain related obstacles and (iii) challenges related to consumer demand and marketability.

Barriers at production stage

On-farm challenges in fruit production can prevent farmers from adopting certain crops, thereby limiting the diversification of agroforestry systems. These challenges are often related to a lack of knowledge or capacity. In a study on native Amazonian fruit species selected for their high economic, social and ecological benefits, the identified barriers to adoption, whether for home consumption or as cash crops, were primarily socio-technical rather than market or profit oriented (Lagneaux et al., 2021). The most common limitations were, in order of frequency: lack of knowledge about the species, skepticism about productive potential given soil conditions, challenges related to harvests (e.g., in the case of tall palm species) and limited access to seeds and seedlings. This suggests that the factors keeping farmers from integrating more NUS fruit species are diverse and not exclusively economic.

For greater uptake of NUS it is helpful to understand which species are compatible with important commercial crops. Coffee in Central America is often

TABLE 7.1 Selection of 25 high-potential Latin American fruit-tree NUS along with their possible uses, current geographical distribution and recommended scale of market

Scientific name	Family	Common name(s)	Primary product	Other products	Current geographical distribution in Latin America	Recommended scale of market	Description of primary barrier(s) to adoption
Acca sellowiana	Myrtaceae	Feijoa	Fresh fruit	Processed fruit products (frozen pulp, puree, jams)	South America	International	Peculiar subtropical ecology of the species, which requires cold temperatures to fruit; sensitive to soil and microclimatic conditions of degraded drylands (e.g., dry tropical forest and highlands deforested to desert); strict cold-chain requirement
Annona cherimola	Annonaceae	Cherimoya, Custard apple	Fresh fruit	Processed fruit product (frozen pulp, puree)	South America (tropical highlands: Andean valleys)	International	Lack of public knowledge on its nutritional properties; relative fragility of the fruit itself
Annona muricata	Annonaceae	Soursop, Guanábana, Graviola	Fresh fruit	Processed fruit products (frozen pulp, puree); leaf powder (medicinal)	Caribbean, Central America, South America	International	Lack of public knowledge on its nutritional properties; underdevelopment of medicinal products (leaf powders); relative fragility of the fruit itself
Annona squamosa, A. squamosa x A. cherimola	Annonaceae	Sugar Apple; Atemoya	Fresh fruit	Processed fruit products (frozen pulp, puree)	Caribbean, Central America, South America	International	Lack of broad productive base; relative fragility of the fruit itself

Bertholletia excelsa	Lecythidaceae	Brazil nut, Castaña	Edible seeds	Seed oil (edible & cosmetics); timber (illegal throughout most of the species' range); medicinal resin	Southwestern Amazonia	International	Long-term duration of growth before maturity and fruiting can be a disincentive for planting; predation on planted individuals is frequent; most current production is from wild populations, limiting growth opportunities
Brosimum alicastrum	Moraceae	Manchinga, Ramón, Maya Nut	Edible seeds	Leaves (rich fodder in silvipastoral systems); timber; medicinal resin	Central and South America	National	Lack of awareness of the products' nutritional benefits – seeds for human consumption and leaves for animal fodder; lack of diffusion of silvopastoral systems
Caryodendron orinocense	Euphorbiaceae	Metohuayo, Cacay, Inchi, Tacay, Nogal	Seed kernel oil (cosmetics)	Edible seed; seed kernel oil (edible)	Western Amazonia	International	Underdevelopment of markets for this promising product (seed kernel oil)
Erythrina edulis	Fabaceae	Pisonay, Porotón	Edible seeds	Ornamental; living fence	Western South America, and Panama	International	Lack of harvesting technologies, possible issues with insect pests affecting per hectare yields, lack of experiences with intensive cultivation
Eugenia stipitata	Myrtaceae	Arazá	Processed fruit product (frozen pulp, puree)	Fresh fruit	South America	National	Relatively small yield of individual trees leads to understanding of species as "non-cash crop"; fast ripening, perishability, and presence of insects in pre-harvest fruit
Eugenia uniflora	Myrtaceae	Suriname Cherry, Pitanga	Processed fruit product (frozen pulp, puree)	Fresh fruit	South America (central and southern)	International	Relatively small yield of individual trees leads to understanding of species as "non-cash crop"; lack of marketing of fruit product

(Continued)

Scientific name	Family	Common name(s)	Primary product	Other products	Current geographical distribution in Latin America	Recommended scale of market	Description of primary barrier(s) to adoption
Euterpe precatoria	Arecaceae	Huasaí	Processed fruit product (frozen pulp)	Palm heart "cabbage"; timber (for crafts and construction); seeds for handicrafts; medicine (roots)	Amazonia	National	Lack of access to value-added processing equipment (cold-chain), complexity of harvesting (high canopy), unfamiliarity of the product
Garcinia humilis, G. madruno, G. macrophylla, Garcinia sp.	Clusiaceae	Charichuelo, Achachairú	Fresh fruit	Processed fruit product (frozen pulp)	Amazonia	National	Perishable fruits; lack of development of high value processed fruit products; lack of marketing emphasizing nutritional benefits, similar to its relative mangosteen
Grias peruviana	Lecythidaceae	Sacha Mangua	Fresh fruit	Nut milk or similar product; seed oil	Amazonia	National	Need for detailed nutritional analysis of oily coconut-like fruit pulp; lack of marketing and investment in development of novel products (nut milk, etc.)
Inga ilta	Fabaceae	Inga, Guaba Ilta	Fresh fruit pulp	Edible seeds; firewood; mulch and biomass	South America (Peru, Ecuador)	Regional	Germinated seeds and insects in pods make the fruit less attractive for direct consumption (and overall perishability); lack of value-added product attractive to consumers
Mauritia flexuosa	Arecaceae	Aguaje, Burití	Fresh fruit; fresh juice	Processed fruit product (frozen pulp, puree); seed kernel oil; pulp oil	South America (northern)	International	Limited amounts of fruit pulp per weight of fruit; consumer unfamiliarity with the fruit and derived products; traditional supply-chains associated with deforestation of the species for

Species	Family	Common name	Fresh product	Processed product	Geography	Market	Challenges
Myrciaria dubia	Myrtaceae	Camu Camu	Fresh juice	Processed fruit product (frozen pulp, puree); vitamin C supplements	South America (northern and western)	International	Premature harvest leads to low quality end products; Incomplete domestication leads to irregular, heterogeneous products; relatively small yield of individual trees leads to understanding of species as "non-cash crop"
Oenocarpus bataua	Arecaceae	Ungurahui, Patauá	Fresh fruit; fresh juice	Processed fruit product (frozen pulp, puree); seed kernel oil; pulp oil	South America, Panama, and Trinidad (Caribbean)	International	Slow growth; complexity of harvesting (high canopy); perishability of value-added product (pulp); lack of consumer knowledge about added-value products
Pouteria lucuma	Sapotaceae	Lucuma	Fresh fruit	Processed fruit product (flour, frozen pulp, puree)	Andean valleys	National; International (flour)	Low visibility on the international market; unwillingness of large players to promote fruits with underdeveloped supply-chains
Prosopis flexuosa, P. pallida	Fabaceae	Algarrobo, Huarango, Peruvian Mesquite	Processed fruit product (fruit syrup *algarrobina*)	Dried fruit pods; fruit pods for cattle feed in silvopastoral systems	Arid and semi-arid Andean and coastal valleys	International	Unfamiliarity of product for international consumers; sensitivity to soil and microclimatic conditions of degraded drylands (e.g., dry tropical forest and deforested highlands)
Theobroma grandiflorum	Malvaceae	Copoazú	Processed fruit product (frozen pulp)	Seeds for chocolate-like product; seed oil (butter)	Amazonia	International	Lack of processing equipment at farm-level; insufficient communication between key actors (farmers, producers, transporters, etc.)

TABLE 7.2 List of 25 barriers to adoption for Latin American fruit-tree NUS

Barriers to adoption of NUS fruit trees of Latin America	Solution 1	Solution 2	Solution 3
Availability: Barriers at the production stage			
Properties of fruit itself			
Perishability	Value-added on-farm processing (Frozen fruit pulp, sun-drying)	Other value-added processing (freeze-drying, making of jams)	Community-level education on nutritional benefits to increase the local consumption of fruits
	Example Species: *Theobroma grandiflorum*	Example Species: *Eugenia stipitata*	Example Species: *Perebea guianensis, Casimiroa edulis*
Presence of insects in pre-harvest fruit	Training farmers to harvest fruits at a proper stage of ripeness	Application of natural repellents: neem, hot pepper, etc.	Value-added processing methods (frozen fruit pulp, sun-drying, freeze-drying, jams)
	Example Species: *Chrysophyllum cainito, Pouteria caimito*	Example Species: *Anacardium excelsum*	Example Species: *Psidium guajava, Spondias bahiensis*
Incomplete domestication leading to irregular / heterogeneous products	Product transformation to hide irregularities of the raw fruit (value-added processing methods such as pasteurization and pures)	Plant breeding to complete selection of varieties for domestication, as well as grafting, to achieve more uniform, determinant fruiting	
	Example Species: *Eugenia involucrata*	Example Species: *Myrciaria dubia*	
Challenges on the farm (productive ecology)			

Long duration (10+ years) vegetative growth period before maturity and fruiting	Grafting of productive scions onto adequately managed rootstock results in precocious fruiting. Frequently seen in Latin America for cacao, mango, and avocado; the same technology can significantly increase the attractiveness to the farmer of a variety of species.	The creation of successional agroforestry systems provides short-term crops alongside a second tier of crops productive after two to eight years. Long-term crops are much more likely to be successful when planted among crops offering more immediate returns. The productivity of the parcel in time incentivizes and facilitates the establishment of long-term crops.	Seed selection of precocious individuals.
	Example Species: *Bertholletia excelsa, Eugenia uniflora*	Example Species: *Bertholletia excelsa, Dipteryx sp., Pseudolmedia macrophylla*	Example Species: *Hymenaea courbaril, Juglans neotropica*
Predation of planted seedlings by common local fauna (especially rodents), including peeling or ringing of bark, breaking of principle root or main growth leader, etc.	Implement herbivore protection mechanisms: physical barriers, such as stakes set in a dense "fence" around seedling; placement of netting or shade cloth; etc.	Implement herbivore protection mechanisms: biological deterrents or repellents (such as fermented hot pepper, black pepper, and garlic mix, application of neem or copper sulfate, and vinegar or sugar water fly traps; etc.).	
	Example Species: *Bertholletia excelsa*	Example Species: *Juglans neotropica*	
Pests and diseases	Understanding pest/disease ecology and apply biological control and/or targeted chemical mechanisms	Physical barriers for fruit protection such as bagging of banana racemes	
	Example Species: *Annona muricata, Solanum betaceum*	Example Species: *Eugenia stipitata, Annona muricata*	

(*Continued*)

Barriers to adoption of NUS fruit trees of Latin America	Solution 1	Solution 2	Solution 3
Complexity of harvesting (high canopy, thorns, irregular fruiting, etc.)	Development and/or implementation of harvesting tool technology (e.g., telescopic-cutting device). For palms in particular: encouraging the harvest without falling the palm by climbing or usage of harvest tool technology	Development of an adequate pruning regiment for more compact architecture	Grafting and artificial selection for more compact mature individuals
	Example Species: *Astrocaryum murumuru, Euterpe precatoria, Bactris gasipaes*	Example Species: *Matisia cordata, Theobroma grandiflorum*	Example Species: *Matisia cordata, Helycostilis tomentosa*
Relatively small yield of individual trees, or limited seasonality, leads to understanding of species as "non-cash crop"	Crop is treated as just one productive component in a diversified agroforestry system, including either a primary cash crop or a broad diversity of species that each produce modestly. Non-fruit trees (such as for timber and medicine) are included. Demo/model farm plots for farmer capacitation visits are established (selected or created) by corporate, NGO, or public actors.	Establishment of publicly accessible demonstration plots for various diverse agroforestry systems that yield profitable production while including otherwise individually-neglected species. Proper signage and/or in depth educational tours accompany the areas themselves.	Interventions at a level of plant breeding and selection. Participatory selection and breeding of most productive genotypes; management of pollinators and consideration of hand pollination (where economically viable).
	Example Species: *Eugenia biflora, Lecointia amazonica, Theobroma bicolor*	Example Species: *Platonia insignis, Myrciaria tenella*	Example Species: *Grias peruviana, Myrciaria dubia*

Lack of access to productive ecologic knowledge (How, when and where do you plant which tree? How productive is it at what age?) due to limited knowledge transfer among farmers, practitioners, and research institutes	Knowledge-spreading platforms: seed-sharing events, farmer field schools, farmer horizontal knowledge exchange; events oriented toward farmer's associations or cooperatives as well as individuals; working groups on specific underutilized species that bring together actors from all sectors. Example Species: *Theobroma bicolor, Myrciaria dubia*	Production of more practitioner-friendly learning resources (e.g., species technical sheets on productive ecology; adapting existing tools such as "Agroforestree" database from ICRAF, ECHO Community, Agroforestry.net species sheet, etc.); and making them available to farmers (e.g., free and accessible online, sharing them through outreach programs). Example Species: *Oenocarpus bacaba, Brosimum alicastrum*	Creation of bridges across national research institutions (e.g., IIAP in Peru, Embrapa in Brazil, etc.), research organizations (e.g., NGOs CINCIA in Peru, Herencia in Bolivia) and local/regional governments in order to transfer evidence-based tools for the development of diversity-friendly policies and interventions. Example Species: *Euterpe precatoria, Mammea americana*
Products primarily derived from wild populations rather than plantations, making harvest costly and challenging	Development of new approved management plans for harmless harvest of Non-Timber Forest Products (NTFPs) in forestry concessions and protected areas Example Species: *Euterpe precatoria*	Enrichment planting in primary and secondary forest areas; in agroforestry system orchards Example Species: *Bertholletia excelsa, Carapa guianensis*	Identification of wild productive forests in the territories of native or *campesino* communities, harvest by and for community members Example Species: *Oenocarpus bataua, Mauritia flexuosa*
Poor plantation performance due to degraded, contaminated and/or poor soils	Planting design considerations: extra high planting density, successionality, basic earthworks (swales, large planting holes), cover crops, tutor species, irrigation and shade, desert and drylands reforestation technologies (e.g., polypropylene plant cocoons and other self-irrigation mechanisms), etc. Example Species: *Prosopis flexuosa*	Agroforestry systems including organic matter-producing and nitrogen-fixing fruit-tree species. Example Species: *Acca sellowiana*	Public and private farmer-facing programs emphasizing on-farm fertility: compost production, breeding of beneficial indigenous microorganisms, locally available organic agricultural inputs such as rock phosphate and bird or bat guano, etc. Example Species: *Annona muricata*

(Continued)

Barriers to adoption of NUS fruit trees of Latin America | *Solution 1* | *Solution 2* | *Solution 3*

On-farm infrastructure

	Solution 1	Solution 2	Solution 3
Difficulty of germination or propagation for certain species, meaning farmers have limited access to seedlings and are unable to successfully propagate their own seedlings without specialist nurseries	Dissemination of knowledge on how to establish nurseries, ideally with the use of local materials and familiar or appropriate design. Community- and family-level nursery can be developed by NGO, governmental, or corporate initiative. The cost of widespread nursery establishment may require financial mechanisms designed to recognize the value of seedlings produced in the nurseries over time. Example Species: *Bertholletia excelsa, Juglans neotropica*	Establishment of public and private nurseries operating at adequate volume to produce seedlings at low cost. Public programs oriented toward making nursery products more accessible to farmers. Example Species: *Grias peruviana*	Corporate supply-chain programs, in which farmers supplying raw materials are incentivized to enhance on-farm diversity, including the delivery of seedlings for species that are challenging or costly to propagate. Example Species: *Bertholletia excelsa*
Lack of accessibility to farm sites by vehicles, impacting farm and market access for farm products, especially perishable ones, and ease of delivering seedlings to farm	Supply-chain investment in innovative alternative means for scaling delivery and distribution of seedlings and reception of farm products. In some rural areas, motorcycles and beasts of burden can provide solutions. Examples in the Amazon region include investment in boats to purchase farm products directly from inaccessible producer communities. Example Species: *Annona sp.*	Public investment in appropriate transportation systems, such as public boat transportation systems in regions with navigable rivers and few roads. Construction of alternative roads, such as motorcycle-accessible gravel paths that avoid some of the pitfalls of the environmental and social impacts of large road construction, including increased deforestation and delinquency. Example Species: *Euterpe sp, Pourouma cecropiifolia*	Empowerment and financing of cooperatives to allow efficient accumulation and transportation of the goods of various farmers. With greater scale of production, short-term investments such as truck hire for transport of goods to market are justifiable and financially feasible. Example Species: *Euterpe sp, Theobroma sp.*

Lack of on-farm infrastructure that would allow the extension of species' geographical range, such as irrigation systems; and related financial constraints	Public investment in infrastructure, such as agricultural water access via pipelines that extend urban grids into rural areas; and preferably decentralized solutions such as motor pumps and hoses for individual farmers in areas with river or stream access. Example Species: *Myrciaria dubia*	Dissemination of knowledge and appropriate seed varieties for dry farming. Use of grafting with naturalized and native rootstock. Example Species: *Prosopis flexuosa*	Planting design considerations: extra high planting density, basic earthworks (swales, large planting holes), cover crops, tutor species, irrigation and shade. Example Species: *Vasconcellea pubescens*

Accessibility: Barriers at the value–chain stage
Supply-Chain Side

Current lack of a promising, known value-added product that is attractive to consumers and is shelf-stable	Innovation in product development (e.g, flour, oil extraction, pulp) Example Species: *Inga sp, Perebea guianensis*	Innovation in value-added processing (cold-chain for frozen pulp, jams, freeze-dried products) Example Species: *Mammea americana, Solanum betaceum*	Creation of regional innovation/think hubs connecting local people and creating and also financing innovative activities (startups, knowledge exchange, etc.) Example Species: *Ugni molinae, Theobroma speciosum*
Lack of access to value-added processing knowledge and equipment (e.g., cold-chain infrastructure)	Farmer outreach programs that support the development of value-added products that can be processed on-farm simply and inexpensively: purees, marmalades, dried fruit, jams Example Species: *Theobroma grandiflorum, Eugenia stipitata*	Producer associations with robust governance that manage processing equipment, infrastructure, vehicles for transport to market, and other capital considerations. These association can be directly linked with the buyers of specific products. Example Species: *Myrciaria dubia*	Knowledge transfer of cost-effective and available methods for fruit transformation to transportable and non-perishable products; from technical institutions to farmers, e.g., through increased collaboration between farmer association and local institutions Example Species: *Theobroma bicolor*

(Continued)

Barriers to adoption of NUS fruit trees of Latin America	Solution 1	Solution 2	Solution 3
Prohibitive costs of production	Improved collaboration (e.g., cooperatives) between farmers to organize and finance processing equipment, materials, ingredients and increase processing capacity.	Access to public or private investment (loan, funding, etc.), productive and technological support by governmental institutions	
	Example Species: *Astrocaryum murumuru, Carapa guianensis*	Example Species: *Euterpe precatoria*	
Informal and inadequate transportation from farm to market (e.g., for fruit pulp)	Coordination between farmers to organize and finance transport to village hub locations with cold-chain adequate means of transport (e.g., truck with cold chamber).	Access to funding for cold-chain adequate means of transport.	
	Example Species: *Euterpe precatoria, Spondias bahiensis*	Example Species: *Poraqueiba sericea, Pourouma cecropiifolia*	
Market side			
Lack of a sufficient productive base to support scaling to satisfy a larger market; challenges of promoting new species without a strong existing market	Municipal nurseries, private sector investment in propagation and diffusion of seedlings; connection of these efforts directly to buyers and processors of the raw materials	Large promotion campaigns a large company or many SMEs	
	Example Species: *Pouteria lucuma, Sambucus peruviana*	Example Species: *Theobroma speciosum*	
Acceptability: Barriers in consumer demand and marketability			

Barrier	Strategy	Strategy	Strategy
Unfamiliarity of product and product's nutritional benefits to local/regional consumers	Public campaigns: diffusion of seedlings; knowledge transfer from science to markets. Example Species: *Poraqueiba sericea*	Engagement with public health and education institutions (e.g., Ministry of Health, Ministry of Education) for the distribution of publicity/educational materials around the fruit's benefits in municipal markets and schools. Example Species: *Myrciaria dubia, Euterpe precatoria*	Form a growers' association that works in collective to penetrate new markets; and promote processing to frozen pulp for placement in school lunch and other public food programs as well as private sector sales. Example Species: *Oenocarpus bataua, Euterpe precatoria*
Popular perception of traditional crops as incompatible with economic or cultural progress; preference for introduced competing fruit species considered more "prestigious"	Public campaigns promoting local products. Example Species: *Spondias bahiensis*	Use of new marketing tools, e.g., promotion by influencers, adoption by celebrity chef. Example Species: *Poraqueiba sericea*	Governmental support to increase production volume, reach competitive market prices and facilitate market access. Example Species: *Mauritia flexuosa*
Unfamiliarity of product to international consumers	Promote familiarity on local market as a basis for international promotion. Example Species: *Mauritia flexuosa, Oenocarpus bataua*	Flagship brand communicating heavily about nutritional benefits (like "superfoods" and Sambazon for Açaí). Example Species: *Pouteria lucuma, Theobroma bicolor*	Adoption by recognized company/brand(s) (e.g., Coca-Cola for açaí). Example Species: *Theobroma speciosum*
Competition with other exotic fruits with an existing reputation (Mango, Passion fruit, etc.)	Promotion campaigns. Example Species: *Schinus molle, Vasconcellea pubescens*	Find niche markets, where customers are attracted to unknown tastes (e.g., Latin American restaurants). Example Species: *Ugni molinae*	Inform customers about the nutritional value of the fruit. Example Species: *Eugenia uniflora, Mauritia flexuosa*
Preference to locally produced or regionally existing fruit products, rather than import of novel and unfamiliar products from other regions	Raising awareness of the social and ecological value related to NTFP produced in Latin America. Example Species: *Euterpe precatoria*	Ensure energy efficient transport and low carbon footprint (ship freight) and inform customers about it. Example Species: *Pouteria lucuma*	
Short time window between fruit harvest in tropics and summer market peak in US and EU	Promotion of less season-specific end products. Example: Salty dishes, desserts others than ice cream. **Example Species:** *Eugenia uniflora*	Irrigation systems. Example Species: *Myrciaria dubia*	

interplanted with fruit species that are complementary in terms of size and ecology, such as guava (*Psidium guajava*) and hog plum (*Spondias mombin*) (de Sousa et al., 2019). However, in many cases, in-field compatibility among fruit species may be limited; therefore, optimizing the designs of farm parcels is crucial to achieve successful outcomes. The lack of agronomic knowledge is also a significant barrier to the adoption and success of NUS – albeit one that can be overcome with training and horizontal knowledge sharing among farmers, such as in farmer field schools (see section 3).

Decision-making on which species to plant is sometimes influenced by the time it takes for a species to become productive. The Amazonian Brazil-nut tree (*Bertholletia excelsa*) for example is a hyper-dominant species, delivering multiple ecosystem services (Thomas et al., 2018). However, the trees only become profitable after 15–20 years or more and concerns in the region about its future market viability are increasing despite some enrichment planting (Bronzini, 2019). The investment required to bring this species to productive age may discourage its use when other species can deliver returns quicker. This serves to illustrate how a species' ecology – and the level of knowledge thereof – can act as barriers to adoption, in this case by smallholders in the Amazon.

Barriers at the value-chain stage – from farm to market

Two of the most significant challenges faced by smallholders in rural areas are post-harvest storage and the transport of fresh fruit to market. Given the perishability of many tropical fruits, processing them into value-added products with longer shelf-life is crucial for producers to fully benefit from the economic potential of these species. Farmers often face basic infrastructural limitations such as lack of electricity or potable water, restricting fruit processing options (Smith et al., 2007). Many fruits are never folded into even local market supply-chains due to the simple lack of transportation access to those markets, despite the fruits being delicious and highly nutritious. Fruits are typically moved at the farmer's expense by boat, motorcycle or truck to intermediate buyers, leading to less profitable transactions and an increased risk of damage to the product, resulting from a lack of operational cold chains. Consequently, market opportunities for fresh fruits are often extremely limited, and improvements to on-farm processing are needed for the development of viable fruit value-chains. Fresh copoazú (*Theobroma grandiflorum*), a relative of cacao, has a rather short shelf-life, perishing only four days after fruit-drop if it is not cold stored or processed. Other fruits are even faster to lose their color or flavor after harvest. For such species, the constant monitoring of trees, quick harvests and rapid transport to the closest local market are required in the absence of on-farm infrastructure.

Another Amazonian fruit, camu camu (*Myrciaria dubia*), prized for its outstanding nutritional value (Rodrigues et al., 2001), requires careful handling for optimum quality. If harvested semi-ripe, the fruit contains high amounts of vitamin C – the highest of any fruit – but its color and flavor are less attractive.

Harvested ripe, it possesses an impressive color and flavor, but the vitamin C content is less stable. Promoting fruits such as camu camu requires research and investment in more sophisticated fruit processing capacity, which, in this example, would enable the stabilization of vitamin C while also preserving organoleptic qualities.

Barriers in consumer demand and marketability

Many native fruits are unknown to urban consumers, hampering their inclusion in national diets. In big cities, fruit consumption is often limited to conventional, familiar fruits. Even though copoazú and Brazil-nuts are major income generators in the Peruvian Amazon, they are almost unknown in the capital Lima, where their consumption is surpassed by strawberries and almonds. This was also historically the case for camu camu – until 2019 when the Aje Group, Peru's largest soft-drink company, carried out a national publicity campaign and established the fruit as the main ingredient in a beverage on the market. This example demonstrates the impact that effective awareness-raising and marketing can have on increasing the visibility of NUS fruits. Public institutions can also play an important role in educating the masses on the benefits of a variety of locally produced fruits – for example, municipalities organizing cultural and gastronomic events.

On the other hand, consumer preference for local fruits in the Global North can amount to an obstacle to the promotion of underutilized Latin American fruits in international markets. This is mainly because of concerns about sustainability, the carbon footprint of airfreight and the lack of transparency of supply chains in remote countries. Novel tropical-fruit products can benefit from expanding conscientious markets (Jansen et al., 2020); consumers are increasingly demanding ethically sourced products that demonstrate traceability and provide guarantees of ecological sustainability and social fairness. Value-chains in which brands and retailers are committed to tracing fruits back to farmers can provide powerful incentives to continue growing native fruit species. A new product is mainly successful when it offers great taste, high quality and competitive pricing, but marketing can also break through by focusing on nutritional, social or environmental benefits.

The removal of barriers to adoption of native fruit-tree species in Latin American food systems.

For any particular NUS, multiple barriers to adoption are often present. The removal of one significant barrier can sometimes be sufficient to unlock the potential of a species. We do not necessarily need to confront all barriers to adoption to effectively address crucial bottlenecks.

Farm-facing solutions: In general, the removal of on-farm barriers requires engagement, working closely with farmers at each stage in the production process.

This is especially the case when scaling-up diverse farming systems, where long-term species are intercropped with short- and medium-term income-generating species – successional agroforestry. Government and other programs can help reduce the risk of traditional cultivars being replaced with monocultures by promoting diverse agroforestry systems with multiple perennial species, such as in the case of cacao cultivation in Central America (de Sousa et al., 2019). Diversity fortifies resilience, both on the farm and in the market, yet does require effective decentralized platforms for delivery of technical support, capacity building and knowledge transfer on species selection, planting design, management and harvesting strategies. Agronomic knowledge on the compatibility and complementarity of different native fruit species in agroforestry systems also needs to be more widely available and horizontally shared.

Finally, gaps in infrastructure and know-how for propagation (nurseries) and distribution in rural, remote areas with limited accessibility must be addressed if promising species are to be widely adopted. In the example of cherimoya, access to selected cultivars in combination with improved management (e.g., introduction of new pruning approaches) has resulted in increased incomes for Andean farmers in Peru, Ecuador and Bolivia (Vanhove and Van Dammel, 2013).

Another example is ramón (*Brosimum alicastrum*), a hyper-dominant forest species that is now embraced by local communities in the Maya Biosphere Reserve in Guatemala. Community- and farmer-facing outreach focused on the economic and nutritional benefits of new fruits is fundamental to the success of such non-profit programs. At the same time, the organization of fruit harvesters in associations – in this case with the collaboration of associations such as Asociación de Comunidades Forestales de Petén and NGOs like Rainforest Alliance (Izabela et al., 2019) – can be key to hitting export volumes and for successful co-investing in necessary infrastructure.

Value-chain development: A critical aspect of fomenting the adoption of native fruit species is the provision of economic incentives to farmers through reliable and equitable value-chains. In parallel, ensuring a consistent product through processing is critical to resilient value-chains. In Latin America, numerous examples of successful fruit processing exist, including pulp, dried fruit, flour, jams and freeze- or spray-dried powders (e.g., Moraes et al., 1994). All these added-value mechanisms help remove fruit transport barriers, though each has its own disadvantages. For example, the production of freeze-dried powder requires expensive equipment and fruit pulps need strict cold chains; meanwhile, jams are niche products with low overall consumption and demand. Opportunities exist for private and public sector actors to innovate. One alternative for overcoming fruit durability and transport barriers while preserving most of the fruit's natural features is the production of purees, a pasteurized form of fruit pulp that is highly versatile and can be used as an ingredient in a variety of final products. Puree is simple and cost-efficient to process and it can be stored and transported without a strict cold-chain. This is a solution that can be managed at the farm-level, keeping the added value of fruit processing in the hands of farmers.

Many of Latin America's most economically significant trees provide other commodities such as seed oil for cosmetic or edible use. Brazil-nut, murumuru (*Astrocaryum murumuru*), and andiroba (*Carapa guianensis*) are just a few of the many native seed oils that have been successfully commercialized in Brazil (Plowden, 2004; Campos et al., 2015; Smith, 2015). Basic equipment – in this case oil presses – is all that is needed to enable small producers or cooperatives to offer value-added products to national markets. When this barrier – lack of access to capital or know-how for value-added equipment – is removed, fruits that previously were left to decompose on the forest floor can incentivize sustainable management of natural forest resources and the augmentation of wild production via agroforestry planting. In the example of Brazil, accessible financing terms are offered to farmers for investment in equipment and infrastructure (Machado de Moraes, 2014). Economic policies that promote the establishment of village-level infrastructure and development of local market-facing supply chains play a key role in successfully lifting a species from NUS status. The establishment of robust cooperatives or producer associations can help secure financing for equipment and strengthen producers' negotiating power with buyers and is, therefore, a vital component of fair trade certification standards – ensuring consumer satisfaction while also making a compelling argument for the consumption of unique fruits from distant regions.

Up-scaling NUS access to consumers: The Amazon is home to a rich diversity of palms that are extremely abundant producers of nutritious and culturally significant fruits. The pulp of açaí (*Euterpe oleracea*) is now a globally consumed, antioxidant-rich "superfood" that was traditionally prominent in the forest economy and food security of the lower Amazon basin (Muñiz-Miret et al., 1996). This product became known beyond Brazil's borders thanks both to well-positioned flagship brands in foreign markets (such as Sambazon in the US) and to a range of policies that support farm- and village-level producers (OECD, 2011).

In addition to açaí, acerola (*Malpighia glabra*), burití (*Mauritia flexuosa*) and copoazú have reached markets beyond the Amazon basin, but the lesson learned from Brazil suggests that establishing domestic as well as international markets is important for the creation of resilient value-chains. Brazil boasts a tremendous diversity of effectively domestically marketed products from the Amazon bioregion that remain underutilized in neighboring Peru, Bolivia, Ecuador and Colombia, presumably due to a lack of similar national economic policies. Species that remain underutilized throughout their native range but which have been developed in Brazil include peach palm (*Bactris gasipaes*), pama (*Pseudolmedia macrophylla*), pitanga (*Eugenia uniflora*), patauá (*Oenocarpus bataua*) and charichuelo (*Garcinia madruno*).

Governmental organizations, NGOs and Brazilian agricultural research institutes have promoted the consumption of native fruit species, resulting in the publication of a national ordinance that officially recognizes the nutritional value of more than 60 native food plants (Beltrame et al., 2016). With this policy

106 Robin Van Loon et al.

support, federal and local governments have been able to achieve the inclusion of several fruits in subnational and local school lunch programs. Farmers who supply these programs are able to diversify their farms with nutritious food plants because the officially mediated procurement market incentivizes them to do so (Wittman and Blesh, 2017). Significant secondary benefits are achieved by introducing NUS in such school lunch programs, enhancing cultural acceptance and leading the way to the development of more robust local economies for these fruits.

The promotion of new fruit products in national or international markets requires investment. For successful adoption, knowledge about the fruit diversity of a region must be shared among producers, practitioners and distributors and with food businesses, including restaurants and end-consumers. Food fairs such as Lima's massive Mistura Food Festival, academic presentations and social events are just some of the many platforms that can be used to promote new tastes and raise awareness of the nutritional, social and ecological values of fruits. A key component is the involvement of adequate first-market adopters including tastemakers and celebrity chefs such as Virgilio Martinez and Pedro Miguel Schiaffino, two of Latin America's most important voices currently at the intersection of gastronomy and biodiversity. Successfully promoting new and exotic tastes involves customers that are curious and open, for example, to exotic and innovative cuisine, as found in some of Latin America's top-ranked restaurants. To reach these customers and further broaden these fruits' clientele, creative cultural solutions are needed – a challenge that is multifaceted yet of great importance if broader acceptance of unfamiliar fruits is to be achieved.

Conclusion

The barriers to adoption for underutilized fruit species in Latin America are as complex and diverse as the fruits themselves. The challenges in mainstreaming these fruits in food systems span from on-farm barriers, such as a lack of infrastructure or knowledge, to gaps in the value-chain, such as inability to process, stock or transport raw material, to lack of consumer demand because some products are not affordable or accepted. Heterogeneous and unfamiliar products struggle to scale up market access and consumer awareness. There are burgeoning examples of native fruit species that were, until recently, underutilized, but that now provide much-needed sustainable livelihood options for rural communities across the region. As demonstrated in this chapter, innovative processing systems, social institutions, ethical consumption and awareness-raising all play a role in lifting barriers (Table 7.2), to establish more resilient agroecological systems using the extraordinary genetic diversity of Latin American fruit-tree species. To synergistically contribute to income generation, diet diversification and resilient landscapes, poly-cultural food production using a greater diversity of fruit trees generates benefits from the local to the global scale. Simultaneously, better social outcomes are achieved, such as the creation of diverse diets with increased fruit consumption and associated public health improvement, development of more resilient farming systems, enriched livelihoods and the strengthening of local food security in cases of catastrophes. The diversity of fruits in Latin America is considerable and, while often overlooked, has the potential to help farmers overcome many of the biggest challenges they face. In order to unlock this transformative potential, practitioners from all sectors – governments, academia, businesses and NGOs – must collaborate on generating and sharing the practical knowledge of successes and failures that represents our most viable strategy for diversifying the food system.

References

Beltrame, D. M. O., Oliveira, C. N. S., Borelli, T., Hunter, D., Neves, C., De Andrade, R., et al. (2016). Diversifying institutional food procurement–opportunities and barriers for integrating biodiversity for food and nutrition in Brazil. *Raizes*, 36, 55–69.

Bioversity (2004) New World Fruits Database. Retrieved from https://www.bioversity international.org/e-library/databases/new-world-fruits/ on the 9/11/2020.

Bronzini, L. (2019) *A Decision-Supporting Monetary Framework for Brazil Nut Plantation and Enrichment in Madre de Dios, Peru*. ETH Zürich, Switzerland.

Campos, J.L.A., da Silva, T.L.L., Albuquerque, U.P., Peroni, N. and Araujo, E.L. (2015) Knowledge, use, and management of the babassu palm (*Attalea speciosa* Mart. ex Spreng) in the Araripe Region (Northeastern Brazil). *Economic Botany*, 69(3), 240–250.

Coimbra, C.E., Santos, R.V., Welch, J.R., et al. (2013) The first national survey of indigenous people's health and nutrition in Brazil: Rationale, methodology, and overview of results. *BMC Public Health*, 13, 52. https://doi.org/10.1186/1471-2458-13-52

de Sousa, K., van Zonneveld, M., Holmgren, M., et al. (2019) The future of coffee and cocoa agroforestry in a warmer Mesoamerica. *Scientific Report*, 9, 8828. https://doi.org/10.1038/s41598-019-45491-7

Dobrovolski, R., Diniz-Filho, J.A.F., Loyola, R.D. and De Marco, J.P. (2011) Agricultural expansion and the fate of global conservation priorities. *Biodiversity Conservation*, 20, 2445–2459

Izabela, A., Barrientos, F., Muir, G., Madrid, J. J., Baumanns, E. and Vanderwegen, L. (2019) Guatemala's nutritious green gold from the "tree of life". In: Pullanikkatil D., Shackleton C. (eds.), *Poverty Reduction through Non-Timber Forest Products* (pp. 59–64). Springer, Cham.

Jansen, M., Guariguata, M. R., Raneri, J. E., Ickowitz, A., Chiriboga-Arroyo, F., Quaedvlieg, J. and Kettle, C. J. (2020) Food for thought: The underutilized potential of tropical tree-sourced foods for 21st century sustainable food systems. *People and Nature*, 00, 0–15.

Kermath, B. M., Bennett, B. C. and Pulsipher, L. M. (2014). *Food Plants in the Americas: A Survey of the Domesticated, Cultivated, and Wild Plants Used for Human Food in North, Central and South America and the Caribbean*, Unpubl. Manuscript, University of Wisconsin Oshkosh, Oshkosh.

Lagneaux, E.; Jansen, M.; Quaedvlieg, J.; Zuidema, P.; Anten, N.; García Roca, M.R.; Corvera-Gomringer, R.; Kettle, C. (2021). Diversity bears fruit: Evaluating the economic potential of undervalued fruits for an agroecological restoration approach in the Peruvian Amazon. *Sustainability,* 13(8), 4582. doi:10.3390/su13084582

Leporati, M., et al. (2014) La agricultura familiar en cifras. In S. Salcedo, and L. Guzmán (eds.) *Agricultura Familiar em América Latina y El Caribe: recomendaciones de política*, Part 1, Chapter 2 (pp. 35–56). FAO, Santiago, Chile.

Levis, C., Flores, B.M., Moreira, P.A., Luize, B.G., Alves, R.P., Franco-Moraes, J., Lins, J., Konings, E., Peña-Claros, M., Bongers, F. and Costa, F.R. (2018) How people domesticated Amazonian forests. *Frontiers in Ecology and Evolution*, 5, 171.

Maas, Bea, Thomas, E., Ocampo-Ariza, C., Vansynghel, J. Steffan-dewenter, I. and Tscharntke, T. (2020) Transforming tropical agroforestry towards high socio-ecological standards. *Trends in Ecology & Evolution*, xx(xx), 1–4. https://doi.org/10.1016/j.tree.2020.09.002.

Machado de Moraes, A.L. (2014) Evolução da política agrícola brasileira. *Revista de Política Agrícola*, 23(3), 55–64.

Moraes, V.D.F., Müller, C.H., De Souza, A.G.C. and Antonio, I.C. (1994) Native fruit species of economic potential from the Brazilian Amazon. Embrapa Amazônia Ocidental-Artigo em periódico indexado (ALICE).

Muñiz-Miret, N., Vamos, R., Hiraoka, M., Montagnini, F. and Mendelsohn, R.O. (1996) The economic value of managing the açaí palm (*Euterpe oleracea* Mart.) in the floodplains of the Amazon estuary, Pará, Brazil. *Forest Ecology and Management*, 87 (1–3), 163–173.

OECD (2011) Agricultural Policy Monitoring and Evaluation (2011) OECD Countries and Emerging Economies. Retrieved from http://www.oecd.org/brazil/brazil-agriculturalpolicymonitoringandevaluation.htm#more on the 17/06/2020

Plowden, C. (2004) The ecology and harvest of andiroba seeds for oil production in the Brazilian Amazon. *Conservation and Society*, 2, 251–272.

Rodrigues, R.B., De Menezes, H.C., Cabral, L.M., Dornier, M. and Reynes, M. (2001) An Amazonian fruit with a high potential as a natural source of vitamin C: The camu-camu (*Myrciaria dubia*). *Fruits*, 56, 345–354. https://agritrop.cirad.fr/485634/1/document_485634.pdf

Santos, L. M. P. (2005) Nutritional and ecological aspects of buriti or aguaje (*Mauritia flexuosa* Linnaeus filius): A carotene-rich palm fruit from Latin America. *Ecology of Food and Nutrition*, 44(5), 345–358. doi: 10.1080/03670240500253369.

Sthapit, B., Lamers, H.A. and Rao, V.R. (2016) On-farm and in situ conservation of tropical fruit tree diversity: context and conceptual framework. In: Sthapit, B., et al. (eds.) *Tropical Fruit Tree Diversity: Good Practices for in Situ and On-farm Conservation* (pp. 3–22). Routledge, Abingdon. https://cgspace.cgiar.org/handle/10568/75612

Smith, N. (2015) *Astrocaryum murumuru*. In *Palms and People in the Amazon* (pp. 61–72). Geobotany Studies (Basics, Methods and Case Studies). Springer, Cham.

Smith, N., Vasquez, R. and Wust, W. (2007). 'Amazon River Fruits: Flavors for Conservation'. Amazon Conservation Association. ISBN: 978–9972-2974-2–7

Thomas, E., Atkinson, R. and Kettle, C. (2018) Fine-scale processes shape ecosystem service provision by an Amazonian hyperdominant tree species. *Scientific Reports*, 8(1), 0–11. https://doi.org/10.1038/s41598-018-29886-6

van Zonneveld M, Turmel M-S, and Hellin J. (2020) Decision-making to diversify farm systems for climate change adaptation. *Frontiers in Sustainable Food Systems*, 4, 32. https://doi.org/10.3389/fsufs.2020.00032

Vanhove, W. and Van Damme, P. (2013) Value chains of cherimoya (*Annona cherimola* Mill.) in a centre of diversity and its on-farm conservation implications. *Tropical Conservation Science*, 6(2), pp.158–180.

Wittman, H. and Blesh, J. (2017) Food Sovereignty and Fome Zero: Connecting Public Food Procurement Programmes to Sustainable Rural Development in Brazil. *Journal of Agrarian Change*, 17(1), 81–105.

8

PROMOTING BUCKWHEAT IN CHINA

Zongwen Zhang

Introduction

Buckwheat is an annual crop native to China. There are two main cultivated types, namely, common buckwheat and tartary buckwheat. The former is also called sweet buckwheat, while the latter is also called bitter buckwheat. Buckwheat is used for multiple purposes. The flour from the seed can be used to make a variety of foods. The straw and leaves can be used as fodder and medical raw material. Buckwheat grain is rich in protein, vitamins, minerals, plant cellulose and other nutrients, and contains flavonoids, especially rutin, which has pharmacological effects for reducing blood fat and cholesterol and preventing cardiovascular diseases (Lin, 2013). Buckwheat is very adaptable to poor soil and cool weather conditions. The growth period of buckwheat is short – it generally takes about 60–90 days to mature. The annual planting area for the crop in China is 0.6–0.8 million hectares, and the total output is 0.5–0.6 million tons (Zhang and Wu, 2010). Buckwheat is a minor crop in China, as its cultivation area is small compared to staple crops such as rice, wheat and corn. However, buckwheat is considered an advanced food due to its high nutritional and medical values. The market demand for it at home and abroad is constantly expanding, which brings opportunities for strengthening buckwheat research and development (R&D) in China. This chapter aims to systematically review the current status of buckwheat biodiversity and its R&D in China, as well as identify research gaps and needs, and put forward action plans for improving the sustainable development of the buckwheat industry.

Buckwheat biodiversity

Species diversity

Buckwheat is a pseudo-cereal crop belonging to the genus *Fagopyrum*, Family Polygonaceae. *Fagopyrum* species are widely distributed in Eurasia, with their

DOI: 10.4324/9781003044802-9

110 Zongwen Zhang

primary diversity center being China. Before 1998, there were nine species and two varieties of *Fagopyrum* recorded in China (Li, 1998). Over the past 20 years, however, both Chinese and foreign scholars have discovered and reported more than 20 new wild species of *Fagopyrum* (Xia et al., 2007; Fan et al., 2019), particularly the ancestor species of buckwheat, namely, *F.esculentum* ssp. *ancestrale* (Ohnishi, 1998a) and *F. tataricum* ssp. *potanini* (Ohnishi, 1998b), were identified in the southwest of China. There are two main cultivated species, namely, *F. esculentum* Moench and *F. tataricum* (L.) Gaertn in China. Due to its medicinal use, *F. cymosum* (Trev.) Meisn. is being domesticated and some cultivated forms have been developed in China (Yuan et al., 2019).

Landrace diversity

During the long history of its cultivation and domestication in China, two major types of buckwheat formed, namely, the cross-pollinated common buckwheat with high genetic diversity within its population, and the self-pollinated tartary buckwheat with high genetic diversity between populations. With years of effort, China has collected 3,085 accessions of buckwheat germplasm, including 2869 landraces (Zhang and Wu, 2010), accounting for 93% of the total. Landraces of tartary buckwheat were mainly from southwest regions, including Yunnan, Sichuan, Guizhou and Tibet, while those of common buckwheat were from north, northeast and northwest, including Shanxi, Shaanxi, Hebei, Inner Mongolia, Gansu and Jilin.

Current status of buckwheat R&D in China

Conservation of buckwheat biodiversity

The *ex situ* conservation of buckwheat biodiversity is managed by the National Crop Genebank (NCG) at the Chinese Academy of Agricultural Sciences (CAAS). Buckwheat accessions were identified, multiplied, catalogued and stored for the long term in a temperature of −18°C. At the same time, samples of these accessions were also stored under mid-term conditions. Researchers from national organizations can access the accessions stored in NCG for research purposes (Zhang and Wu, 2010).

As *F. cymosum* (Trev.) Meisn. has very important medicinal value, the Ministry of Agriculture and Rural Affairs established *in situ* conservation sites in the Hubei and Hunan provinces to prevent important populations of *F. cymosum* (Trev.) Meisn. by using the way of physical isolation (Zheng et al., 2019). For on-farm management, Bioversity International and a partner of the Chinese Academy of Sciences initiated a study on *in situ* conservation of buckwheat in Liangshan, Sichuan province. The investigation showed that tartary buckwheat is a staple food crop of the Yi ethnic group that is native to Lianshan. A variety of landraces of tartary buckwheat were managed by local farmers. The living customs in the

Yi ethnic group and the ecological conditions in Liangshan were all conducive to the on-farm management of buckwheat biodiversity (Zhao et al., 1998).

Assessment of genetic diversity

The characterization of agronomic traits of buckwheat biodiversity is mainly carried out by the relevant national research organizations coordinated by NCG at CAAS. Traits of plants, flowers and grains were measured; the growth period, lodging property and seed-shattering were accounted; and the contents of protein, fat, amino acids, vitamins and trace elements were analyzed. The analysis showed that buckwheat accessions had a rich diversity of agronomic traits, which was not only reflected in the traits themselves, but also showed great variation among materials from different regions. For example, the plant height ranged from 34 cm–205 cm, averaging 98.7 cm. The 1,000-grain weight of common buckwheat (26.68g) was significantly higher than that of tartary buckwheat (19.30g) (Zhang and Wu, 2010).

With the rapid development of biotechnology, DNA molecular marker technology has been widely used in the study of the genetic diversity of buckwheat. Bioversity International, in cooperation with the CAAS and other relevant partners in China, analyzed the genetic diversity of buckwheat by using ISSR (inter-simple sequence repeat) markers. The results showed that there were significant genetic differences among local buckwheat cultivars in Yunnan, and there were significant genetic differences among local cultivars in Guizhou, Hubei and Yunnan (Zhao et al., 2006). A genetic linkage map of buckwheat was constructed by combining morphological and molecular markers, laying a foundation for gene mining in buckwheat (Du et al., 2013). Recently, Bioversity and CAAS carried out the transcriptome analysis with a focus on rutin accumulation at filling stage seeds among three buckwheat species (Gao et al., 2017)

Improvement of varieties

China attaches great importance to the improvement of buckwheat varieties. With the support of relevant national and local projects, many research institutions and universities have participated in breeding new buckwheat varieties, aiming to improve the yield per unit area, lodging resistance and quality, so as to meet the farmers' needs for more income generation by growing buckwheat. Breeding technology was mainly based on systematic selection. With the mechanism of cross-pollination of common buckwheat, natural hybridization was the source of variants for selection. With mutagenesis technology to enhance the germplasm, a batch of new buckwheat varieties were bred (Ma et al., 2015). Marker-assisted selection technique is used in buckwheat breeding, by identifying useful diversity, constructing genetic linkage map, and developing molecular indicators such as SSR (simple sequence repeat) markers for identifying the heterozygotes of interspecific hybridization (Yang et al., 2019).

Production and farming system

In the 1950s, the annual planting area of buckwheat in China crossed two million hectares. Later, it was squeezed by high-yielding crops, and the annual sown area has been 0.6–0.8 million hectares in current years. At present, buckwheat is grown in more than 20 provinces and regions, in which tartary buckwheat is mainly present in the southwest and common buckwheat in the north, northeast and northwest. According to buckwheat distribution, the production area can be divided into four ecological regions:

1   Northern spring buckwheat region: This area includes the plateau and mountainous areas along the Great Wall and to the north, including parts of Heilongjiang, Jilin, Liaoning, Inner Mongolia, Hebei, northern Shanxi, Gansu, Ningxia and Qinghai. This area is also the main producing area of common buckwheat.
2   Northern summer buckwheat region: This area is centered on the Yellow River Basin, bounded to the south by Qinling mountain and Huaihe River, west by the Loess Plateau and east by the Yellow Sea, where is also a traditional winter wheat region. Buckwheat is planted as a second crop, generally seeded in June–July.
3   Southern autumn and winter buckwheat region: This region includes the south of Huaihe River, the middle and low reaches of, and the south of the Yangtze River. This region has a wide area, warm climate, long frost-free period and abundant rainfall. It is dominated by rice with buckwheat used to fill vacant fields, and is sowed in September–November.
4   Southwest high land spring and autumn buckwheat region: This region includes Qinghai-Tibet Plateau, Yunnan-Guizhou Plateau and the hills of Sichuan, Hubei, Hunan and Guizhou. It has a high altitude, as well as complicated ecological and geographical environments, with more clouds, less sunshine and large temperature differences between day and night. Tartary buckwheat is mainly planted in this region.

Process and value-chain

The processing of buckwheat products is an important part of the value-chain. Early enterprises engaged in buckwheat processing were very small and were mainly based in the areas of buckwheat production. The processing products were mostly primary products, such as flour and noodles from common buckwheat, and flour and tea from tartary buckwheat. In recent years, buckwheat-processing technologies have developed rapidly, and the product–market chain has extended dramatically. New products include kernels, roasted buckwheat rice, instant noodles, biscuits, beer, liquor, flavonoids, shell pillows, leaf powder, etc. Some traditional buckwheat food such as nest head, vinegar and powdered jelly also have been revitalized and popularized for driving the consumption of buckwheat, and achieving good economic and social benefits from the crop.

Buckwheat is largely consumed by farmers themselves. However, considerable quantities are also made available to local and international markets. Bioversity and the Shanxi Academy of Agricultural Sciences worked on buckwheat value-chains with farmers, processors, traders and consumers to promote its production, processing and marketing, and contribute to income generation for farmers by adding the value to buckwheat produce and linking farmers to markets. However, great changes are have been taking place in buckwheat markets in China in recent years. With the increase of domestic consumption demand driven by its health value, China, an exporter of buckwheat, began to import buckwheat in large quantities – for example, 270,000 tons were imported in 2018 and 365,000 tons in 2019.

National plans and policies related to buckwheat

China attaches importance to the R&D of buckwheat. In 2011, the Ministry of Agriculture and Rural Affairs (MARA) integrated buckwheat into the modern agriculture technology system by establishing the national buckwheat R&D center, including laboratories and experimental stations tasked with the roles of organizing and coordinating the efforts in buckwheat R&D in the country, involving germplasm and breeding, cultivation, pest and disease control and processing. In 2017, the MARA, together with other relevant ministries and commissions, formulated the Outline of Action Plan on Advancing Special Agricultural Products, including buckwheat.

Some local governments in buckwheat-producing provinces also issued relevant policies related to buckwheat R&D. In 2014, the government of Liangshan Prefecture of Sichuan Province promulgated the "Suggestions to Promote the Development of Tartary Buckwheat Industry;" In Guizhou Province, buckwheat was regarded as a green agricultural product in its "Green Agricultural Products Project Work Plan (2017–2020)." In 2019, the government of Shanxi Province formulated the "Suggestions on Accelerating the Development of the Whole Industrial Chain of Miscellaneous Cereals" including buckwheat as one of the eight miscellaneous crops prioritized.

Gaps and opportunities for promoting buckwheat

Rich biodiversity needs more efforts for conservation and identification

Due to the extension of high-yield crops such as corn and potato, the acreage of buckwheat is unstable, and many local varieties are at risk of being lost. In addition, the identification and evaluation of buckwheat germplasm resources are not

114 Zongwen Zhang

sufficient, and many social, ecological and economic values of buckwheat remain unrevealed and underutilized.

The benefit of growing buckwheat needs to be improved for increasing farmer's incomes

Buckwheat yield-per-hectare is low, and generally is below than one ton per hectare. There is a significant gap in productivity, specifically, a lower benefit from growing buckwheat than that from other crops such as potato. Therefore, the benefit of planting buckwheat has become the main concern of farmers and local governments, and is also a key problem that needs to be solved through research on improving yield and market value.

The nutritional and health value has huge potential for marketing

The annual consumption of buckwheat in China is about 0.4 million tons, less than a half kilogram per capita. As buckwheat is recognized as a nutritional and healthy food, a diet centering on it has entered the public's cookbooks. Therefore, there is a huge potential for expanding the consumption of buckwheat in nutritional and healthy diets in China.

Supporting policies are critical to the sustainable development of buckwheat in China

Although the central and local governments have issued some support policies for buckwheat R&D, there is still a big gap compared to those for staple crops, including insufficient funds for research, weak in personnel capacity, lack of seed subsidies and fewer opportunities to secure loans from state-owned banks. Therefore, the sustainable development of buckwheat needs more favorable policies at the local and national levels.

Future action plans for buckwheat R&D

To strengthen the conservation and improvement of buckwheat biodiversity

Efforts should be made to effectively evaluate and use buckwheat biodiversity for food, nutrition and income generation. More landraces and improved varieties should be made available for responding to consumer demand for nutritional and healthy foods and managing climate risks under changing environments. To achieve this objective, the following research activities should be undertaken:

- Strengthening integrated conservation of buckwheat biodiversity through linking *ex situ* collections in genebank to *in situ* conservation in the wild and field.

Promoting buckwheat in China **115**

- Evaluating and identifying buckwheat biodiversity for useful traits with integrated morphological and molecular approaches.
- Using local buckwheat biodiversity to improve yield and nutrition quality for new varieties and meet the needs of farmers, processers and consumers through integrated modern and participatory breeding methods.

To promote farming practices for advancing productive benefits from buckwheat for farmers

Planting mode and farming practices are critical for productivity and quality in buckwheat production. Strengthening the research on cultivation technology of buckwheat will contribute to increasing the yield potential, highlighting its nutritional characteristics and improving benefits from planting buckwheat. To achieve this, the following research activities should be strengthened:

- Promoting green production of buckwheat with low inputs to maintain the nutritional and health value.
- Promoting multiple cropping with buckwheat, by filling seasonal gaps to increase the efficiency of land use as well as farmers' incomes.
- Advancing the mechanization for planting and harvesting to address labor shortage issues.

To strengthen processing and the value-chain of nutritional and healthy diets

Processing and marketing is an important part of the buckwheat value-chain and an effective way to realize the nutritional value of the crop. Only by making buckwheat popular with the masses, will consumers realize its nutritional and health value. To achieve this, the following research activities should be strengthened:

- Promoting the value-chain of buckwheat by linking farmers to processors, traders and consumers.
- Understanding and being aware of the nutritional and health value of buckwheat diets.
- Promoting communication to change consumer behavior for dietary diversity.

To enhance the external environment for buckwheat R&D

The external environment of buckwheat R&D includes relevant supporting policies and research programmes. Through strengthening the partnership with research institutions, enterprises and farmers to understand their need for support the relevant policy suggestions should be put forward to relevant government departments. To achieve this objective, the following policies must be promoted:

- Subsidizing buckwheat seed production and distribution.
- Financial support to support buckwheat activities, including funds for research, loans for businesses and small loans for farmers.
- Enhancing national action in mainstreaming buckwheat for nutritious and healthy diets.

Summary

Buckwheat is an underutilized crop, mainly used for food, feed and medical raw materials. Buckwheat biodiversity is the material basis for its improvement and options for farmers to improve their livelihoods. With the support of relevant national programmes and international collaboration, great progress has been made in buckwheat research on biodiversity conservation, variety improvement, production practice and food processing and marketing. The gaps and opportunities for further development of buckwheat have been identified in this chapter, and key action plans were put forward to promote the conservation of biodiversity for variety improvement, farming practices for high productivity, the value-chain for income generation and relevant policies for the revival and sustainable development of the buckwheat industry.

References

Du, X., Zhang, Z., Wu, B., Li, Y. and Wang, A. (2013) 'Construction and analysis of genetic linkage map in tartary buckwheat (*Fagopyrum tataricum*) using SSR', *Chinese Agricultural Science Bulletin*, vol 29, no 21, pp61–65.
Fan, Y., Ding, M., Zhang, K., Yang, K., Tang, Y., Zhang, Z., Fang, W., Yan, J. and Zhou, M. (2019) 'Germplasm resource of the genus Fagopyrum mill', *Journal of Plant Genetic Resources*, vol 20, no 4, pp813–828.
Gao, J., Wang, T., Liu, M., Liu, J. and Zhang, Z. (2017). 'Transcriptome analysis of filling stage seeds among three buckwheat species with emphasis on rutin accumulation', *PLoS ONE*, vol 12, no 12, ppe0189672.
Li, A. (1998) *Flora of China*. Science Press, Beijing.
Lin, R. (2013) *Tartary Buckwheat*. China Agricultural Science and Technology Press, Beijing.
Ma, M., Liu, L., Zhang, L., Zhou, J. and Cui, L. (2015). 'Research progress in buckwheat breeding', *Journal of Shanxi Agricultural Science*, vol 43, no 2, pp240–243.
Ohnishi, O. (1998a) 'Search for the wild ancestor of buckwheat I. Description of new Fagopyrum (Polygonacea) species and their distribution in China', *Fagopyrum*, 15, pp18–28.
Ohnishi O. (1998b) 'Search for the wild ancestor of buckwheat III. The wild ancestor of cultivated common buckwheat, and of tartary buckwheat', *Economic Botany*, vol 52, no 2, pp123–133.
Xia, M., Wang, A., Cai, G., Yang, P. and Liu, J. (2007) *Studies on Wild Buckwheat Resources*. China Agricultural Press, Beijing.

Yang, L., Yao, Q., Li, H., Wang, Y., Ran, P., Cui, Y. and Chen, Q. (2019) 'Identification of Perennial Interspecific Hybrids on Genus Fagopyrum', *Journal of Plant Genetic Resources*, vol 20, no 2, pp387–395.

Yuan, J., Lu, Y. and Tian, Y. (2019) 'Variety development and cultivation technology of Fagopyrum cymsome', *Chinese Journal of Ethnomedicine and Ethnopharmacy*, vol 28, no 9, pp23–26.

Zhang, Z. and Wu, B. (2010) 'Conservation and utilization of buckwheat' in Z. Zhang, D. Zheng and R. Lin (eds) *Research and Development of Oat and Buckwheat – Proceedings of the First and Second National Symposium on Oat and Buckwheat*. China Agricultural Science and Technology Press, Beijing, pp41–50

Zhao, L., Zhang, Z., Li, Y. and Wang, T. (2006). 'Analysis of genetic diversity in tartary buckwheat based on ISSR markers', *Journal of Plant Genetic Resources*, vol 7, no 2, pp159–164.

Zhao, Z., Zhou, M., Luo, D., Li, F. and Cao, J. (1998) 'Ethnic investigation of in-situ conservation of tartary buckwheat in China', *Chinese Journal of Applied and Environmental Biology*, vol 4, no 4, pp320–326.

Zheng, X., Chen, B., Song, Y., Li, F., Wang, J., Qiao, W., Zhang, L., Cheng, Y., Sun, Y. and Yang, Q. (2019) 'In-situ conservation of wild relatives of crops', *Journal of Plant Genetic Resources*, vol 20, no 5, pp1103–1109.

9

NEGLECTED AND UNDERUTILIZED SPECIES OF MEXICO

Status and priorities for their conservation and sustainable use

Tiziana Ulian, Hugh W. Pritchard, Alejandro Casas, Efisio Mattana, Udayangani Liu, Elena Castillo-Lorenzo, Michael Way, Patricia Dávila Aranda, and Rafael Lira

Introduction

Mexico is a megadiverse country, harbouring at least 23,314 species of vascular plants, around half of which are endemic (Villaseñor, 2016) and distributed in ecosystems ranging from arid to wet and tropical to temperate types (Sarukhán et al., 2009). This high biodiversity is complemented by considerable cultural richness, encompassing 62 ethnic groups speaking 290 languages and over 360 linguistic variants (Eberhard, 2020), making Mexico among the top-ten countries for linguistic diversity and cultural richness (Sarukhán et al., 2009; Casas et al., 2016a) and home to one of the most important biocultural legacies of the world (Sarukhán et al., 2009).

Such tremendous biocultural diversity has resulted in more than 9,000 plants with known uses (Casas et al., 2016a), including 2,168 documented as edible (Mapes and Basurto, 2016). From prehistoric times, southern Mexico together with Central America has been a centre of origin of plant domestication (Harlan, 1971; MacNeish, 1992; Vavilov et al., 1992), including for many current main crops, such as maize (*Zea mays*), beans (*Phaseolus vulgaris*, *P. coccineus*, *P. acutifolius*), pumpkins (*Cucurbita argyrosperma*, *C. pepo*), and chilies (*Capsicum annuum*). These crops contributed to shaping the agricultural system known as 'milpa' – meaning cultivated field in Nahuatl – which gave origin to and sustained Mesoamerican cultures (Moreno-Calles et al., 2010).

In Mexico, there are nearly 200 native species of plants bearing clear signs of domestication. Moreover, nearly 1,200 species are under some form of management, such as let standing and planting enhancement, or special care dedicated to desirable plants. Such interventions may affect vegetation composition and promote particular 'types' (phenotypes of some species with favourable traits) and incipient domestication processes (Casas et al., 2007, 2016b).

DOI: 10.4324/9781003044802-10

NUS in Mexico **119**

Within the edible plants basket, plentiful are the neglected and underutilized species (NUS) relatively unknown outside Mesoamerica. These are of importance locally for both food security and people's livelihoods. Being highly adapted to marginal environments, they contribute significantly to agroecosystem resilience and are crucial in our efforts to fight food and nutritional insecurity and help communities cope with pervasive climate change (Padulosi et al., 2011; Ulian et al., 2020). This chapter reviews NUS in Mexico, in relation to the urgent need to secure their conservation and promote their sustainable use.

Overview of NUS in Mexico

From the great diversity of edible plants occurring in Mexico, it is possible to cluster species in three main groups:

1   **Traditional vegetables, such as the** *quelites* (plural for 'quilitl' in Nahuatl, meaning 'greens'). These are herbs from several plant families, mainly used as food, but also include some species of trees and shrubs with edible flowers (Bye and Linares, 2016). Examples include *Anoda cristata* (alache), *Crotalaria pumila* (chipil or chipilin), *Amaranthus hybridus*, *A. hypochondriacus*, *A. cruentus* (quelites or amarantos), *Dysphania ambrosioides* (epazote), *Chenopodium berlandieri* (huauzontle), and *Porophyllum ruderale* subsp. *macrocephalum* (pápalo or pápaloquelite) whose young leaves – and in some cases inflorescences or seeds – are used as food (leaves from this taxon and *D. ambrosioides* are used as condiment as well as in anthelmintic treatments [Frei et al., 1998]). Young leaves and flowers from trees, such as *Leucaena* spp. (in náhuatl 'guaxquilitl' meaning *quelite de guaje*) are eaten as vegetables (Bye and Linares, 2016). It is very common that *quelites* are gathered in the wild for family consumption, grown in marginal cultivations, or sold in traditional and regional markets (Bretting, 1982; Casas et al., 2007; Blanckaert et al., 2012; Casas et al., 2016a,b; Mapes and Basurto, 2016; Vibrans, 2016).

2   **Edible flowers and/or fruits of agave and cacti.** Since prehistoric times, the leaves, flowers, and shortened stem (corm) of numerous *Agave* species have been used for fibres, food, and beverages. Alcoholic drinks such as *pulque* are at least 3,000 years old, while distilled beverages (mescal, bacanora, sotol, and tequila) are more recent innovations. Some species have been used also for medicinal and ceremonial purposes. Incipient or advanced domestication has been documented only in few species (i.e., *A. fourcroydes*, *A. salmiana*, *A. americana*, *A. mapisaga*, *A. tequilana*, *A. angustifolia*, *A. hookeri*) (Figueredo-Urbina et al., 2017; Álvarez-Ríos et al., 2020). Demographic and genetic studies indicate that around 20 species (including *A. marmorata*, *A. potatorum*, *A. kerchovei*, *A. tequilana* blue variety, *A. inaequidens*) are at a high risk of extinction due to the degradation of their habitat, low genetic variation and/or overexploitation (Colunga García-Marín et al., 1996; Figueredo-Urbina et al., 2017; Aguirre-Planter et al., 2020).

120 Tiziana Ulian et al.

Among the cacti, the *Opuntia* species have been used since prehistory. This group of around 80 species and 200 local varieties native to Mexico (Reyes-Aguero et al., 2005) are used as living fences, to enhance soil conservation and/or combat desertification. Its cladodes (modified stems called *nopales* or *nopalitos* in Spanish) and fruits (grouped into two types, tunas – sweet – and xoconostles – sour) of about 10 species, are appreciated as sources of food, forage, medicinal remedies, additives for cosmetics, and used for feeding cochineal insects from which a carmin pigment is obtained (widely used as colourant in the food and cosmetic industries). Artificial selection (e.g., spineless cladodes), traditional management (*Opuntia streptacantha*), or partial domestication (*O. hyptiacantha*) or full domestication (*O. ficus-indica*) are known. Several food products, juices, and alcoholic beverages are commercially produced from cladodes and fruits. For example, 'colonche' (also referred to as *coloche* and *nochoctli*) is fermented from fruits of different cactus species, including columnar cacti (Ojeda-Linares et al, 2020). More recently, there has been an improvement of quality and safety, reduction of production costs, and development of marketing strategies for products of *Opuntia* (Illoldi-Rangel et al., 2012; López-Palacios et al., 2012).

Several species of columnar cacti are used as food (e.g., *Escontria chiotilla* [Jiotilla], *Polaskia chende* [chilmoxtli, chende o chinoa], *P. chichipe*, *Myrtillocactus geometrizans* [garambullo], *M. schenckii* [garambullo negro], *Stenocereus stellatus* [pitaya de agosto] and *S. pruinosus* [pitaya de mayo]) (Casas and Barbera, 2002; Casas et al., 2007, 2016b). These have great economic and cultural importance in several regions of Mexico (including in the Tehuacán Valley and the Balsas River Basin in the south-central region). The fruits of wild-cultivated *Cephalocereus tetetzo* (tetecho) and *Pachycereus weberi* (cardón) are highly appreciated by local people and are commonly left standing in agroforestry systems.

3　**Fruits of trees and shrubs.** Mexico has at least 2,885 native tree species (Tellez et al., 2020), many of which produce fruits that have been consumed by native populations since pre-Hispanic times (González and DelAmo, 2012). Four species of well-documented management practices and domestication processes are *Brosimum alicastrum* (ramón), *Leucaena esculenta* (guaje), *Sideroxylon palmeri* (tempesquistle), and *Ceiba aesculifolia* ssp. *brevifolia* (pochote) (Casas and Caballero, 1996; González-Soberanis and Casas, 2004; Avendaño et al., 2006). *Brosimum alicastrum* is one dominant tree species that grows in perennial and sub-deciduous tropical forests along both coasts of Mexico. The fresh pulp of the fruit and the seeds are edible, while the leaves provide good fodder for animals. Its seeds, roasted and ground, are used as a coffee substitute and for preparing dark-coloured dough used for making bread and tortillas. They are rich in carbohydrates and protein (12% in dry weight), tryptophan (an amino acid poorly represented in maize), and also contain calcium, phosphorus, iron, and vitamins A, B, and C. The abundance of this tree in Mayan archaeological sites is an indication of its management and cultivation by this ethnic group (Mapes and Basurto 2016). *Leucaena esculenta* [guaje] is mainly used as food by the Mixtec, Nahua, and Tlapanec peoples in the state of Guerrero, as well as by the Popoloca people in the Tehuacán Valley, Central Mexico. Its leaf buds, flowers, seeds, and young pods are eaten, and local gatherers have ensured morphological divergence between managed and unmanaged wild populations (Casas and Caballero, 1996; Casas et al., 2007; Blancas et al., 2010). *Sideroxylon palmeri* fruits are consumed in the Tehuacán Valley, where the species is widely commercialised (González-Soberanis and Casas, 2004). Flowers, roots, and seeds of *C. aesculifolia* subsp. *parvifolia* are consumed by local people, while stems and branches are used as firewood. Occasionally, the bark is used as a medicinal remedy to cure kidney disorders and skin infections, as well as to decrease blood sugar levels. In addition, locals protect individual trees and can favour varieties with purple-reddish fruits (Avendaño et al., 2006).

In Mesoamerica, the Sapotaceae is among the most important fruit tree families (González and DelAmo, 2012). *Manilkara zapota* (zapote or chicozapote) grows in tropical forests in southern Mexico, where its fruits are consumed raw, used for making preserves or preparing sherbets and ice cream. At the global level, this species is better known for its latex, which is used in the production of the chewing gum. *Pouteria campechiana* (zapote amarillo, canistel), whose fruits are consumed raw, also thrives in the tropical forests of Mexico and is also commonly found in home gardens. Similarly, *P. sapota* (mamey), also occurring in tropical forests, is much appreciated for its fruits, which are consumed raw, or used in the preparation of sherbets, ice cream, and refreshing beverages. Mix cropping of these trees with other species is known, e.g citrus, ornamentals, and timber species, and they can be found in cocoa and coffee plantations to provide shade. *Chrysophyllum cainito* (caimito or cayumito) is a common tree of tropical forests with two variants of its fruit (purple and yellow/green) which are edible (Mapes and Basurto, 2016). Finally, about 13 species of *Pinus* (pinos piñoneros) are valued for their edible nuts (the most important being *Pinus cembroides*, *P. nelsonii*, *P. maximartinezii* and *P. culminicola*). These are distributed mainly in northern Mexico, and their seeds, rich in lipids and proteins, are collected from the wild and are used in desserts and snacks, and marketed all over the country (Mapes and Basurto, 2016).

Conservation of NUS in Mexico

Traditional forms of plant management in Mexico range from the cultivation and selection of plant varieties to different types of management *in situ*, and to the gathering of plant products directly from the wild (Blancas et al., 2010). The degree of artificial selection and divergence between managed and wild populations is guided by farmers' needs (Casas et al., 2007; Blancas et al., 2010) and multiple approaches might be applied in the same field (Blancas et al., 2010;

122 Tiziana Ulian et al.

Moreno-Calles and Casas, 2010). When plants are cultivated, they are sown by seeds or planted in agriculture fields or home gardens (Blanckaert et al., 2004). Some useful plants (wild or weedy forms) that already grow there may be spared by farmers and may be protected in various ways against herbivores, competitors, cold spells, intense solar radiation, and drought (Blancas et al., 2010). Their presence can be also increased through the dispersal of sexual or vegetative propagules. Among those plants maintained through seed dispersal, there are many *quelites*, such as *A. hybridus*, *C. pumila*, and *A. cristata* (Mapes and Basurto 2016). In contrast, many species of columnar cacti, several *Opuntia* species, agaves, and trees (e.g., *L. esculenta* and *S. palmeri*) are multiplied through vegetative propagation (Blancas et al., 2010). Nonetheless, many plant products are still extracted directly from the wild, based on sound ecological understanding and following sustainable practices, as recorded in the Guerrero Mountains, the Tehuacán Valley (Casas et al., 2007; Blancas et al., 2010), and the Sierra Tarahumara of the Chihuahua State (LaRochelle and Berkes, 2003).

Since February 2002, the Royal Botanic Gardens, Kew (RBG Kew), and the Facultad de Estudios Superiores, Iztacala of the National Autonomous University of Mexico (Fes-I UNAM) have been banking seeds of wild plant species, including crop wild relatives from the arid and semiarid areas of Mexico (León-Lobos et al., 2012). Seeds have been collected and deposited at Fes-I UNAM, with duplicates stored at RBG Kew's Millennium Seed Bank (MSB), under an Access and Benefit Sharing Agreement. This has been achieved in the framework of the Millennium Seed Bank Project (now Partnership), which acts globally to safeguard plant diversity, focusing in particular on plants most at risk and most useful for the future of humankind (Liu et al., 2018). The partnership in Mexico served as a catalyst also for two further projects, viz. 'The MGU – Useful Plant Project' and the 'Science-based conservation of tree species in Mexico'. Overall, 2,598 wild plant species (more than 10 % of the Mexican flora) have so far been conserved *ex situ* as seeds in Mexico, with duplicates stored in the MSB. However, only 62 seed samples of 21 species from the most important groups of NUS mentioned here have been safeguarded through seed banking, and it is hoped that such a significant gap in the long-term conservation of NUS in Mexico can be filled in the near future.

Needs, challenges, and opportunities

In situ and on-farm conservation realised by local communities in Mexico is an important facet of integrated plant conservation (Blanckaert et al., 2004, 2007; Rodríguez-Arévalo et al., 2017). Traditional management systems have been recognised, in fact, to have higher capacity for conserving biodiversity and ensuring the resilience of socioecological systems than industrial systems (Berkes et al., 2000; Blancas et al., 2010; Chappell and LaValle, 2011).

Many NUS in Mexico do occur in a wide range of traditional agroecosystems, such as *milpa* (corn-bean-squash field), *chilar* (chili pepper field), *frijolar* (bean field), and *cafetal* (coffee groves) (Mapes and Basurto, 2016), or are grown as crops in forests and agroforestry systems (Moreno-Calles et al., 2013). These are crucial places for the conservation of biodiversity and represent important reservoirs of plant genetic diversity (Blanckaert et al., 2004; Larios et al., 2013).

Seed banking is highly complementary to these traditional management systems. However, seed conservation needs to be better targeted to adequately capture the great genetic diversity of NUS present in Mexico. More research on seed germination requirements and dormancy is also needed to support large-scale species propagation. Moreover, seed conservation is still highly challenging for numerous species with recalcitrant, desiccation-sensitive seeds, therefore alternative conservation measures (e.g., cryopreservation of embryos and embryonic axes) should also be deployed (Li and Pritchard, 2009; Wyse et al., 2018). In addition, there is a need to promote the 'cross pollination' of ideas between agriculturalists focusing on major crops and farmers working on traditional forms of management (Turner et al., 2011), if we are to really strengthen the development of sustainable agriculture practices.

Although NUS are mainly collected for family consumption, some species are exchanged among households (Arellanes et al., 2013; Blancas et al., 2013) or are traded in markets (tianquiste), including those in big cities (Arellanes et al., 2013; Farfán-Heredia et al., 2018). Local and regional media, schools, NGOs, social organisations, and some governmental institutions promote the conservation and sustainable use of native plant resources and their associated traditional knowledge. However, greater mainstreaming and inclusion of NUS in policy and legal frameworks is much needed, as noted in 'The National Report on the Status of The Phyto-genetic Resources for Agriculture and Food Supply' (Molina-Moreno and Córdova-Téllez, 2006). In this regard, the Mexican Commission for the Knowledge and Use of Biodiversity (CONABIO), serves as an important bridge between academia, governmental agencies, and other sectors of society, by compiling and sharing knowledge on biological diversity, and promoting its conservation and sustainable use.[1] CONABIO now leads the project 'Securing the Future of Global Agriculture in the Face of Climate Change by Conserving the Genetic Diversity of the Traditional Agro-ecosystems of Mexico,' that seeks to support agrobiodiversity conservation, sustainable use, and resilience,[2] in line with the Nagoya Protocol of the Convention on Biological Diversity (https://www.cbd.int/abs/). However, strong participatory-driven decision-making (Padulosi et al., 2011; Noorani et al., 2015; FAO, 2019), economic incentives, and subsidies (Padulosi et al., 2019) are needed for such a process to be truly successful.

References

Aguirre-Planter, E., Parra-Leyva, J. G., Ramírez-Barahona, S., Scheinvar, E., Lira-Saade, R. and Eguiarte, L. E. (2020) Phylogeography and genetic diversity in a southern North American desert: Agave kerchovei from the Tehuacán-Cuicatlán Valley, Mexico. *Frontiers in Plant Science*, 11, pp. 863.

124 Tiziana Ulian et al.

Álvarez-Ríos, G. D., Pacheco-Torres, F., Figueredo-Urbina, C. J. and Casas, A. (2020) Management, morphological and genetic diversity of domesticated agaves in Michoacán, México. *Journal of Ethnobiology and Ethnomedicine*, 16(1), pp. 3.

Arellanes, Y., Casas, A., Arellanes, A., Vega, E., Blancas, J., Vallejo, M., Torres, I., Rangel-Landa, S., Moreno, A. I. and Solís, L. (2013) Influence of traditional markets on plant management in the Tehuacán Valley. *Journal of Ethnobiology and Ethnomedicine*, 9(1), pp. 38.

Avendaño, A., Casas, A., Dávila, P. and Lira, R. (2006) Use forms, management and commercialization of "pochote" Ceiba aesculifolia (HB & K.) Britten & Baker f. subsp. parvifolia (Rose) PE Gibbs & Semir (Bombacaceae) in the Tehuacán Valley, Central Mexico. *Journal of Arid Environments*, 67(1), pp. 15–35.

Berkes, F., Colding, J. and Folke, C. (2000) Rediscovery of traditional ecological knowledge as adaptive management. *Ecological Applications*, 10(5), pp. 1251–1262.

Blancas, J., Casas, A., Pérez-Salicrup, D., Caballero, J. and Vega, E. (2013) Ecological and socio-cultural factors influencing plant management in Náhuatl communities of the Tehuacán Valley, Mexico. *Journal of Ethnobiology and Ethnomedicine*, 9(1), pp. 39.

Blancas, J., Casas, A., Rangel-Landa, S., Moreno-Calles, A., Torres, I., Pérez-Negrón, E., Solís, L., Delgado-Lemus, A., Parra, F., Arellanes, Y., Caballero, J., Cortés, L., Lira, R. and Dávila, P. (2010) Plant Management in the Tehuacán-Cuicatlán Valley, Mexico1. *Economic Botany*, 64(4), pp. 287–302.

Blanckaert, I., Paredes-Flores, M., Espinosa-García, F. J., Piñero, D. and Lira, R. (2012) Ethnobotanical, morphological, phytochemical and molecular evidence for the incipient domestication of Epazote (Chenopodium ambrosioides L.: Chenopodiaceae) in a semi-arid region of Mexico. *Genetic resources and crop evolution*, 59(4), pp. 557–573.

Blanckaert, I., Swennen, R. L., Flores, M. P., López, R. R. and Saade, R. L. (2004) Floristic composition, plant uses and management practices in homegardens of San Rafael Coxcatlán, Valley of Tehuacán-Cuicatlán, Mexico. *Journal of Arid Environments*, 57(2), pp. 179–202.

Blanckaert, I., Vancraeynest, K., Swennen, R. L., Espinosa-Garcia, F. J., Pinero, D. and Lira-Saade, R. (2007) Non-crop resources and the role of indigenous knowledge in semi-arid production of Mexico. *Agriculture, Ecosystems & Environment*, 119(1–2), pp. 39–48.

Bretting, P. (1982) Papaloquelite y la etnobotánica de las especies Porophyllum en México. (No. 04; CP, FOLLETO 42).

Bye, R. and Linares, E. (2016) Ethnobotany and ethnohistorical sources of Mesoamerica. in *Ethnobotany of Mexico*. Springer. pp. 41–65.

Casas, A. and Barbera, G. (2002) Mesoamerican domestication and diffusion. in P. S. Nobel (Ed.), *Cacti: Biology and uses*. University of California Press. pp. 62.

Casas, A., Blancas, J. and Lira, R. (2016a) Mexican ethnobotany: Interactions of people and plants in Mesoamerica. in R. Lira, A. Casas, J. Blancas (Eds.), *Ethnobotany of Mexico*. Springer. pp. 1–19.

Casas, A., Blancas, J., Otero-Arnaiz, A., Cruse-Sanders, J., Lira, R., Avendaño, A., Parra, F., Guillén, S., Figueredo, C. J. and Torres, I. (2016b) Evolutionary ethnobotanical studies of incipient domestication of plants in Mesoamerica. in R. Lira, A. Casas, J. Blancas (Eds.), *Ethnobotany of Mexico*. Springer. pp. 257–285.

Casas, A. and Caballero, J. (1996) Traditional management and morphological variation in Leucaena esculenta (Fabaceae: Mimosoideae) in the Mixtec Region of Guerrero, Mexico. *Economic Botany*, 50(2), pp. 167–181.

Casas, A., Otero-Arnaiz, A., Pérez-Negrón, E. and Valiente-Banuet, A. (2007) In situ management and domestication of plants in Mesoamerica. *Annals of Botany*, 100(5), pp. 1101–1115.

Chappell, M. J. and LaValle, L. A. (2011) Food security and biodiversity: Can we have both? An agroecological analysis. *Agriculture and Human Values*, 28(1), pp. 3–26.

Colunga‑García Marín, P., Estrada‑Loera, E. and May‑Pat, F. (1996) Patterns of morphological variation, diversity, and domestication of wild and cultivated populations of Agave in Yucatan, Mexico. *American Journal of Botany*, 83(8), pp. 1069–1082.

Eberhard, D. M., Gary F. Simons, and Charles D. Fennig (eds.) (2020) *Ethnologue: Languages of the World*. 23rd edition. Dallas, TX: SIL International. Online version: http://www.ethnologue.com

FAO (2019) The state of the world's biodiversity for food and agriculture, FAO Commission on Genetic Resources for Food and Agriculture Assessments.

Farfán-Heredia, B., Casas, A., Moreno-Calles, A. I., García-Frapolli, E. and Castilleja, A. (2018) Ethnoecology of the interchange of wild and weedy plants and mushrooms in Phurépecha markets of Mexico: economic motives of biotic resources management. *Journal of Ethnobiology and Ethnomedicine*, 14(1), pp. 5.

Figueredo-Urbina, C. J., Casas, A. and Torres-García, I. (2017) Morphological and genetic divergence between Agave inaequidens, A. cupreata and the domesticated A. hookeri. Analysis of their evolutionary relationships. *PloS One*, 12(11), pp. e0187260.

Frei, B., Baltisberger, M., Sticher, O. and Heinrich, M. (1998) Medical ethnobotany of the Zapotecs of the Isthmus-Sierra (Oaxaca, Mexico): Documentation and assessment of indigenous uses. *Journal of Ethnopharmacology*, 62(2), pp. 149–165.

González, R. and DelAmo, S. (2012) Frutos mesoamericanos: Breve historia de sabores y sin sabores. *Biodiversitas*, 103, pp. 6–11.

González-Soberanis, C. and Casas, A. (2004) Traditional management and domestication of tempesquistle, Sideroxylon palmeri (Sapotaceae) in the Tehuacán-Cuicatlán Valley, Central Mexico. *Journal of Arid Environments*, 59(2), pp. 245–258.

Harlan, J. R. (1971) Agricultural origins: Centers and noncenters. *Science*, 174(4008), pp. 468–474.

Illoldi-Rangel, P., Ciarleglio, M., Sheinvar, L., Linaje, M., Sánchez-Cordero, V. and Sarkar, S. (2012) Opuntia in Mexico: Identifying priority areas for conserving biodiversity in a multi-use landscape. *PLoS One*, 7(5), pp. e36650.

Larios, C., Casas, A., Vallejo, M., Moreno-Calles, A. I. and Blancas, J. (2013) Plant management and biodiversity conservation in Náhuatl homegardens of the Tehuacán Valley, Mexico. *Journal of Ethnobiology and Ethnomedicine*, 9(1), pp. 74.

LaRochelle, S. and Berkes, F. (2003) Traditional Ecological Knowledge and Practice for Edible Wild Plants: Biodiversity Use by the Rarámuri, in the Sierra Tarahumara, Mexico. *International Journal of Sustainable Development and World Ecology*, 10(4), pp. 361–375.

León-Lobos, P., Way, M., Aranda, P. D. and Lima-Junior, M. (2012) The role of ex situ seed banks in the conservation of plant diversity and in ecological restoration in Latin America. *Plant Ecology and Diversity*, 5(2), pp. 245–258.

Li, D.-Z. and Pritchard, H. W. (2009) The science and economics of ex situ plant conservation. *Trends in Plant Science*, 14(11), pp. 614–621.

Liu, U., Breman, E., Cossu, T. A. and Kenney, S. (2018) The conservation value of germplasm stored at the Millennium Seed Bank, Royal Botanic Gardens, Kew, UK. *Biodiversity and Conservation*, 27(6), pp. 1347–1386.

López-Palacios, C., Peña-Valdivia, C. B., Reyes-Agüero, J. A. and Rodríguez-Hernández, A. I. (2012) Effects of domestication on structural polysaccharides and dietary fiber in nopalitos (Opuntia spp.). *Genetic Resources and Crop Evolution*, 59(6), pp. 1015–1026.

MacNeish, R. S. (1992) The origins of agriculture and settled life. (No. 04; GN799. A4, M3.).

126 Tiziana Ulian et al.

Mapes, C. and Basurto, F. (2016) Biodiversity and edible plants of Mexico. in *Ethnobotany of Mexico*. Springer. pp. 83–131.

Molina-Moreno, J. and Córdova-Téllez, L. (2006) Recursos fitogenéticos de México para la alimentación y la agricultura: Informe Nacional 2006. Secretaría de Agricultura, Ganadería, Desarrollo Rural, Pesca y Alimentación y Sociedad Mexicana de Fitogenética, A.C. Chapingo, México. 172 p.

Moreno-Calles, A. I. and Casas, A. (2010) Agroforestry systems: Restoration of semiarid zones in the Tehuacán Valley, Central Mexico. *Ecological Restoration*, 28(3), pp. 361–368.

Moreno-Calles, A. I., Casas, A., Blancas, J., Torres, I., Masera, O., Caballero, J., García-Barrios, L., Pérez-Negrón, E. and Rangel-Landa, S. (2010) Agroforestry systems and biodiversity conservation in arid zones: The case of the Tehuacán Valley, Central México. *Agroforestry Systems*, 80(3), pp. 315–331.

Moreno-Calles, A. I., Toledo, V. M. and Casas, A. (2013) Los sistemas agroforestales tradicionales de México: una aproximación biocultural. *Botanical Sciences*, 91(4), pp. 375–398.

Noorani, A., Bazile, D., Diulgheroff, S., Kahane, R. and Nono-Womdim, R. (2015) Promoting neglected and underutilized species through policies and legal frameworks. in *EUCARPIA International Symposium on Protein Crops, V Meeting AEL* [V Jornadas de la AEL]. Spanish Association for Legumes (AEL), Pontevedra, Spain, pp. 107–111.

Ojeda-Linares, C. I., Vallejo, M., Lappe-Oliveras, P. and Casas, A. (2020) Traditional management of microorganisms in fermented beverages from cactus fruits in Mexico: An ethnobiological approach. *Journal of Ethnobiology and Ethnomedicine*, 16(1), pp. 1–12.

Padulosi, S., Cawthorn, D., M., Meldrum, G., Flore, R., Halloran, A. and Mattei, F. (2019) Leveraging neglected and underutilized plant, fungi, and animal species for more nutrition sensitive and sustainable food systems. in P. Ferranti, E. M. Berry, and J. R. Anderson (Eds.), *Encyclopedia of Food Security and Sustainability*. Elsevier. pp. 361–370.

Padulosi, S., Heywood, V., Hunter, D. and Jarvis, A. (2011) Underutilized species and climate change: Current status and outlook. *Crop Adaptation to Climate Change*, Chapter 26, pp. 507–521.

Reyes-Aguero, J., Rivera, J. and Flores, J. (2005) Morphological variation of Opuntia (Cactaceae) in connection with its domestication in Meridional Highland Plateau of Mexico. *INTERCIENCIA*, 30(8), pp. 476–484.

Rodríguez-Arévalo, I., Mattana, E., García, L., Liu, U., Lira, R., Dávila, P., Hudson, A., Pritchard, H. W. and Ulian, T. (2017) Conserving seeds of useful wild plants in Mexico: Main issues and recommendations. *Genetic Resources and Crop Evolution*, 64(6), pp. 1141–1190.

Sarukhán, J., Koleff, P., Carabias, J., Soberón, J., Dirzo, R., Llorente-Bousquets, J., Halffter, G., González, R., March, I. and Mohar, A. (2009) Capital natural de México. Síntesis: Conocimiento actual, evaluación y perspectivas de sustentabilidad. Comisión Nacional para el Conocimiento y Uso de la Biodiversidad, México. p. 104.

Tellez, O., Mattana, E., Diazgranados, M., Kühn, N., Castillo-Lorenzo, E., Lira, R., Montes-Leyva, L., Rodriguez, I., Way, M., Dávila, P. and Ulian, T (2020) Native trees of Mexico: Diversity, distribution, uses and conservation. *PeerJ*, 8, e9898. doi:10.7717/peerj.9898

Turner, N. J., Łuczaj, Ł. J., Migliorini, P., Pieroni, A., Dreon, A. L., Sacchetti, L. E. and Paoletti, M. G. (2011) Edible and tended wild plants, traditional ecological knowledge and agroecology. *Critical Reviews in Plant Sciences*, 30(1–2), pp. 198–225.

Ulian, T., Diazgranados, M., Pironon, S., Padulosi, S., Liu, U., Davies, L., Howes, M. J. R., Borrell, J. S., Ondo, I. and Pérez‑Escobar, O. A. (2020) Unlocking plant resources to support food security and promote sustainable agriculture. *Plants, People, Planet*, 2(5), pp. 421–445.

Vavilov, N. I., Vavylov, M. I., Vavílov, N. Í. and Dorofeev, V. F. (1992) *Origin and Geography of Cultivated Plants* Cambridge University Press, Cambridge. ISBN 978-0-521-40427-3.

Vibrans, H. (2016) Ethnobotany of Mexican weeds. in *Ethnobotany of Mexico*. Springer. pp. 287–317.

Villaseñor, J. L. (2016) Checklist of the native vascular plants of Mexico. *Revista Mexicana de Biodiversidad*, 87(3), pp. 559–902.

Wyse, S. V., Dickie, J. B. and Willis, K. J. (2018) Seed banking not an option for many threatened plants. *Nature Plants*, 4(11), pp. 848–850.

Notes

1 https://www.biodiversidad.gob.mx/v_ingles/
2 http://www.fao.org/gef/projects/detail/en/c/1113211/

10

NEGLECTED AND UNDERUTILIZED SPECIES IN BRAZIL

From obscurity to non-conventional edible plants

Nuno Rodrigo Madeira, Valdely Ferreira Kinupp and Lidio Coradin

Introduction

Like other elements of nature such as minerals and fossil fuels, plant genetic resources are not dispersed evenly around the globe, being found in greater abundance in certain regions, especially in tropical countries (Shulman, 1986). The process of evolution, adaptation and migration of human beings across the planet and the exchange of germplasm was and continues to be decisive for the great variety of grains, fruits, vegetables, roots and tubers available, even if the majority of this diversity is still neglected and underutilized (NUS). During the period of the world's great navigations, between the fifteenth and seventeenth centuries, there was a substantial exchange of species (Hue, 2008) and some acclimatized on different continents, which became the basis of the world economy, resulting in strong global interdependence. Another important contingent of species, native to each region, was slowly being replaced by exotic species that were being introduced and, over time, these species stopped being used and fell into relative obscurity, though pockets of this biodiversity have been maintained by many different communities. Thus, thousands of native species have not received their due attention or support from research, despite the fact that their economic value and nutritional advantages have already been demonstrated in many instances (Coradin et al., 2018). With the greatest biodiversity on the planet, Brazil seeks, on the one hand, to increase the production of exotic commercial species that form the basis of most mainstream agriculture and food production in the country. On the other hand, also seeks to draw increasing attention to the wealth of NUS, native or naturalized, which can become new cultivation options, diversifying the agricultural and food portfolio, and bringing economic and socio-environmental advantages while also addressing many of the increasing malnutrition, health and environmental concerns of the country.

DOI: 10.4324/9781003044802-11

Many initiatives from different sectors of the federal and state governments, non-governmental organizations and the private sector have emerged to expand knowledge of NUS in the country. By valuing these species, which are more resilient and adaptable to adverse soil and climate conditions, thus requiring fewer external inputs, people's perception of their potential is increased and there is now greater awareness of the opportunities and possibilities of using such a vast diversity.

Historical context

Brazil presents a great variety of biomes and climates. Its geographical, environmental, climatic and cultural diversity is reflected in its food diversity, one of the greatest on the planet, as shown by the several sources of carbohydrates (rice, beans, corn, wheat, cassava, potatoes, sweet potatoes, yams) and meats (cattle, poultry, pigs, fish, goats and sheep) consumed.

When the Portuguese arrived in Brazil, in the year 1500, most Indigenous Peoples (around 300 ethnic groups at the time, 180 nowadays) had a diet based on cassava, sweet potatoes and corn, complemented by hunting and fishing. However, as hunters and gathers, there is also evidence of the use of many other species (Lopes, 2017).

During the colonization process, intense cultural miscegenation took place, with the introduction of exotic ingredients that joined those already present in Brazilian territory. Cascudo (1983) concluded that Europeans introduced technique and sophistication in taste to existing Indigenous and African elements. Madeira et al. (2008) mention the exchange of vegetables by the Portuguese, directly and intentionally but also indirectly, through the enslavement of Africans.

After the 1960s, with intense urbanization and globalization, there was a loss of the reference of the home gardens production and a strong influence of market power in food choices and in the strengthening of large supermarket chains, driving dietary simplification and prioritizing species with structured production chains, resulting in the reduced use of local species with subsequent social, economic and environmental impacts (Brazil, 2013).

However, some NUS with significant cultural importance in certain regions of Brazil continued to be used, mainly by local people (Brasil, 2015). Table 10.1 presents examples of these NUS, including assaí, cupuaçu, peach palm, Brazilian nut, jambu and waterleaf in the north; buriti, babassu, uricuri, umbu, hibiscus and yam in the northeast; Brazilian grape, pereskia, taro, arracacha, taioba and Malabar spinach in the southeast; feijoa, pindo, horseradish, Brazilian pine and mate in the south; and pequi, baru, Brazilian copal, cagaita and guariroba in the mid-west (Coradin et al., 2018). Other species that were once part of daily life are now largely forgotten, including arrowroot, mangarito, yam bean and sweetcorn root (Table 10.1), in a process termed "food extinction" by Madeira and Botrel (2019).

In Brazil, there is some confusion regarding the use of the term "NUS" and which species call under this terminology proposed by IPGRI (2002). Other terminologies also used in Brazil are "traditional vegetables" (Madeira et al., 2013), "unconventional vegetables" (Cardoso et al., 1998) and PANC – "non-conventional edible plants" in Portuguese (Kinupp and Lorenzi, 2014), the last of which is nowadays the most accepted term in Portuguese for NUS.

Kinupp and Lorenzi (2014) defined PANC as "unconventional and non-ordinary plants that have one or more parts that can be used as food, or even unconventional parts of conventional plants, as banana navel or papaya tree marrow." Madeira and Kinupp (2016) add that PANC don't have a structured production chain. This term being very expressive and euphonic in Brazil (PANC in Portuguese sounds exactly like "punk," giving it a curious double meaning) seeks to promote and disseminate the use of these plants that are still considered unusual by the vast majority of the Brazilian population.

Initiatives for the conservation and promotion of NUS in Brazil

In Brazil, a considerable number of studies on individual species can be found, but academic studies on the broader themes of NUS or PANC are rare or few in number. The lack of studies regarding the cultivation and promotion of NUS use is a real concern and requires special attention being given to further research, as well as development that might stimulate their greater conservation, production and consumption. In this context, a series of initiatives related to the promotion of NUS use in Brazil have been developed and continue to be encouraged. One classic work is the *Dictionary of Useful Plants of Brazil and Cultivated Exotic Plants*, written by Pio Corrêa and published by the Ministry of Agriculture between 1926 and 1978 in six volumes, with the collaboration of Penna for volumes 3–6, and reissued in 1984. With the objective of promoting regional food culture, Cardoso (1997) released the book *Non-conventional vegetables from the Amazon*, emphasizing their full adaptability to high temperatures and humidity and low soil fertility.

The sustainable use of the Brazilian genetic resources is limited to areas of natural occurrence and involves complex activities, from bioprospecting and research to the transformation, production and creation of new markets and commercialization.

To tackle this issue, the Ministry of the Environment has developed, since the 2000s, the "Plants for the Future" initiative, which aims to identify native species that can be used for crop diversification to expand opportunities for investment by the private and public sectors and to reduce the vulnerability of the agri-food system.

As biodiversity is used and valued, its conservation is enhanced and the number of species used is expanded, together with its contribution to climate change adaptation. Through partnerships with governmental and non-governmental institutions, the academic and business sectors and civil society, the initiative

FIGURE 10.1 Plants for the Future books. Brasília, 2020.
Source: Lidio Coradin (used with permission).

has been building strategies that also contribute to the rescue and valorization of knowledge and flavors of traditional and regional Brazilian cuisine (Santiago and Coradin, 2018).

More than 800 species have been prioritized in the five geopolitical regions of Brazil. In each of these regions, the species were organized into different groups of use: food, aromatic, medicinal, oilseed and ornamental, among others. A richly illustrated and technically comprehensive book has been published for each region (Figure 10.1). In 2011, the first book, *Native species of Brazilian flora with actual or potential economic value: Plants for the Future − South Region*, was published. In 2017, the book for the mid-west region was released; in 2018, the one for the northeast region; and the books for the north and southeast regions are set to be launched in 2021, concluding this series, which has involved the contributions of hundreds of experts.

These books collectively and for the first time present to the wider Brazilian society quality information about the value of native biodiversity for health, food security, wellbeing, employment opportunities and income generation − all of which are key objectives of the Plants for the Future initiative. It's the first effort of this kind carried out in a megadiverse country. It also represents an important step for the implementation, in Brazil, of the Aichi Biodiversity Targets of the Convention on Biological Diversity (CBD), as well as of the Targets of the Global

FIGURE 10.2 Biodiversity for Food and Nutrition project team visiting the traditional vegetable collection at Embrapa, highlighting *Pereskia aculeata* crop. Brasília, 2017.
Source: Lidio Coradin (used with permission).

Strategy for Plant Conservation (GSPC) and will no doubt be an important platform for the country to enhance the conservation and use of NUS in the Convention on Biological Diversity (CBD) post-2020 global biodiversity framework.

Furthermore, the partnership between Plants for the Future and the Biodiversity for Food and Nutrition project[1] has enabled greater collaborations with different sectors of society, in addition to creating new opportunities for mainstreaming biodiversity into already existing federal initiatives related to food and nutrition (Figure 10.2) (Hunter et al., 2020). This facilitated the capacity building of many technicians and students and the publication of recipe book – *Brazilian Biodiversity: tastes and flavours*. With 335 classic and contemporary recipes, the book has revived traditional flavors and techniques, presenting new tastes and textures (Santiago & Coradin, 2018). This partnership also facilitated the organization of a list of native Brazilian socio-biodiversity species of nutritional value (Ordinance MMA/MDS n° 284/2018) to guide governmental actions, while also creating incentives for family farmers who produce and commercialize these species (Coradin et al., 2018).

In 2007, Kinupp defended his doctoral thesis on native plants known as bushes and weeds in the urban area of Porto Alegre, Brazil, whose economic potential was then little-known (Kinupp, 2007). He estimated the local floristic wealth of 1,500 species, out of which 311 (21%) could be used for food, proving the undeniable potential of NUS for enriching the human diet. As of 2020, this document had been accessed online more than 54,000 times.

Later, Kinupp and Lorenzi (2014) released the book *Non-conventional Edible Plants (PANC) in Brazil: identification guide, nutritional aspects and illustrated recipes*, which is a significant reference volume in Brazil about NUS. The book is richly

TABLE 10.1 Some NUS traditionally used in Brazil cited in this chapter

English common names	Portuguese common names	Scientific names	Family
'Açaí'	Açaí	*Euterpe oleracea*	Arecaceae
'Arracacha'	Mandioquinha	*Arracacia xanthorrhiza*	Apiaceae
Arrowroot	Araruta	*Maranta arundinacea*	Marantaceae
'Babaçu'	Babaçu	*Attalea speciosa*	Arecaceae
Barbados shrub, pereskia	Ora-pro-nóbis	*Pereskia aculeata*	Cactaceae
'Baru'	Baru	*Dipteryx alata*	Fabaceae
Brazilian copal	Jatobá	*Hymenaea spp*	Fabaceae
Brazilian grape	Jabuticaba	*Plinia* spp.	Myrtaceae
Brazilian nut	Castanha-do-brasil	*Bertholletia excelsa*	Lecythidaceae
Brazilian pine	Pinhão	*Araucaria angustifolia*	Araucariaceae
'Buriti'	Buriti	*Mauritia flexuosa*	Arecaceae
'Cagaita'	Cagaita	*Eugenia dysenterica*	Myrtaceae
'Chaya', spinach tree	Chaya	*Cnidoscolus aconitifolius*	Euphorbiaceae
'Cupuaçu'	Cupuaçu	*Theobroma grandiflorum*	Malvaceae
'Feijoa'	Goiaba-serrana, feijoa	*Acca sellowiana*	Myrtaceae
'Guariroba'	Gueroba, guariroba	*Syagrus oleracea*	Arecaceae
Hibiscus	Vinagreira, cuxá	*Hibiscus sabdariffa*	Malvaceae
Horse radish	Crem, raiz-forte	*Armoracia rusticana*	Brassicaceae
Jambu	Jambu	*Acmella oleracea*	Asteraceae
Malabar spinach	Bertalha	*Basella alba*	Basellaceae
'Mangarito'	Mangarito	*Xanthosoma riedelianum*	Araceae
Mate	Erva-mate	*Ilex paraguariensis*	Aquifoliaceae
'Moringa'	Moringa	*Moringa ovalifolia*	Moringaceae
Peach palm	Pupunha	*Bactris gasipaes*	Arecaceae
'Pequi'	Pequi	*Caryocar brasiliense*	Caryocaraceae
Pindo, 'butiá'	Butiá	*Butia odorata*	Arecaceae
Purslane	Beldroega	*Portulaca oleracea*	Portulacaceae
Sweet corn root	Ariá	*Goeppertia allouia*	Marantaceae
'Taioba'	Taioba	*Xanthosoma taioba*	Araceae
Taro	Inhame	*Colocasia esculenta*	Araceae
'Umbu'	Umbu	*Spondias tuberosa*	Anacardiaceae
'Uricuri'	Licuri, ouricuri	*Syagrus coronata*	Arecaceae
Waterleaf	Cariru	*Talinum fruticosum*	Talinaceae
Yam	Cará, inhame	*Dioscorea* spp.	Dioscoreaceae
Yam bean	Jacatupé	*Pachyrhizus tuberosus*	Fabaceae

illustrated (with 2,510 photos) and presents data on 351 species (205 of which are native), including a brief botanical characterization, agronomic recommendations and three culinary recipes for each species, totaling 1,053 recipes. This book and the associated research that has arisen from it has led to considerable advances in the promotion of NUS in Brazil. Through interaction with traditional and social media, this work has given fresh impetus to the topic and has helped shine a light on some of the uses of NUS in daily life of Brazilian society, especially for those in search of healthy diets.

Embrapa Vegetables, located in the Federal District, maintains an important collection of traditional vegetables (Figure 10.3), with about 400 accessions and 80 NUS. It is responsible for research focusing on the rescue of traditional vegetables, of which many are native while others are considered "naturalized" thanks to their full climatic adaptation. More recently, some NUS new to Brazil that have traditionally been used in other countries, such as moringa and chaya (Table 10.1), have been incorporated into the Embrapa collection due to their resilience and adaptability. This work began officially in 2006 and has been responsible for building partnerships with different institutions working on traditional vegetables, and for encouraging activities in different regions of the country. These vegetables were once maintained by farmers, which is why the term "traditional vegetables" is used to refer to them. Besides the maintenance of the collection and conservation of the species, the studies encourage the cultivation and consumption of these vegetables, improving production systems, diversifying production and increasing the income of family farmers (Madeira et al., 2013).

FIGURE 10.3 Traditional vegetable collection, all NUS, at Embrapa Vegetables. Brasília, 2017.

Source: Lidio Coradin (used with permission).

In a partnership between Embrapa, the Ministry of Agriculture, Livestock and Supply, and the Technical Assistance and Rural Extension Agency of Minas Gerais State, including family farmers' organizations, the cultivation and consumption of traditional vegetables was encouraged in Minas Gerais through the establishment of community seed-banks, the organization of events (technical lectures, field days, culinary and agronomic workshops, etc.) and the publication of booklets and books (Brasil, 2013; Pedrosa, 2013), benefiting more than 50 municipalities and 5,000 family farmers. Other important partners of this work are the Agricultural Research Company of Minas Gerais, Viçosa Federal University and Lavras Federal University, all of which have established collections of traditional vegetables and have carried out important research on the topic. These actions were also expanded to the states of São Paulo and Santa Catarina, through partnerships with research and rural extension state agencies (Castro and Devide, 2013; Calegari and Matos Filho, 2017), also influencing the states of Goiás, Paraná and Bahia.

Many NUS have recognized functional effects as nutraceuticals. Important nutritional studies were carried out by the VP Functional Nutrition Center, analyzing the nutritional value of several NUS, showing that they contained high levels of proteins, minerals and functional compounds (Paschoal et al., 2017).

The Ministry of Health, in partnership with several institutions, released *Dietary Guidelines for the Brazilian Population* (Brasil, 2014) with ten steps for establishing an adequate and healthy diet. They also published the book *Brazilian Regional Foods* (Brasil, 2015) with 191 foods (123 NUS) emphasizing that healthy eating involves, besides the choice of suitable foods, biodiversity conservation and the recognition of cultural heritage.

The Slow Food organization has described 186 food communities in Brazil, out of which 110 are food plant species and 98 are NUS (Slow Food Brasil, 2020).

Another important aspect of the promotion of NUS in Brazil is the organization of technical and scientific events in the country. In 2007, in Porto Seguro, Bahia, the 47th Brazilian Horticulture Congress prioritized the theme "Rescuing and Valuing Underutilized Vegetables." In 2018, Brazilian biomes and their NUS were the themes of the 14th International Congress of Functional Nutrition in São Paulo.

In 2017, Embrapa Vegetables and its partners organized, in Brasília, the 1st HortPANC – National Meeting for Non-conventional Vegetables, focused on agriculture (Figure 10.4), but also promoting dialogue about nutrition and cooking and linking producers with cooks and chefs. The second HortPANC was held in São Paulo in 2018; the third in Curitiba, Paraná State, in 2019; and the fourth was postponed to 2021 in Salvador, Bahia State, due to the COVID-19 pandemic. It is important to highlight the growing number of researchers, and related initiatives, in Brazil that are developing studies with NUS. Today more than 1,000 researchers, professors and students use the acronym PANC in their curriculums, as registered in the Brazilian National Research Council (CNPq) platform.

FIGURE 10.4 Field day highlighting non-conventional vegetables, all NUS, at Embrapa Vegetables on the 1st HortPANC – National Meeting of Non-conventional Vegetables. Brasília, 2017.
Source: Nuno Rodrigo Madeira (used with permission).

Finally, it's noteworthy that the interest of the population and the use of NUS are growing in Brazil, which can be verified by the large number of social media communities that are using the acronym PANC on platforms such as WhatsApp, Facebook and Instagram. And, as a response to that, there is a growing number of farmers and entrepreneurs investing in the production of NUS.

Final considerations

Actions to promote the valorization, production and the consumption of NUS, which are part of Brazil's socio-cultural heritage, are very important not only for food security but also for food and nutritional sovereignty, since they contribute to increasing diversity and richness in diets, in addition to promoting good eating habits that encompass cultural, economic and social aspects. Such actions also bring to light more resilient species with a high capacity to mitigate the negative effects of climate change.

References

Brasil (2013) 'Manual de Hortaliças Não Convencionais', Ministério da Agricultura, Pecuária e Abastecimento, Brasília, 99p.

Brasil (2014) 'Guia alimentar para a população brasileira', 2ª ed., Ministério da Saúde, Brasília, 156p.

Brasil (2015) 'Alimentos regionais brasileiros', 2ª ed., Ministério da Saúde, Brasília, 484p.

Calegari, C. R. and Matos Filho, A. M. (2017) 'Plantas alimentícias não convencionais – PANCs', Epagri, 53p.

Cardoso, M. O. (1997) 'Hortaliças não-convencionais da Amazônia', Embrapa-SPI/CPAA, Brasília, 150p.

Cascudo, L. C. (1983) 'História da alimentação no Brasil', Itatiaia, Belo Horizonte, Udesp, São Paulo, 926p.

Castro, C. M. e Devide, A. C. P. (2013) 'Plantas alimentícias não-convencionais: Sazonalidade no Vale do Paraíba', Agência Paulista de Tecnologia dos Agronegócios, Pindamonhangaba-SP, 66p.

Coradin, L., Camillo, J. and Oliveira, C. N. S. (2018) 'A Iniciativa Plantas para o Futuro', in L. Coradin, J. Camillo e F. G. C. Pareyn (eds). *Espécies nativas da flora brasileira de valor econômico atual ou potencial: plantas para o futuro: região Nordeste, Série Biodiversidade 51*, Ministério do Meio Ambiente, Brasília, pp31–72.

Hue, S. M. (2008) 'Delícias do descobrimento: a gastronomia brasileira do século XVI', Jorge Zahar, Rio de Janeiro, 208p.

Hunter, D., Borelli, T. and Gee, E. (2020) 'Biodiversity, food and nutrition: A new agenda for sustainable food systems', Issues in Agricultural Biodiversity, Earthscan/Routledge, London, 334p.

IPGRI (2002) 'Neglected and underutilized plant species: Strategic action plan of the international plant genetic resources institute', IPGRI, Rome, Italy, 28p.

Kinupp, V. F. (2007) 'Plantas alimentícias não-convencionais da região metropolitana de Porto Alegre-RS', UFRGS, Porto Alegre-RS, 562p.

Kinupp, V. F. and Lorenzi, H. (2014) 'Plantas alimentícias não convencionais (PANC) no Brasil: guia de identificação, aspectos nutricionais e receitas ilustradas', Plantarum, São Paulo, 768p.

Lopes, R. J. (2017) '1499: o Brasil antes de Cabral', Harper Collins, Rio de Janeiro, 246p.

Madeira, N. R., Silva, P. C., Botrel, N., Mendonça, J. L., Silveira, G. S. R. and Pedrosa, M. W. (2013) 'Manual de produção de Hortaliças Tradicionais', Embrapa, Brasília-DF, 155p.

Madeira, N. R. e Botrel, N. (2019) 'Contextualizando e resgatando a produção e o consumo das hortaliças tradicionais da biodiversidade brasileira', *Revista Brasileira de Nutrição Funcional*, vol 43, no 78, pp27–32.

Madeira, N. R. and Kinupp, V. F. (2016) 'Experiências com as Plantas Alimentícias Não Convencionais no Brasil', *Informe Agropecuário, Belo Horizonte*, vol 37, no 295, pp7–11.

Madeira, N. R., Reifschneider, F. J. B. and Giordano, L. B. (2008) 'Contribuição portuguesa à produção e ao consumo de hortaliças no Brasil: uma revisão histórica', *Horticultura Brasileira, Brasília*, vol 26, no 4, pp428–432.

Paschoal, V., Baptistella, A. B. and Souza, N. S. (2017) 'Nutrição funcional e sustentabilidade', Valéria Paschoal Ed. Ltda., São Paulo, 384p.

Pedrosa, M. W. (2013) 'Hortaliças não-convencionais', Empresa de Pesquisa Agropecuária de Minas Gerais (Epamig), Prudente de Morais-MG, 23p.

Santiago, R. A. C. and Coradin, L. (2018) 'Sabores e Aromas do Brasil', In R. A. C. Santiago e L. Coradin (eds), *Biodiversidade brasileira: sabores e aromas. Série Biodiversidade 52*, Ministério do Meio Ambiente, Brasília, pp23–44, www.mma.gov.br/publicacoes/biodiversidade/category/54-agrobiodiversidade.html

Shulman, S. (1986) 'Seeds of controversy: Nations square off over who will control plant genetic resources', *BioScience*, vol 36, no 10, pp647–651.

SlowFood Brasil (2020) 'Produtos do Brasil na arca do gosto', http://www.slowfoodbrasil.com/arca-do-gosto/produtos-do-brasil. Accessed 06 May 2020.

11

UNDERUTILIZED GENETIC RESOURCES AND CROP DIVERSIFICATION IN EUROPE

Ambrogio Costanzo

Introduction

Current European agriculture results from half a century of dramatic transformation, mostly driven by the Common Agricultural Policy, which, after being launched in 1962, pursued food security and self-sufficiency through increased productivity and farm restructuring at the expense of crop diversity. Despite policy shifts from a productivist to what Ward et al. (2008) call a post-productivist paradigm in the 1970s and 1980s (mostly aiming at managing produce surplus, and balancing agricultural production with environmental protection and rural development), the picture of European cropland is today far from being biodiverse. The development of European agriculture has been characterised by high concentration and specialisation, which led to a remarkable hegemony of a small number of crops, both in cultivated areas and in breeding efforts. In fact, Europe is no exception to the dominance of the "big three" crops in food systems – rice, maize and wheat – that cover 40% of worldwide arable land and provide 55% of humankind's total caloric intake (Stamp et al. 2012).

Looking at the EU's combinable crops area – the sector where biodiversity loss is most severe (Stamp et al., 2012), and which this chapter mostly focuses on – there is a striking dominance of cereal crops (82%), followed by a 16% of oilseed crops and just 2% of protein crops. As of 2020, 64% of the total combinable area is covered by just three species: common wheat, grain maize and barley. The European common catalogue of plant varieties shows a similar picture: only two species (common wheat and maize) have more than 2,000 registered varieties, and three (barley, oilseed rape and sunflower) have more than 1,000, whereas all protein grain crops comprise 633 varieties, of which 411 for just one species (*Pisum sativum*). Both the area partitioning and the number of registered varieties show the remarkable dominance of major cereal (common wheat, maize) and

DOI: 10.4324/9781003044802-12

NUS in Europe **139**

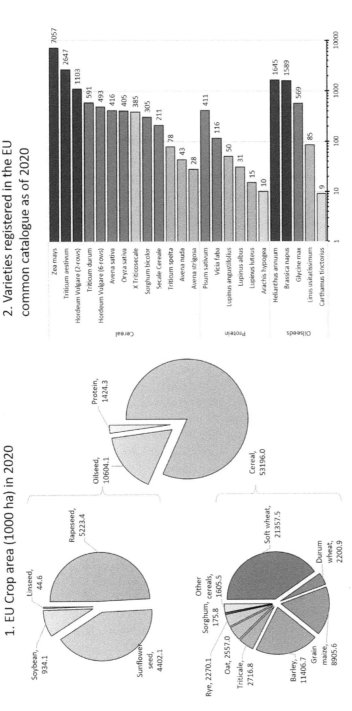

FIGURE 11.1 Area occupied by major arable crops in the European Union as of 2020 and number of registered varieties in the EU common catalogue in 2020 (data from EUROSTAT[1] and the EU database[2] of registered plant varieties, both retrieved on 28 June 2020).

140 Ambrogio Costanzo

oilseed (sunflower, oilseed rape) crops, and the alarming trend of disappearing grain legumes (Zander et al., 2016) in the agricultural as well as in the research and development landscape. It is against this extremely hegemonic landscape that the diversification of the seed and crops basket should be framed, as part of an overall transition towards more sustainable agricultural system (Chable et al., 2020).

What are underutilized crops in the European context?

With awareness that long-term stable and sustainable development through locally adapted agricultural systems will depend on increasing crop biodiversity, the EU H2020 Framework Programme (2014–2020) targeted, under the Societal Challenges 2 "Food security, sustainable agriculture and forestry", the rediscovery and valorisation of underutilized genetic resources for food and agriculture. As part of this effort, the project "Embedding crop diversity and networking for local high-quality food systems – DIVERSIFOOD (2014–2019)" has explored a wide range of species, both major and those actually neglected and underutilized, and the potential of diversification in the context of communities of users. This project aimed to go beyond mere conservation of diversity, emphasising innovation and appreciating these genetic resources as a key asset to restore agricultural diversity. In a continuous debate over the meaning of "orphan", "neglected", "underutilized" and "alternative" crops, the DIVERSIFOOD project formulated a working definition according to which an underutilized crop is classified as "a plant genetic resource with limited current use and potential to improve and diversify cropping systems and supply chains in a given context". According to this conceptualisation, underutilized crops are not to be looked at in isolation. On the contrary, the focus needs to shift from the plants to the processes of their (re)introduction as related to agricultural diversification in specific contexts. In the European context, at least three interconnected processes are worth exploring to purse this goal: (i) shifting cultivation areas; (ii) reintroducing "minor" species and (iii) diversifying the genetic structure of major and minor species.

Shifting cultivation areas

Shifts in cultivation areas of cultivated plant are a process that has characterised human history for millennia. As demonstrated by Vavilov (1935), cultivated plants were domesticated in relatively small foci, the "centres of origin", from which their cultivation expanded into larger areas. In several instances, species have crossed geographical barriers, like the acquisition of a huge variety of species in Europe following the colonisation of the Americas, sometimes generating further diversification centres far from the centre of origin, like for maize in Europe (Thenaillot and Charcosset, 2011). Shifting cultivation areas is a continuous process and is still ongoing (Khoury et al., 2016).

A first notable trend is that some species start to be cultivated commercially in a region whose climate is not too dissimilar from their home region, as e.g., for quinoa (*Chenopodium quinoa* Wild.). This annual species, a staple food for Andean Amerindian people going back millennia, is currently cultivated in 95 countries worldwide, notably in Europe, as a result of significant research efforts in the 1990s (Bazile et al., 2016). Quinoa is possibly a "model species" of underutilized crop, for its success as well as for the problematic implications of such a success. The renaissance of its demand and production was due to the discovery of its interesting nutritional properties (Repo-Carrasco et al., 2003) as well as to its high level of genetic diversity, which facilitates its adaptation (Ruiz et al., 2014). However, the sudden explosion in interest in this species has also roused non-negligible and still unsolved concerns of equity and fairness regarding the exploitation of the value added to its genetics by generations of Andean farmers (Chevarria-Lazo et al., 2015).

A second trend is plants following human migrations: "exotic" crops introduced on a non-commercial basis, grown for family consumption as part of growers' cultural heritage, and that are rarely surveyed or subjected to formal documentation and conservation. A notable example is the "Sowing New Seeds" work conducted in the early 2010s on the plants cultivated by overseas immigrants in their allotments in the English Midlands, which highlighted the potential of non-professional cultivation and selection of exotic species for more general resilience and the diversification of agriculture (Kell et al., 2013).

A third trend is expanding cultivation areas following, or anticipating, changes in climate. A remarkable example is the northwards expansion of field crops in northern Europe, due to milder winters and potentially longer growing seasons, and the subsequent opportunities (Peltonen-Sainio et al., 2009) as well as challenges of pests and diseases expansion (Hakala et al., 2011).

Rediscovering "forgotten" species

During the process of the modernisation and standardisation of European agriculture, which led to massive concentration on a few species, a large number of species has been abandoned. This is true both in dominant and currently shrinking sectors, like cereals and protein crops, respectively. In cereals, among the most widespread and most intensively bred crops in Europe is wheat, mostly common wheat (*Triticum aestivum*) with a small fraction of durum wheat (*Triticum durum*) in Mediterranean regions. However, the genus *Triticum* comprises many species that were once more widely cultivated but that are now relegated to the status of "minor cereals" and cultivated in limited amounts, often in marginal areas. The hulled wheats are typical examples of this: the hexaploid spelt (*Triticum spelta*), the tetraploid emmer (*Triticum dicoccon*) and the diploid einkorn (*Triticum monococcum*), in order of decreasing acreage, have seen increasing market interest due to their nutritional properties as well as to their rusticity and adaptation

142 Ambrogio Costanzo

to stressful environments (Cubadda and Marconi, 2002; Zaharieva et al., 2010; Zaharieva and Monneveux, 2014).

Amongst *Triticum* species, of particular interest is *Triticum turgidum* L., known as "Rivet wheat" in the British Isles or "Poulard wheat" in France. It is a close relative of durum wheat, particularly common in mountainous areas and Atlantic climates, the object of some breeding efforts in England during the 1930s (Zeven et al., 1990), but then abandoned and nowadays virtually non-existent. The European Search Catalogue for Plant Genetic Resources (EURISCO) database[3] records more than 14,000 accessions of *T. turgidum* conserved in European genebanks, 38% of which are in Italy. A large representation of these accessions was tested during the DIVERSIFOOD project, highlighting a great variability in physical properties of the grain, which makes it a good candidate for both bread- and pasta-making, also in regions (like north-western France, the Netherlands and the United Kingdom) that are not suitable for durum wheat (Chable, 2018; Costanzo et al., 2019).

Diversifying the genetic structure of major and minor species

The shrinking number of species is only part of the loss of diversity resulting in modern European agriculture. In fact, for many crops – notably the most widespread and those with the largest number of registered varieties – crop phenotypic diversity is drastically reduced to extremely homogeneous crop types, such as pure line varieties (e.g., wheat and barley) and F1-bybrids (e.g., maize), to the detriment of historic varieties, landraces and genetically diverse populations. Increasing genetic and phenotypic diversity within species is, however, a key process to enhance sustainability and resilience through functional diversity (Newton et al., 2009) and a key priority in "neglected" as well as mainstream species. For instance, whilst testing einkorn entries in view of exploring their potential for UK organic farming, a striking yield advantage of landraces over a modern, dwarf commercial einkorn variety (Costanzo et al., 2019) was a reminder that the reintroduction of forgotten species cannot be separated from a rethinking of the relationships between breeding and target cropping systems. In fact, marginal environments, with low input use and high exposure to climatic extremes, need, and can be added value by, more resilient/rustic phenotypes for both minor (like einkorn) and major crops (like wheat and barley). However, diversification of the genetic structure of crops goes well beyond the "rediscovery of the old", towards the development of decentralised, evolutionary (Phillips and Wolfe, 2005) and participatory breeding methods, that have extensively been applied in wheat (Van Frank et al., 2020), showing clear advantages in terms of performance, robustness and stability (Goldringer et al., 2020).

It is important to point out that, whilst the rediscovery of forgotten species mainly faces practical and economic constraints, genetic diversification within species encounters further obstacles in the legislative structure developed by and for mainstream agriculture, such as the UPOV (International Union for the Protection Of New Varieties of Plants) convention and national and international seed legislation (European catalogue and seed certification for seed marketing). In fact, the official conditions for cereal seed certification, namely the "sufficient identity and varietal purity" (Art. 1, 66/402/EEC of 1966), effectively ruled out any marketing of cereal seed that was not distinct, uniform and stable in appearance, until a "temporary experiment", under the Commission Implementing Decision 2014/150/EU, allowed the marketing of "populations" of wheat, oats, barley and maize, under quantitative restrictions. The legality of genetically heterogeneous seeds and of genetic structures not covered by the above "temporary experiment" is currently under discussion in the context of the Revised Organic Regulation that will come into force in 2021 (Costanzo and Bickler, 2019).

Where are underutilized genetic resources and how to assess them

In the DIVERSIFOOD project, nine species have been tested with hundreds of accessions in 33 different trials across Europe. Seeds have been sourced from a variety of both informal and formal, *ex situ* and *in situ* collections, and evaluated in participatory experiments with multi-actor communities of researchers, farmers and potential users. Four main outcomes have been noted:

- Certain traits that during modern breeding had been lost have reappeared, including undesirable traits that might have played a role in the abandonment of certain species or phenotypes, like the brittle rachis in einkorn that can generate huge grain losses, or the extreme straw height of certain winter cereals like rivet wheat, which creates problems of lodging. Moreover, many genetic resources show considerable within-crop phenotypic heterogeneity, either linked to their genetic structure, them being landraces or resulting from intentional or even accidental mixtures occurred during *in situ* informal conservation.
- Agroecosystem performance of the same genetic resource can vary greatly depending on where it is grown and must, therefore, be looked at on a very local scale. As obvious as it might seem, this reinforces the importance of deploying and testing genetic resources in multiple farms rather than on research stations, to identify the potential for local adaptation.
- The yield of "underutilized genetic resources" can be a serious limiting factor, as the tested material can be either low-yielding or difficult to harvest, but, in many cases, can be a relief for marginal conditions. Species such as einkorn, emmer or rivet wheat can thrive where their commonly grown closest relatives (e.g., durum or bread wheat) are not a viable option. This is one of the key benefits expected from underutilized crops: they can be a valuable resource for more marginal land.
- A diversity of crops triggers a diversity of products that, in turn, need adaptation in both processing and the methods and concepts used to assess their quality. Grains from minor cereals are not necessarily suited to industrial

milling but provide an opportunity for artisanal millers and bakers, whose processing methods are more flexible and adaptable to the raw material, to add value to highly nutritious grains.

A series of booklets covering practical aspects of research, farming, marketing and policy around increasing crops diversity on the farm and in the supply chain has been produced by the DIVERSIFOOD project, including a guide to participatory experiments on underutilized genetic resources (Costanzo and Serpolay, 2019) and case studies of the marketing of products from newly bred lines and underutilized crops (Padel et al., 2018).

Conclusion

European agriculture faces the urgent need to diversify its crop basket, to overcome the dominance of a very limited number of crops and the related socio-technical lock-ins that prevent easy diversification (Meynard et al., 2018). Neglected and underutilized crops have a huge strategic importance in the transition towards more sustainable agricultural systems. However, diversification is a complex process, not limited to the (re)introduction of a few species, but encompassing all aspects of cropping systems' design, connection between genetic resource conservation and use, more inclusive breeding systems and stronger connections between producers and consumers.

Notes

1 https://ec.europa.eu/eurostat/web/agriculture/data/database
2 https://ec.europa.eu/food/plant/plant_propagation_material/plant_variety_catalogues_databases/search/public/index.cfm
3 https://eurisco.ipk-gatersleben.de/apex/f?p=103:30:9540465718747, retrieved 28 June 2020

References

Bazile, D., Jacobsen, S.-E. and Verniau, A. (2016). 'The global expansion of Quinoa: trends and limits. *Frontiers in Plant Sciences*. DOI: 10.3389/fpls.2016.00622

Chable, V. (2018) 'The rivet wheat'. Innovation factsheet no. 5. DIVERSIFOOD. http://www.diversifood.eu/wp-content/uploads/2018/06/Diversifood_innovation_factsheet5-Rivet-Wheat.pdf (retrieved 16 January 2020).

Chable, V., Nuijten, E., Costanzo, A., et al. (2020) 'Embedding cultivated diversity in society for agro-ecological transition'. *Sustainability* 12(3), 784. DOI: 10.3390/su12030784

Costanzo, A., Amos, D., Dinelli, G., et al. (2019) 'Performance and nutritional properties of einkorn, emmer and rivet wheat in response to different Rotational position and Soil tillage'. *Sustainability* 11(22), 6304. DOI: 10.3390/su11226304

Costanzo, A. and Bickler, C. (2019) 'Proposal for a toolbox for identification and description of organic heterogeneous material'. *Deliverable*, 2(8), LIVESEED project: www.liveseed.eu

Costanzo, A. and Serpolay, E. (2019) 'A guide to participatory experiments with underutilized genetic resources'. DIVERSIFOOD Booklet, no. #2. ITAB. https://orgprints.org/35259/

Chevarria-Lazo, M., Bazile, D., Dessauw, D., Louafi, S., Trommetter, M. and Hocdé, H. (2015) 'Quinoa and the exchange of genetic resources: Improving the regulation systems'. In: Bazile D., Bertero H.D. and Nieto C. (eds.) *State of the Art Report on Quinoa Around the World in 2013*. Rome: FAO and CIRAD, pp. 83–105. http://www.fao.org/3/a-i4042e.pdf (retrieved 28 June 2020).

Cubadda, R. and Marconi, E. (2002) 'Spelt wheat'. In Belton, P.S. and Taylor, J.R.N. (eds.) *Pseudocereals and Less Common Cereals: Grain Properties and Utilization Potential*. Springer Science & Business Media. Springer, Berlin, Heidelberg. ISBN 9783540429395. pp. 153–175

Goldringer, I., Van Frank, G., d'Yvoire, C.B., et al. (2020) 'Agronomic evaluation of bread wheat varieties from participatory breeding: A combination of performance and robustness'. *Sustainability* 12(1), 128. DOI: 10.3390/su12010128

Hakala, K., Hannukkala, A.O., Huusela-Veistola, E., Jalli, M. and Peltonen-Sainio, P. (2011) 'Pests and diseases in a changing climate: A major challenge for Finnish crop production' *Agriculture and Food Science* 20, 3–14

Kell, S., Rosenfeld, A., Cunningham, S., Dobbie, S. and Maxted, N. (2013) *Benefits of Non-Traditional Crops Grown by Small Scale Growers in the Midlands*. Garden Organic: Coventry. https://www.gardenorganic.org.uk/sites/www.gardenorganic.org.uk/files/sns/SNSReportFinal.pdf (retrieved 28 June 2020).

Khoury, C.K., Achicanoy, H.A., Bjorkman, A.D., Navarro-Racines, C., et al. (2016) 'Origins of food crops connect countries worldwide'. *Proceedings of the Royal Society B*. DOI: 10.1098/rspb.2016.0792

Meynard, J.-M., Charrier, F., et al. (2018) 'Socio-technical lock-in hinders crop diversification in France'. *Agronomy for Sustainable Development* 38. DOI: 10.1007/s13593-018-0535-1

Newton, A.C., Begg, G.S. and Swanston, J.S. (2009) 'Deployment of diversity for enhanced crop function'. *Annals of Applied Biology*. DOI: 10.1111/j.1744-7348.2008.00303.x

Padel, S., Rossi, A., D'Amico, S., Sellars, A. and Oehen, B. (2018) 'Case studies of the marketing of products from newly bred lines and underutilized crops'. DIVERSIFOOD project; https://orgprints.org/34456/

Peltonen-Sainio, P., Jauhiainen, L., Hakala, K. and Ojanen, H. (2009). 'Climate change and prolongation of growing season: Changes in regional potential for field crop production in Finland'. *Agricultural and Food Science* 18(3), 171–190. DOI: 10.2137/145960609790059479

Phillips, S.L. and Wolfe, M.S. (2005) 'Evolutionary plant breeding for low input system'. *Journal of Agricultural Science*. DOI: 10.1017/S0021859605005009

Repo-Carrasco, R., Espinoza, C. and Jacobsen, S.E. (2003) 'Nutritional value anf use of the Andean crop Quinoa (Chenopodium quinoa) and Kañiwa (Chenopodium pallidicaule)'. *Food Reviews International*. DOI: 10.1081/FRI-120018884

Ruiz, K.B., Biondi, S., Martinez, E., et al. (2014) 'Quinoa biodiversity and sustainability for food security under climate change. A review'. *Agronomy for Sustainable Development*. DOI: 10.1007/s1359-013-0195-0

Stamp, P., Messmer, R. and Walter, A. (2012) 'Competitive underutilized crops will depend on the state funding of breeding programmes: An opinion on the example of Europe'. *Plant Breeding* 131, 461–464. DOI: 10.1111/j.1439-0523.2012.01990.x

Thenaillot, M.I. and Charcosset, A. (2011) 'A European perspective on maize history'. *Comptes Rendus Biologies*. DOI: 10.1016/j.crvi.2010.12.015

Van Frank, G., Riviere, P., Pin, I., et al. (2020) 'Genetic diversity and stability of performance of wheat population varieties developed by participatory breeding'. *Sustainability* 12(1), 384. DOI: 10.3390/su12010384

Vavilov, N.I. (1935) *The Phytogeographical Basis for Plant Breeding* (D. Love, transl.). Vol. 1: pp. 316–366. 1992. Cambridge University Press, Cambridge.

Ward, N., Jackson, P., Russel, P. and Wilkinson, K. (2008) 'Productivism, post-productivism and European agricultural reform: the case of sugar'. *Sociologia Ruralis* 48(2), 118–132. DOI: 10.1111/j.1467-9523.2008.00455.x

Zaharieva, M. and Monneveux, P. (2014) 'Cultivated einkorn wheat (*Triticum monococcum* L. subsp. *monococcum*): The long life of a founder crop of agriculture'. *Genetic Resources and Crop Evolution*. DOI: 10.1007/s10722-014-0084-7

Zaharieva, M., Ayana, N.G., Hakimi, A., Misra, S.C. and Monneveux, P. (2010) 'Cultivated emmer wheat (*Triticum dicoccon* Schrank), an old crop with promising future: A review'. *Genetic Resources and Crop Evolution* 57, 937–962. DOI: 10.1007/s10722-010-9572-6

Zander, P., Amjat-Babu, T. S., Preissel S., et al. (2016) 'Grain legume decline and potential recovery in European agriculture: A review'. *Agronomy for Sustainable Development* 36, 26. DOI: 10.1007/s13593-016-0365-y

Zeven, A.C. (1990) 'Classification of landraces and improved cultivars of rivet wheat (*Triticum turgidum*) and bread wheat (*T. aestivum*) from Great Britain and described in 1934'. *Euphytica* 1990(47), 249. DOI: 10.1007/BF00024248

12

NEGLECTED AND UNDERUTILIZED SPECIES AND INDIGENOUS FOODWAYS OF OCEANIA

Danny Hunter, Nick Roskruge, Simon Apang Semese, Philip Clarke and Gerry Turpin

Introduction

Oceania represents a considerable array of agro-ecosystems comprising an equal diversity of wild and cultivated plant species that, at one time or another, have been used for food and other purposes. This diversity of crops and trees has played an important role in the lives and development of populations throughout the islands that comprise Oceania. However, today only a small fraction of this diversity is regularly used or found in markets and value-chains. Bringing back these foods to the tables of consumers will play a strategic role in supporting local production systems in adapting to pervasive climate change, and will also help re-establish healthy food habits to tackle a growing public-health nutrition problem and safeguard cultural identity. For the purpose of this chapter, we define Oceania broadly as to include the islands of the Pacific, including Melanesia, Polynesia and Micronesia, as well as New Zealand and Australia. Communities in Oceania have historically created an enduring relationship with the landscape and its compliment of flora and fauna. With 60,000 years of settlement, Indigenous Australians have developed an intimate relationship with and knowledge of a diverse range of ecosystems (Mulvaney and Golson, 1971; Clarke, 2003a). Today, this reliance on traditional foods is still evident in some communities, such as those in Arnhem Land, Cape York and the Torres Strait, where hunting, gathering and fishing supplement significantly modern store-bought diets. But in most instances these communities have been disconnected from their traditional foods and food systems. Around 30,000 years ago, people started to move into the western part of the Pacific, including Papua New Guinea (PNG), the Solomon Islands and northern Australia, taking with them their traditional foods such as taro, yams, banana, coconut, breadfruit pandanus, sago and arrowroot, before moving further east into the more scattered islands of the Pacific Ocean

DOI: 10.4324/9781003044802-13

148 Danny Hunter et al.

(Denham, 2008; Denham et al., 2009a, 2009b; Haden, 2009). Pacific Islanders have developed farming and food systems that have exploited a unique range of ecosystems in high and low islands throughout the region and which employ a diversity of plant and animal species. Many of these crops, such as taro, have adapted to range of agro-ecologies, resulting in hundreds of unique cultivars (Iosefa et al., 2013). Significant and unique genetic diversity exists in many other traditional Pacific Island crops, and this is true also for nutritional diversity (Englberger et al., 2009). A thousand years ago, the indigenous Māori, on settling Aotearoa-New Zealand from the more tropical Pacific Islands, had to adapt their horticultural practices to a new environment of unpredictable and limiting climate with extremes from sub-tropical features in the north to sub-Antarctic in the south, thereby shaping a lifestyle that was based on a subsistence approach, including both cultivated and uncultivated plants and the seasonal harvesting of birds and fish.

While this chapter can only provide a very general and brief survey of NUS from Oceania, it also looks at those factors that have undermined traditional food systems and diversity, particularly colonisation and globalisation, and the impacts this has had on food systems, diets and nutrition. The chapter highlights recent efforts to promote and reinvigorate traditional food systems and NUS, and highlights the challenges and opportunities for their promotion.

Dietary and lifestyle change, loss of food diversity and human health

The nutrition transition – the process by which development, globalisation, urbanisation, poverty and subsequent changes in lifestyle have led to excessive calorific intake, poor-quality diets and low physical activity – is particularly prevalent among communities of Indigenous People in Australia, New Zealand and the Pacific. This has been exacerbated by inappropriate development, health, food and trade policies that support imports of less healthy foods, as in the case of the Pacific, or the availability of culturally inappropriate and nutritionally poor foods in remote stores, as in Australia (Lee et al., 2009; Brimblecombe et al., 2013). Urbanisation and lifestyle changes, including in eating habits, also seriously threaten the very existence of many traditional food crops. An alarming dietary shift from traditional foods and healthy diets towards the consumption of poor-quality processed foods and diets has taken place, which has led to the dramatic emergence of obesity and associated chronic diseases, especially among Indigenous compared to non-Indigenous groups. In addition to over-nutrition and obesity, there are problems of deficiency of vital micronutrients (hidden hunger) among communities of Indigenous People in Oceania, which can – in extreme situations – contribute to early death or to impaired physical and intellectual development, especially of children. For example, the Federated States of Micronesia (FSM), recently ranked among the most obese countries in the world, is also a country where pockets of Vitamin A deficiency (VAD) are among the highest in the world (Englberger et al., 2003, 2011). Studies of Indigenous peoples' diets in Australia, New Zealand and the Pacific consistently report low intake of fruit and vegetables; high intake of soft drinks, refined cereals and sugars; excessive sodium intake and limited availability of several key micronutrients. This is contributing to a rapidly growing health burden, including both physical and psychological illnesses, as countries try to deal with the associated healthcare costs, reduced productivity and shorter lifespans.

The great irony is that few of these health problems existed prior to European contact, colonisation or development. Indigenous foodways contained a diversity of animals, plants and trees that were nutritionally rich and that provided ample levels of both macro- and micronutrients for a healthy diet. Englberger et al. (2003, 2011) highlight reports from early visitors that suggest that island communities consumed a diversity of foods and appeared healthy. There was little evidence of malnutrition, diabetes or hypertension before the 1940s, with VAD not documented until 1998. Many of the traditional foods of these Indigenous foodways have now been marginalised or forgotten, and the traditional ecological knowledge associated with them lost or threatened. Furthermore, inappropriate policy and regulatory frameworks have, especially in the Pacific, increased dependency on food imports through the promotion of culturally unsuitable foods, which have resulted in changes in the pattern of agricultural production towards cash- and export-oriented cropping, ignoring the potential of nutritionally rich local foods (Englberger et al., 2003).

Indigenous foodways of Māori and Aboriginal Australia have also fared poorly, been marginalised and been forgotten (Haden, 2009). Davy (2016) reports that the much higher rate of illness and disease among Aboriginal populations in Australia compared to non-Indigenous ones is directly related to food insecurity and poor diet, including through poorer access to nutritious foods. Bussey (2013) highlights that, prior to European contact, levels of diet-related illness among Indigenous Australians were non-existent. National health surveys in New Zealand consistently reveal the high prevalence of obesity and non-communicable diseases (i.e., cardiovascular disease, cancer and type 2 diabetes) among the Māori – the indigenous people of Aotearoa-New Zealand. These trends have been mostly attributed to sedentary lifestyles and poor dietary choices, as the Māori move to urban areas and away from their lands and highly adapted food production systems. With the shift to urban settings, much of the traditional knowledge surrounding the production, processing and storing of local foods is being lost and agricultural/cultural systems are reportedly being further threatened by biosecurity risks and corporate ownership (Roskruge, BFN website[1]).

Increasingly there are calls to revitalise these Indigenous foodways and the nutrient-rich foods they contain (Kuhnlein et al., 2009; FAO, 2021). The access, availability and use of traditional foods have the potential to recover, improve and safeguard food security, especially among remote Indigenous communities, now and in the future, and, thus, demands further research (Bussey, 2013).

150 Danny Hunter et al.

Neglected and underutilized species of Oceania: a brief overview

A short chapter of this nature, dealing as it does with such a large geographic area, prevents a comprehensive inventory or description of the neglected and underutilized species (NUS) present in Oceania. Rather, what we aim for here is a brief introduction highlighting some of the key NUS with most potential for development and utilisation. For those interested in exploring the NUS of Oceania in more detail we provide some important references.

Although the land area of the Pacific is small, it is recognised as a centre of diversity and/or origin of a small number of crops, for example, taro (*Colocasia esculenta*), which comprises thousands of distinct cultivars grown and eaten throughout the region (Rao et al., 2010). In 2009, participants from 15 countries representing the three sub-regions of the Pacific (Melanesia, Micronesia and Polynesia) came together for the first time to develop a regional priority list of NUS/species groups, including a strategy for their conservation and sustainable use, addressing: generation and collection of knowledge/research; communication and dissemination; policy advocacy, market development and partnerships; capacity building and institutional strengthening (Taylor et al., 2009). Regional priorities were clearly breadfruit (*Artocarpus altilis*), bananas (*Musa* spp.) of the Fe'i group and/or Pacific plantain, Polynesian chestnut (*Inocarpus fagifer*), and tava (*Pometia pinnata*). Other priority species included bele/aibika (*Abelmoschus manihot*), pandanus (*Pandanus tectoris*), noni (*Morinda citrifolia*), football fruit (*Pangium edule*), pitpit/duruka (*Sacchurum edule*), arrowroot (*Tacca leontopetaloides*), saijan/drumstick (*Moringa oleifera*), *Spondias dulcis*, *Diplazium esculentum*, fig tree (*Ficus tinctoria*), sago (*Metroxylon sagu*), aupa (*Amaranthus* spp.), fern (*Diplazium* spp.), Malay apple (*Syzygium malaccense*) and mangrove (*Bruguiera* spp.). Some locally important underutilized nut species were also identified, including Galip nut (*Canarium indicum*) and nuts from *Terminalia catappa*, *Barringtonia edulis* and *B. procera*. Some of the tree species mentioned here, and others with edible parts, have also been prioritised and described in detail more recently by Thomson et al. (2018). Much of this diversity remains under threat in the Pacific, as a result of urbanisation and lifestyle changes, including changes in eating habits as well as pests and diseases (Box 12.1).

Box 12.1 Highlands pitpit: a knowledge-rich, multifunctional food crop in PNG

Highlands pitpit (*Setaria palmifoliai*) has been a crop of significant value in terms of food, medicine, bride price ceremonies and other traditional uses since ancestral days, matched with a diversity of varieties and traditional names found across the Highlands region (Figure 12.1). For instance, the Imbbonggu people of the Southern Highlands call it *moi* in the local Imbbonggu language, while it is known as *wotani* in the Gahuku language of the Eastern Highlands. The different varieties of highlands pitpit are characterised by colour, leaf shape and other morphology. The local people believe that the geographical location of the crop contributes to its attributes. For example, the Imbbonggu District of the Southern Highlands Province has four main varieties: Moi kapatumbe (sacred colourful variety); Moi arimoka (white stem variety); Moi gene (green variety) and Moi leruku (purple stem variety). There is also considerable nutritional diversity among the many varieties (Semese, 2018). Highlands pitpit is only eaten within Melanesian society, especially in PNG. The inner thickened soft base of the plant is the main edible portion and is cooked as a green vegetable or in a mixed meal with meat and other vegetables. Cooking recipes differ per region and tribe. The most common approach to cooking is the traditional *mumu*, a hermetic form of underground cooking. In addition, highlands pitpit is used as stock-feed across the region, especially for pigs, which have currency value in PNG. Many believe the fibrous nature of the plant helps grind the pigs' teeth and keeps them healthy. It is also a good supplement food for pigs in times when other food crops are limited in supply or during natural disasters such as drought due to its potential to withstand drought periods. Today, introduced crops, urbanisation and climate change all contribute to the declining production and use of highlands pitpit.

FIGURE 12.1 Highlands pitpit (*Setaria palmifolia*).
Source: Simon Apang Semese (used with permission).

152 Danny Hunter et al.

A number of horticultural plants have featured in traditional Māori society in Aotearoa–New Zealand and most are underutilized today. They provide the basis for a body of knowledge or *mātauranga* aligned with traditional horticulture; however, not all plants used by the Māori were cultivated (Best, 1976). There was also the availability of food stores from 'uncultivated' plants such as *aruhe* (fernroot, *Pteridium* spp.) a range of seaweeds including *kārengo* (*pārengo* or sea lettuce, *Porphyra columbina*) and berries or fruit of tree crops such as the *hīnau* (*Elaeocarpus dentatus*) and *miro* (*Prumnopitys ferruginea*), which were often located near settlements and harvested in much the same way as cultivated plants were. While these crops were considered uncultivated, they were no less managed to ensure maximum production of the harvested plant parts, e.g., the timing of harvest or minimising competition between plants for the best quality produce. The primary cultivated crops included *kūmara* (sweet potato, *Ipomoea batatas*), taro (*Colocasis esculenta*), the *uwhi* or yam (*Dioscorea alata*), various species of *pūhā* or *rauriki* (*Sonchus* spp., see Box 12.2), varieties of *tii kouka* (*Cordyline australis* and related species) (Fankhauser, 1990) and *kōkihi* (NZ spinach, *Tetragonia tetragonioides*). Since colonisation, a number of other crops have been introduced in traditional gardens and are now considered traditional Māori foods. These include *kamokamo*, a local selection of *Cucurbita pepo*, *taewa* or *Māori* potatoes and *kānga* or Indian corn. Roskruge (2014, 2015) has worked with the national Māori

---

Box 12.2 *'Ka katokato au i te rau pororua'*, 'I am gathering the bitter leaves of the sow-thistle'

Pūhā (also known as pūwhā, rauriki, kautara, tiotio or pororua) is a generic term for the introduced leafy green plant commonly called sow-thistle (various plants of the *Sonchus* spp.) in other countries, which is now a favourite utility plant for both the food and health needs of the Māori (Figure 12.2). Traditionally not a cultivated crop by early Māori but harvested from the wild, pūhā has previously been described as a culturally important plant to the Māori (Taylor and Smith, 1997; Roskruge, 2013, 2015). The traditional *whakatauaki* or proverb in the subheading above is recognition in itself of the value of pūhā to Māori, which is now considered an uncultivated Indigenous vegetable or food by most New Zealanders. The reliability of pūhā as an available green vegetable plant throughout most of the year has perpetuated its usefulness in the New Zealand diet. Coincidentally, the resurgent interest in traditional crops such as pūhā has enhanced its potential to contribute to the cultural and economic future of the Māori, including as a processed product, similar to new and innovative spinach products that are frozen, canned and/or mixed with other base products. Looking at the diversity of vegetable-based products on supermarkets shelves, the opportunities seem endless: a good example is pre-packaged liquid soups sold in a ready-to-heat pouch.

FIGURE 12.2 Pūhā (*Sonchus oleraceus*).
Source: Nick Roskruge (used with permission).

Historically, pūhā was only ever used as a fresh green vegetable, especially in the spring and early summer, and was provided a relish for meals, primarily fish. For the Māori, there are several ways in which pūhā is utilised in the diet. Primarily, it is a basis for a meal that includes pūhā, meat, potatoes and, sometimes, further relishes such as plain dumplings (colloquially termed dough-buoys). Often substitute green vegetables such as kouka (*Cordyline* spp.), kōkihi (*Tetragonia tetragonioides*), tohetaka (native dandelion, *Taraxacum magellanicum*), watercress (*Nasturtium* spp.), silverbeet (*Beta vulgaris*) or cabbage (*Brassica oleracea* var. *capitata*) are used to replace the pūhā when it is out of season or is in limited supply. A further use for pūhā is as a base for *toroī* or bottled/preserved pūhā and mussels (Dixon, 2007). In this case, the older, more stringent or bitter plants are often preferred. *Penupenu* is a mash of vegetables generally prepared for infants or invalids. In this meal, the pūhā forms the basis of a mix of vegetables usually including taewa (Māori potatoes) and kamokamo (*Cucurbit* spp.).

horticultural collective *Tahuri Whenua* to highlight the diversity of the traditional Māori diet and this has contributed to a resurgence in the cultivation of these foods.

With 60,000 years of continuous settlement of the land and an ongoing intimate relationship with and knowledge of many different ecosystems and the possible array of plants and animals available therein, it is almost impossible to comprehend the extent to which these would have been used for food. A comprehensive database of Australian Aboriginal food-plants (currently containing about 1,400 – and rising – plant/fungi species, including 100 insect-based foods) has been constructed in a project at the University of Adelaide, funded by the Orana Foundation.[2] This inventory is based primarily on historical sources but, by extension, the true number of edible plants used by Aboriginal foragers likely would have been greater when accounting for the many species of *Acacia* etc. that were also probably utilised as food. A few writers and researchers have written generally about some of the more common or key plant-food species that have been used, including Tim Low in his two books *Wild Food Plants of Australia* (1988) and *Bush Tucker: Australia's Wild Food Harvest* (1989) or, more recently, John Newton's *The Oldest Foods on Earth: A*

FIGURE 12.3 Pandanus (*Pandanus tectorius*).
Source: Philip A. Clarke (used with permission).

History of Australian Native Foods with Recipes (2016). Other researchers such as Peter Latz have written in considerable detail about Indigenous plant use in a specific geographical region, such as *Bushfires and Bushtucker: Aboriginal Plant Use in Central Australia* (1995). There is a considerable literature on this topic and the reader is guided to key publications by Clarke (2003a,b, 2007, 2018; Jones and Clarke, 2018) for more information. The list of plant species used for food alone runs in the thousands and covers environments ranging from temperate to arid and tropic. Indigenous food sources include native fruits like wild fig (*Ficus platypoda*), desert quandong (*Santalum acuminatum*), native gooseberry (*Cucumis melo*), native plum (*Santalum lanceolatum*), bush banana (*Leichhardtia australia*), finger lime (*Citrus australasica*), Kakadu plum (*Terminalia ferdinandiana*), bush tomato (*Solanum centrale*), pandanus (*Pandanus tectorius*, Figure 12.3). They also comprise greens (various pigweeds [*Portulaca* spp.], wood sorrels [*Oxalis* spp.]); wild seeds (including grasses, beans and nuts) such as various *Acacia* spp., native millet (*Panicum decompositum*); woolybutt (*Eragrostis eriopoda*), *Oryza* spp., wild sorghums, cycads (*Cycas media*), desert kurrajong (*Brachychiton gregorii*), bunya nut (*Araucaria bidwillii*); roots and tubers (bush potato (*Ipomoea costata*), nalgoo (*Cyperus* spp.) and murnong (*Microseris lanceolata*) to highlight relatively few. It should also be mentioned, that many of these heavily relied-on food sources are highly toxic and labour intensive to prepare, which is a challenge when attempting to reintroduce them into contemporary food pathways.

Safeguarding and promoting NUS in Oceania for sustainable food systems and healthy diets

There are a number of past and ongoing initiatives to promote NUS in Oceania such as bringing back traditional, forgotten foods through links to gastronomic initiatives and chef alliances, food tourism, community-based initiatives to promote improved food security through healthier diets based on local foods and efforts targeting schools and the youth.

Ongoing initiatives in New Zealand aimed at reviving traditional crops and traditional food systems as well as exploring production systems, markets, Indigenous branding, education and research include those led by the National Māori Horticultural Collective known as Tahuri Whenua,[3] which has projects aligned with food crops/plants including:

- 'Ahi kōuka i te ata, he ai i te po – the value of kōuka from a Māori lens'[4] a project looking at the traditional knowledge and use of the Tii Kouka/cabbage tree (Nga Pae o Maramatanga funded).
- E moe tonu ana te tohetaka, kaore ano i kohera – a project to identify and document heritage food-plants (primarily introduced during colonisation); Ministry of Business Innovation and Employment (New Zealand) (MBIE) (Government) funded.

156 Danny Hunter et al.

- Kaore te kumara e korero mo tona ake reka – a project to establish protocols for a germplasm collection of traditional Māori foods; MBIE funded (Tahuri Whenua/Massey University collaboration).

The Tahuri Whenua has also been active in reconnecting young Māori with their traditional food heritage through school gardens (māra) and other school-based activities (Roskruge, 2020).

In the Pacific, probably the most high-profile initiative to promote healthier diets and better nutrition based on local NUS has been the 'Let's go Local' approach (see Box 12.3), implemented in the FSM (Englberger, 2011; Englberger and Johnson, 2013). A more recent similar example to leverage neglected and underutilized food crops for improved nutrition in the Pacific is the Community Food and Health project. Other efforts to promote the utilisation of neglected and underutilized crops in the Pacific include initiatives to link local foods to tourism, gastronomy and chef alliances, e.g., the Pacific Island Food Revolution,[5] and other market opportunities (CTA, 2017; Berno, 2020), as well as through schools and school garden activities (Redfeather and Cole, 2020; Ming Wei and Bikajle, 2020)

Box 12.3 Let's Go Local: promoting nutritious, underutilized crops in the Federated States of Micronesia

The FSM has recently witnessed significant dietary shifts and increasing dependence on imported, often unhealthy, foods. Traditionally, these islands have relied on sustainable agriculture practices producing nutritious staples such as roots and tubers, breadfruit and banana for food security and nutrition. Since the middle of the twentieth century onwards, however, less nutritious, cheaper imported foods have dominated the food supply and have contributed to the poor health of the population. VAD levels in this community are among the highest in the world, as are levels of obesity and being overweight, and non-communicable diseases like diabetes.

These problems occur despite a remarkable biodiversity of underutilized nutritious foods available in the country. Pohnpei state alone has 133 varieties of breadfruit, 55 bananas, 171 yams, 24 giant swamp taros, nine tapiocas and many pandanus varieties documented. To address these health- and food-system problems, efforts were started in 1998 to identify local plant foods that could be promoted to alleviate the VAD problem. Local biodiversity experts highlighted the rare *Karat* and other yellow-fleshed banana varieties as potential options. Subsequent nutrient compositional analyses demonstrated that *Karat*, a variety traditionally given to infants as a weaning food, was rich in beta-carotene, the most important of the provitamin A carotenoids, with amounts much higher than those found in common white-fleshed bananas.

> Further studies also highlighted the many additional yellow-fleshed varieties of banana, giant swamp taro, breadfruit and pandanus rich in beta-carotene and other carotenoids, nutrients and fibre, which could be an important part of food-based approaches to addressing nutritional problems in FSM.
>
> Most of this work was led by the Island Food Community of Pohnpei (IFCP), a national NGO working to promote the production, consumption and marketing of local nutritious plant diversity through it's 'Let's Go Local' national campaign (Englberger, 2011).

A number of the earlier mentioned publications have raised awareness of the importance of traditional foods in Australia, which have largely been neglected and unexplored despite the multiple benefits they might bring. Bussey (2013) stresses that the gaps in literature on the access, availability and use of traditional foods should be examined further, as they have the potential to recover, improve and safeguard food security in remote Aboriginal communities. While there is increasing attention being paid to the importance of traditional foods and Indigenous food systems in Australia – through school-based education (Dawe et al., 2020), improving food security and nutrition in a more sustainable manner, links to sustainable gastronomy initiatives (such as The Orana Foundation[6] and Black Olive Catering[7]) and other niche markets for native foods, as well as other livelihood and income-generating opportunities – there is still some way to go in developing a suitable environment that enables and ensures that the benefits arising from the sustainable use of traditional foods are accessible for Indigenous communities, the long-time custodians of these genetic resources.

Notes

1 Cultivated foods of the Māori http://www.b4fn.org/case-studies/case-studies/cultivated-foods-of-the-maori/
2 https://theoranafoundation.org/projects/indigenous-food-database/, accessed 2 July 2020.
3 Information about the Tahuri Whenua Incorporated Society can be found at https://www.tahuriwhenua.org/and also on the BFN Project website http://www.b4fn.org/case-studies/case-studies/cultivated-foods-of-the-maori/
4 Ahi kōuka i te ata, he ai i te po http://www.maramatanga.co.nz/project/ahi-kouka-i-te-ata-he-ai-i-te-p-value-kouka-m-ori-lens
5 Pacific Island Food Revolution https://www.pacificislandfoodrevolution.com/
6 The Orana Foundation https://thefoundation.org/
7 Black Olive Catering http://www.blackolive.net.au/

References

Berno, T. (2020) Linking food biodiversity and food traditions to food tourism in Small Island Developing States (SIDS). In Hunter, D., Borelli, T. and Gee, E. (eds.)

Biodiversity, Food and Nutrition: A New Agenda for Sustainable Food Systems. Abingdon: Routledge, pp. 236–254.

Best, E. (1976) *Māori agriculture*, Dominion Museum Bulletin 9, Govt Printer, Wellington. (Reprint, repaginated but without textual alteration, of 1925 edition), 315pp.

Brimblecombe, J.K., Ferguson, M.M., Liberato, S.C. and O'Dea, K. (2013). Characteristics of the community-level diet of Aboriginal people in remote northern Australia. *Medical Journal of Australia*, 198(7), 380–384.

Bussey, C. (2013) Food security and traditional foods in remote Aboriginal communities: A review of the literature. *Australian Indigenous Health Bulletin*, 13(2), 1–9.

Clarke, P.A. (2003a). *Where the Ancestors Walked. Australia as an Aboriginal Landscape.* Sydney: Allen & Unwin.

Clarke, P.A. (2003b). Australian ethnobotany: An overview. *Australian Aboriginal Studies*, 2, 21–38.

Clarke, P.A. (2007). *Aboriginal People and Their Plants.* Dural Delivery Centre, NSW: Rosenberg Publishing [Second edition published in February 2011; electronic edition published in December 2011].

Clarke, P.A. (2018). Plant food. In F. Cahir, I. Clark, & P.A. Clarke (eds.) *The Indigenous Bio-cultural Knowledge of Southeastern Australia.* Melbourne: CSIRO Publishing, pp. 55–71.

CTA (2017) *Stories from the Field: Transforming Food Systems in the Pacific.* Technical Centre for Agricultural and Rural Cooperation (CTA). 26pp.

Davy, D. (2016) Australia's efforts to improve food security for Aboriginal and Torres Strait Islander Peoples. *Health and Human Rights Journal*, 18(2), 209–218

Dawe, P., Fawcett, A. and Webb, T. (2020) Learning gardens cultivating health and well-being – stories from Australia. In Hunter, D., et al. (eds.) *Agrobiodiversity, School Gardens and Healthy Diets.* Earthscan, pp. 192–207.

Denham, T. (2008). Traditional forms of plant exploitation in Australia and New Guinea: The search for common ground. *Vegetation History & Archaeobotany*, 17(2), 245–248.

Denham, T., Atchison, J., Austin, J., Bestel, S., Bowdery, D., Crowther, A., Dolby, N., Fairbairn, A., Field, J., Kennedy, A., and Lentfer, C., (2009a). Archaeobotany in Australia and New Guinea: Practice, potential and prospects. *Australian Archaeology*, 68(1), 1–10.

Denham, T., Donohue, M. and Booth, S. (2009b). Horticultural experimentation in northern Australia reconsidered. *Antiquity*, 83(321), 634–648.

Dixon, L.L.B. (2007) *Microbiological Quality of Toroi: A Māori Food Delicacy.* Masters of Philosophy in Biological Science Thesis, University of Waikato, Hamilton, 163pp.

Englberger, L. (2011) *Let's Go Local: Guidelines Promoting Pacific Island Foods.* FAO.

Englberger, L. and Johnson, E. (2013) Traditional foods of the Pacific: Go local, a case study in Pohnpei, Federated States of Micronesia. In Fanzo, J., Hunter, D., Borelli, T. and Mattei, F. (eds.) *Diversifying Food and Diets: Using Agricultural Biodiversity to Improve Nutrition and Health.* Abingdon: Routledge, pp. 231–241.

Englberger, L., Lorens, A., Levendusky, A., Pedrus, P., Albert, K., Hagilmai, W., Paul, Y., Nelber, D., Moses, P., Shaeffer, S. and Gallen, M. (2009) Documentation of the traditional food system of Pohnpei. In Kuhnlein, H.V., Erasmus, B. and Spigelski, D. (eds.) *Indigenous Peoples' Food Systems: The Many Dimensions of Culture, Diversity and Environment for Nutrition and Health.* Rome: FAO and CINE, pp. 109–138.

Englberger, L., Lorens, A., Pretrick, M., Raynor, B., Currie, J., Corsi, A., Kaufer, L., Naik, R.I., Spegal, R. and Kuhnlein, H.V. (2011) Approaches and lessons learned for promoting dietary improvement in Pohnpei, Micronesia. In Thompson, B. and Amoroso, L. (eds.) *Combating Micronutrient Deficiencies: Food-Based Approaches.* CABI Books, pp. 224–253.

Englberger, L., Marks, G.C. and Fitzgerald, M.H. (2003) Insights on food and nutrition in the Federated States of Micronesia: A review of the literature. *Public Health Nutrition*, 6, 5–17.

Fankhauser, B.L. (1990) The Māori use of tii (cabbage trees) for food. In Harris, W. and Kapoor, P (eds.) *Ngā mahi Māori ō te wao nui ā Tane*. Christchurch: Botany Division, DSIR, pp. 43–47.

FAO (2021) *The White/Wiphala Paper on Indigenous Peoples Food Systems*. Rome.

Kuhnlein, H.V., Erasmus, B. and Spigelski, D. (2009) *Indigenous Peoples' Food Systems: The Many Dimensions of Culture, Diversity and Environment for Nutrition and Health*. Rome: FAO and CINE, 339pp.

Haden, R (2009) *Food Culture in the Pacific Islands*. Greenwood Press.

Iosefa, T., Taylor, M., Hunter, D. and Tuia, V (2013). Supporting farmers' access to the global gene pool and participatory selection in taro in the Pacific. In De Boef, W.S., Peroni, N., Subedi, A. and Thijssen, M.H. (eds.) *Community Biodiversity Management: Promoting Resilience and the Conservation of Plant Genetic Resources*. Abingdon: Routledge, pp. 285–289.

Jones, D.S. and Clarke, P.A. (2018). Australian Aboriginal culture and food-landscape relationships: Possibilities of indigenous knowledge for the future Australian landscape. In Zeunert, J. and Waterman, T. (eds.) *Routledge Companion to Landscape and Food*. London: Routledge, pp. 41–60.

Latz, P. (1995) *Bushfires and Bushtucker: Aboriginal Plant Use in Central Australia*. Alice Springs: IAD Press.

Lee, A.J., Leonard, D., Moloney, A.A. and Minniecon, D.L. (2009). Improving Aboriginal and Torres strait islander nutrition and health. *Medical Journal of Australia*, 190(10), 547–548.

Low, T. (1988) *Wild Food Plants of Australia*. Sydney: Angus and Robertson.

Low, T. (1989) *Bush Tucker: Australia's Wild Food Harvest*. Sydney: Angus and Robertson.

Ming Wei, K. and Bikajle, S. (2020) Katakin komman jikin kallib ilo jikuul – Republic of the Marshall Islands school learning garden programme. In Hunter, D., et al. (eds.) *Agrobiodiversity, School Gardens and Healthy Diets*. Earthscan, pp. 257–266.

Mulvaney, D.J. and Golson, J. (eds.) (1971). *Aboriginal Man and Environment in Australia*. Canberra: Australian National University Press.

Newtown, J. (2016) *The Oldest Foods on Earth: A History of Australian Native Foods with Recipes*. Sydney: New South Publishing.

Rao, V.R., Matthews, P.J., Eyzaguire, P.B. and Hunter, D. (2010) *The Global Diversity of Taro: Ethnobotany and Conservation*. Rome: Bioversity International.

Redfeather, N. and Cole, E. (2020) Reviving local food systems in Hawai'i. In Hunter, D., et al. (eds.) *Agrobiodiversity, School Gardens and Healthy Diets*. Abingdon: Routledge, pp. 171–184.

Roskruge, N.R. (2013) The significance of Pūhā or Sow-thistle in Māori society. *He Pukenga Korero*, 12, 26–35.

Roskruge, N.R. (2014) *Rauwaru: Ngaweri, Maori Root Vegetables*. Massey University: Publ. School of Agriculture & Environment

Roskruge, N.R. (2015) *Tahuaroa: Korare, Maori Green Vegetables*. Massey University: Publ. School of Agriculture & Environment.

Roskruge, N. (2020) School gardens (māra): Today's learning spaces for Māori. In Hunter, D., et al. (eds.) *Agrobiodiversity, School Gardens and Healthy Diets*. Abingdon: Routledge, pp. 222–230.

Semese, S.A. (2018) *Traditional Systems Production and Uses of Rungia (Rungia klossii) and Highlands Pitpit (Setaria palmifolia) in Three Villages of the Highlands Region of Papua New Guinea*. Papua New Guinea: University of Goroka.

Taylor, M., Jaenicke, H., Skelton, P. and Mathur, P.N. (2009) Regional consultation on Crops for the Future: towards food, nutritional, economic and environmental security in the Pacific. Secretariat of the Pacific Community, Suva. 40pp.

Taylor, R. and Smith, I. (1997) *The State of New Zealand's Environment: 1997*. Wellington: Publ. Ministry for the Environment.

Thomson, L., Doran, J. and Clarke, B. (2018) Trees for life in Oceania: Conservation and utilization of genetic diversity. ACIAR Monograph No. 201. Australian Centre for International Agricultural Research, Canberra. 278pp.

PART II

Approaches, methods and tools for the use enhancement of NUS

13

THE BFN MAINSTREAMING TOOLKIT. A ROADMAP TO USING NEGLECTED AND UNDERUTILIZED SPECIES FOR FOOD SYSTEM CHANGE

Teresa Borelli, Daniela Beltrame, Victor W. Wasike, Gamini Samarasinghe, Ayfer Tan and Danny Hunter

Introduction

2019 will be remembered as a landmark year for food and nutrition reports, which have laid bare the nutritional and environmental disruptions our global food system is causing (FAO, 2019; IPBES, 2019; IPCC, 2019; Willett et al., 2019). In many of these and in preceding reports, the conservation and sustainable use of biodiversity for food and agriculture features as one critical pathway to diversify diets, improve livelihoods and contribute to environmental outcomes (Mouillé et al., 2010; FAO, 2015; Bioversity International, 2017; FAO, 2019). Meanwhile, the impact of the COVID-19 pandemic in 2020 has exposed the limitations of our food systems and has highlighted the importance of localising food production and consumption as well as the need to re-evaluate local foods including neglected and underutilized species (NUS) (HLPE, 2020).

Despite the existence of guiding principles to support countries in increasing the use of biodiversity in the agriculture, climate change, food security and nutrition and other relevant sectors (FAO, 2016), their implementation constitutes a challenge for many countries (FAO, 2017). Partly to blame are the limited examples of effective cross-sectoral collaboration and multi-disciplinary platforms to make the case for investing in biodiversity for better food and nutrition (Hunter et al., 2016). It is with these challenges in mind that, in 2018, the Biodiversity for Food and Nutrition (BFN)[1] project sought to fill the existing information gaps by putting together the *Biodiversity Mainstreaming for Healthy & Sustainable Food Systems Toolkit* – a compendium of experiences and lessons learned, gained by Brazil, Kenya, Sri Lanka and Turkey during project implementation (Figure 13.1). The BFN toolkit provides concrete examples that can help implement these guidelines and provides a framework to support countries in transitioning toward healthier, more sustainable diets as indicated by the recent WHO Guidance on Mainstreaming Biodiversity (WHO, 2020).

DOI: 10.4324/9781003044802-15

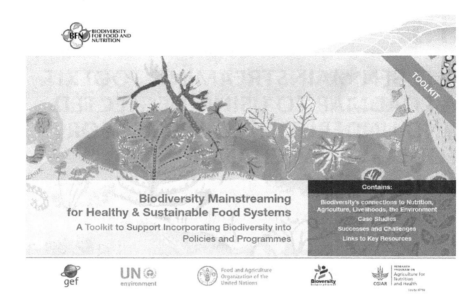

FIGURE 13.1 The *Biodiversity Mainstreaming for Healthy & Sustainable Food Systems Toolkit*.
Source: Hunter et al., 2018. Used with permission.

How the toolkit was developed

Underpinning the BFN project is a comprehensive three-pillar approach for mainstreaming biodiversity for better food and nutrition into policies and practices by: 1) Providing Evidence, 2) Influencing Policy and 3) Raising Awareness (Figure 13.2). Originally conceptualised by the Leveraging Agriculture for Nutrition in South Asia research consortium in its effort to improve understanding of how agriculture and food systems could be better designed to enhance nutrition in the region (Hunter et al., 2016), the approach had never been tested on the ground. With limited examples and literature to guide the countries at the onset of the project, Brazil, Kenya, Sri Lanka and Turkey put the theory into practice and took stock of the experiences gained (and challenges faced) during project implementation to assist fellow practitioners wishing to replicate the approach. In the final stages of the project, implementers from each country came together during a writing workshop to lay the groundwork for the toolkit.

What the toolkit offers

The toolkit lays out the fundamentals of biodiversity mainstreaming, including NUS, into sectoral policies and practices, including how to facilitate and align activities to support achieving national targets linked to biodiversity conservation

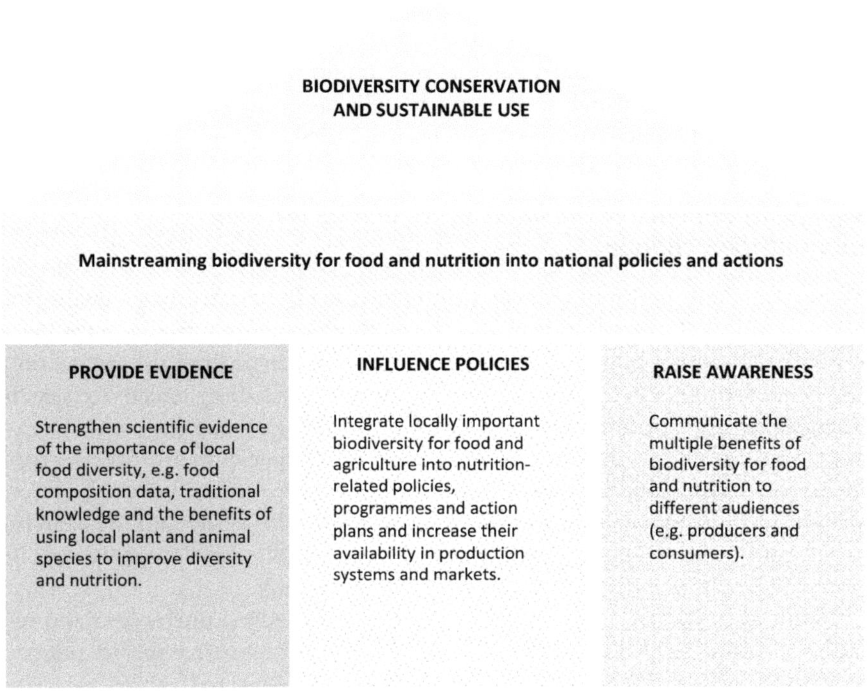

BIODIVERSITY CONSERVATION AND SUSTAINABLE USE

Mainstreaming biodiversity for food and nutrition into national policies and actions

PROVIDE EVIDENCE

Strengthen scientific evidence of the importance of local food diversity, e.g. food composition data, traditional knowledge and the benefits of using local plant and animal species to improve diversity and nutrition.

INFLUENCE POLICIES

Integrate locally important biodiversity for food and agriculture into nutrition-related policies, programmes and action plans and increase their availability in production systems and markets.

RAISE AWARENESS

Communicate the multiple benefits of biodiversity for food and nutrition to different audiences (e.g. producers and consumers).

FIGURE 13.2 The three-pillar approach adopted by the BFN project to establish enabling environments for mainstreaming biodiversity for food and nutrition into local food systems, from production to consumption. (Adapted from Borelli et al., 2020.) Used with permission.

and food and nutrition security. A practical set of steps and examples are clearly outlined, chief among them are the need to:

i   supply factual evidence that justifies NUS mainstreaming for the specific purpose of improving food and nutrition
ii   identify suitable entry points for effective implementation
iii   establish a positive enabling environment and gain financial and institutional support
iv   increase market interest and capacity to produce NUS
v   develop awareness-raising campaigns
vi   set up a monitoring and evaluation framework.

The approaches used to implement activities across the three pillars are not prescriptive and are likely to vary depending on country context and the different social, cultural or economic backgrounds. In implementing the BFN project, for example, it was found that efforts to increase the appreciation of local NUS was easier in some countries and less so in others. Efforts to step up production and consumption of the target species proved easier in Turkey and Sri Lanka than in Kenya, a country where, under colonial rule, traditional crops were displaced by European staples and which were, until recently, associated with backwardness, poverty and/or famine (Raschke and Cheema, 2008).

Supplying evidence of the importance of NUS for food and nutrition change

To reverse these perceptions, scientific evidence of the nutritional quality of NUS is critical. The toolkit dwells on the methodologies that facilitate the generation of nutrition information to gain support for increasing the production and consumption of NUS. Of chief importance is establishing effective research partnerships, determining what information already exists and the country's capacity to fill any research gaps, and identifying which species to prioritise for further research and marketing. In Kenya, selecting the species for food composition analysis was undertaken following a market survey and ensuing ranking exercise carried out by the Kenya Agricultural and Livestock Research Organization in close collaboration with local communities (Figure 13.3).

Similarly, in Brazil, partnerships established with federal universities and research institutes helped build the nutritional profile of 64 native species (mostly fruit and nuts) identified by the project (Figure 13.4) (Gee et al., 2020).

FIGURE 13.3 Indigenous neglected and underutilized species that were targeted by the BFN project in Kenya for food composition analysis. (Credit: BFN Kenya.) Used with permission.

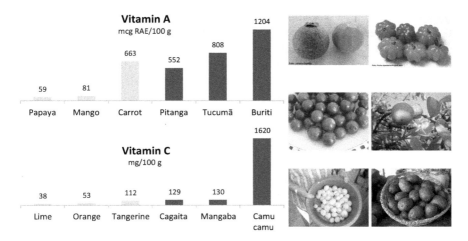

FIGURE 13.4 Vitamin A content (expressed in mcg RAE per 100g of fresh raw pulp/whole fruit, with or without peel) of the native Brazilian fruits pitanga (*Eugenia uniflora*), tucumã (*Astrocaryum aculeatum*) and buriti (*Mauritia flexuosa*); and vitamin C content (expressed in mg per 100g of fresh raw pulp/whole fruit, with or without peel) of cagaita (*Eugenia dysenterica*), mangaba (*Hancornia speciosa*) and camu camu (*Myrciaria dubia*) compared to introduced non-native species. (Adapted from Hunter et al., 2019; Borelli et al., 2020.) Used with permission.

Identifying partners and suitable entry points for NUS mainstreaming

Once data is available, time and resources should be made available to identify and engage the most strategic and like-minded allies to achieve the intended mainstreaming aims as well as to understand the policy landscape in which these partners operate. Pre-existing multi-stakeholder policy platforms that deal with food and nutrition security are an ideal starting point to garner policy support for the greater use of NUS (Gee et al., 2020). However, platforms will often need to be built from scratch. Considerable time can be spent in creating the enabling environment needed for durable mainstreaming to occur and this should carefully be accounted for during the planning process (Gee et al., 2020). An entire section in the toolkit outlines several fruitful strategies for targeting receptive policies, such as the National Biodiversity Strategy and Action Plans (NBSAPs), as well as agricultural and nutrition policies. It also highlights ideal partners to engage. These can range from sectoral ministries to universities and research organisations, as well as civil society actors such as community groups, non-governmental organisations, Indigenous groups and professional associations that have an interest in promoting NUS. However, non-conventional partners such as schools and chefs can also help promote the use of these unfamiliar ingredients.

In Brazil, for example, collaboration with prominent chefs and gastronomy professionals helped raise the profile of identified NUS, which are now used, elevated and transformed, in novel food recipes, some of which are captured in the recipe book *Brazilian Biodiversity: tastes and flavours* (Santiago de Andrade Cardoso and Coradin, 2018). Collaboration with the Ministry of Health and Education has additionally helped integrate these species into the national school meals programme, PNAE, which reaches 40 million students in 150,000 schools on a daily basis (FNDE, 2020). Along with nutritional data, training was provided to the 8,000 plus nutritionists who work with the schools to develop the children's weekly menus in line with national nutrition standards and guidelines, as well as to the school cooks that translate these guidelines into meals. School cooks are generally unaccustomed to transforming or incorporating NUS into flavourful and healthy meals for children, but there is now strong recognition of the important role school 'chefs' can play in nutrition education and in teaching young palates to appreciate native nutritious foods. Healthful school meals competitions, such as the *Melhores Receitas da Alimentação Escolar* (best school meal recipes) launched in Brazil in 2018–2019, can be a powerful way to acknowledge this role and stimulate improvements and dialogue around safe and healthy eating in schools.

How to increase the production, marketing and awareness of NUS

Long-term behavioural change and the acceptance of NUS in food systems and diets requires coordinated action on both the supply and demand side of the food value-chain as well as the public and private institutions working together to develop and implement policies and strategies that support more sustainable and diversified food systems centred on NUS. These actions include:

- Building the capacity of producers to incorporate NUS in their production systems
- Improving or creating market infrastructure for biodiversity
- Raising consumer awareness to increase demand for food biodiversity.

In the toolkit, one example from Kenya illustrates how business training and technical assistance provided to small-scale farmers helped establish market linkages with schools and how this increased the sustainable production of African indigenous vegetables and forgotten pulses such as bambara groundnut (*Vigna subterranea*). Additional examples are provided in Module 2 of the BFN e-learning course and in Hunter et al. (2019), Borelli et al. (2020) and Gee et al. (2020). Naturally, where the aim is increased production of species that are collected or managed in the wild, domestication programmes and/or sustainable collection or management guidelines will need to be considered to avoid overexploitation (Gee et al., 2020).

The toolkit also demonstrates how to select NUS that are more suited for marketing. Most likely biological, environmental, cultural, economic and nutritional considerations, such as the presence of limiting nutrients, will come into play. An interesting example is offered by the *ad hoc* sustainability index developed by Turkey to reduce an initial sample size of 43 species, mostly wild edibles, to three target species – foxtail lily (*Eremurus spectabilis*), golden thistle (*Scolymus hispanicus*) and einkorn wheat (*Triticum monococcum*) – which have since been the object of domestication research as well as post-harvest handling and value-chain analysis (Table 13.1).

Once the target species have been identified, new marketing options can be explored that make it profitable for actors along the food value-chain – from producers to consumers – to engage. Such is the case for the 28 NUS identified by Sri Lanka. Jackfruit (*Artocarpus heterophyllus*), water lily (*Nymphaea pubescens*), traditional cowpea (*Vigna unguiculata*), black gram (*Vigna mungo*) and horse gram (*Macrotyloma uniflorum*) varieties, as well as traditional rice varieties, are among

TABLE 13.1 The sustainability index developed by BFN Turkey to prioritise a selected number of crops for further marketing and use.

Environmental	*Economic*	*Food & nutrition*
1 Conservation • *Ex situ, in situ*/on farm	**1 Collection/production** • Collection/production constraints • Distance from collection/ production site • Collection/production continuity	**1 Iron content** **2 Calcium content** **3 Fibre ratio**
2 Cultivation • Ease of production		**4 Antioxidant ratio**
• Growth rate	**2 Market characteristics**	**5 Vitamin A content**
• High adaptability • Vegetation period • Annual growth	• Recognisable • Easily packed • Suitable for storage	
3 Disappearance/threat	**3 Processing industry available**	
• Habitat destruction and fragmentation	**4 Marketing opportunities available for collected/ produced species**	
• Pollution/exploitation	**5 Distance from market (Km)**	
• Destructive practices		
6 Widespread distribution		
7 Habitat preferences		

the NUS currently on offer at *Helabojun*, 18 BFN-supported food outlets that source their ingredients directly from local communities and which sell freshly prepared local foods. The outlets also employ and empower rural women who receive nutrition, food and business training from the Women Farmers Extension Program of the Department of Agriculture and can earn a reasonable wage by working at the stores.

In terms of awareness-raising, the toolkit illustrates how practitioners can make the most of ongoing activities in the country that revolve around gastronomy and ecotourism, such as the Alaçatı herb festival in Turkey, and national and regional campaigns on healthy eating, to get the messaging across (Figure 13.5). It also discusses the benefits of engaging the health sector to target the development and revision of national dietary guidelines or the advantages of engaging the education sector to include the teaching of biodiversity for food and nutrition in health and nutrition education in schools and vocational training programmes.

FIGURE 13.5 The different marketing strategies and awareness raising events that have helped countries raise the profile of NUS. From top left to right: *Helabojun* outlet, Sri Lanka (Credit: BFN Sri Lanka/NAICC, DoA); BFN gastronomy event within the BFN Symposium, Brazil (Credit: Bioversity/S. Landersz); food stalls at the Alaçatı Herb Festival, Turkey (Credit: Bioversity/S. Landersz); Food and Biodiversity Festival, Busia, Kenya (Credit: BFN Kenya). Used with permission.

The BFN Mainstreaming Toolkit **171**

Monitoring and evaluating the effective mainstreaming of NUS

Careful monitoring of increased use of NUS can help garner additional support for their inclusion in food and nutrition security strategies. For this reason, desired mainstreaming outcomes should be established early in the planning phase and be SMART,[2] tailored to the specific country context as well as the resources that are available (Hunter et al., 2018). SMART objectives can range from community- to policy-level improvements, and enough time should be devoted to establishing a baseline against which the suggested monitoring indicators can be measured. These include, but are not limited to:

- Level of funding/resource mobilisation to support NUS for improving diets/nutrition in research and interventions.
- Extent of NUS mainstreaming for healthy diets and nutrition in relevant national instruments including in national dietary guidelines, NBSAPs or multi-sectoral nutrition action plans and strategies.
- Level of diversification in public food procurement and school feeding programmes.
- Increase in scientific literature focusing on the composition or consumption of NUS.

In conclusion, the BFN Mainstreaming Toolkit provides practitioners with a foundation and much-needed guidance on how to achieve wider appreciation of biodiversity, including NUS, in health, nutrition, agriculture and food security programmes, as envisaged by the FAO *Voluntary Guidelines for Mainstreaming Biodiversity into Policies, Programmes and National and Regional Plans of Action on Nutrition* and the Committee on World Food Security (CFS)*Voluntary Guidelines on Food Systems for Nutrition* of the Committee on World Food Security (CFS). However, in implementing the BFN approach, it has become apparent that there is no one-size-fits-all solution to the challenge of using biodiversity and NUS to transform our current food systems. Without a doubt, there are countless possibilities waiting to be explored. We encourage workers to identify the most suitable and culturally appropriate solutions for their individual countries that can help transition the BFN approach into actionable policies that matter for people and the planet.

Notes

1 Spanning 2021–2019, the GEF 'Mainstreaming biodiversity for nutrition and health' project was led by Brazil, Kenya, Sri Lanka and Turkey and was coordinated by Bioversity International, with implementation support from the United Nations Environment Programme (UNEP) and the Food and Agriculture Organization of the United Nations (FAO). Additional support was provided by the CGIAR Research Program on Agriculture for Nutrition and Health and the Australian Centre for International Agricultural Research.

2 Specific, measurable, achievable, relevant and time-bound (SMART)

References

Bioversity International (2017) *Mainstreaming Agrobiodiversity in Sustainable Food Systems: Scientific Foundations for an Agrobiodiversity Index*. Bioversity International. Rome, Italy.

Borelli, T., et al. (2020) 'Local solutions for sustainable food systems: The contribution of orphan crops and wild edible species', *Agronomy*, 10(2). https://www.mdpi.com/2073-4395/10/2/231

FAO (2015) *Coping with Climate Change – The Roles of Genetic Resources for Food and Agriculture*. Food and Agriculture Organization of the United Nations. Rome, Italy.

FAO (2016) *Voluntary Guidelines for Mainstreaming Biodiversity Into Policies, Programmes and National and Regional Plans of Action on Nutrition*. Food and Agriculture Organization of the United Nations. Commission on Genetic Resources for Food and Agriculture. Rome, Italy

FAO (2017) Sixteenth Regular Session of the Commission on Genetic Resources for Food and Agriculture - CGRFA-16/17/Report Rev.1. Food and Agriculture Organization of the United Nations. Rome, Italy.

FAO (2019) *The State of the World's Biodiversity for Food and Agriculture*. Food and Agriculture Organization of the United Nations. Rome, Italy.

Fundo Nacional de Desenvolvimento da Educação (FNDE) (2020) FNDE repassa R$ 375 milhões para alimentação escolar. 3 June, 2020.

Gee, E., et al. (2020) 'The ABC of mainstreaming biodiversity for food and nutrition. Concepts, theory and practice', in Hunter, D., Borelli, T. and Gee, E. (eds) *Biodiversity Food and Nutrition. A New Agenda for Sustainable Food Systems*. Abingdon. Routledge., pp. 85–186.

HLPE (2020). *Impacts of COVID-19 on Food Security and Nutrition: Developing Effective Policy Responses to Address the Hunger and Malnutrition Pandemic*. Committee on World Food Security. High Level Panel of Experts on Food Security and Nutrition. Rome. Italy.

Hunter, D., et al. (2016) 'Enabled or disabled: Is the environment right for using biodiversity to improve nutrition?' *Frontiers in Nutrition*, 3, pp. 1–6.

Hunter, D., et al. (2018) *Biodiversity Mainstreaming for Healthy and Sustainable Food Systems: A Toolkit to Support Incorporating Biodiversity Into Policies and Programmes*. Bioversity International. Rome, Italy.

Hunter, D., et al. (2019) 'The potential of neglected and underutilized species for improving diets and nutrition', *Planta*, 250(3), pp. 709–729.

IPBES (2019). *Summary for Policymakers of the Global Assessment Report on Biodiversity and Ecosystem Services of the Intergovernmental Science-Policy Platform on Biodiversity and Ecosystem Services*. Díaz, S., et al. (eds.). IPBES Secretariat, Bonn, Germany, 56 pp.

IPCC (2019). *Climate Change and Land: An IPCC Special Report on Climate Change, Desertification, Land Degradation, Sustainable Land Management, Food Security, and Greenhouse Gas Fluxes in Terrestrial Ecosystems* Shukla, P.R., Skea, J., Calvo Buendia, E., Masson-Delmotte, V., et al. (eds.). In press. https://www.ipcc.ch/srccl/cite-report/

Mouillé, B., Charrondière, R. and Burlingame, B. (2010) *The Contribution of Plant Genetic Resources to Health and Dietary Diversity: Thematic Background Study*. Food and Agriculture Organization of the United Nations. Rome, Italy. pp. 1–20.

Raschke, V. and Cheema, B. (2008) 'Colonisation, the new world order, and the eradication of traditional food habits in East Africa: Historical perspective on the nutrition transition', *Public Health Nutrition*, 11(7), pp. 662–674.

Santiago de Andrade Cardoso, R. and Coradin, L. (eds.) (2018) *Biodiversidade Brasileira: sabores e aromas*. Ministerio do Meio Ambiente (MMA). Brasília, DF, Brazil.

Willett, W., Rockström, J., Loken, B., Springmann, M., Lang, T., Vermeulen, S., Garnett, T., Tilman, D., DeClerck, F., Wood, A., et al. (2019) 'Food in the Anthropocene: The EAT–Lancet Commission on healthy diets from sustainable food systems', *The Lancet*, 393, pp. 447–492.

World Health Organization (WHO) (2020) *Guidance on Mainstreaming Biodiversity for Nutrition and Health*. World Health Organization. Geneva, Switzerland.

14

DEVELOPMENT OF SEASONAL CALENDARS FOR SUSTAINABLE DIETS – EXPERIENCES FROM GUATEMALA, MALI AND INDIA

Gaia Lochetti

Introduction

Seasonality is a concept that can often be overlooked, particularly in Western countries where people are accustomed to having constant access to fresh plant foods even out of season. Economic development and consequent urbanisation have produced a shift in food systems, which have become more global and standardised worldwide, particularly in urban areas. It is, in fact, not uncommon to find food products out of season in markets and supermarkets, which have been transported from the other side of the world at a high cost to the environment.

While this might go unnoticed by the inattentive consumer, seasonal patterns and food availability are still very much at the core of the food systems and their efficient functioning, and are important for the sustainable management and use of agrobiodiversity, as well as understanding market fluctuations and the consequent availability of and access to food. Particularly in low- and middle-income countries, food production is dependent on the passing of seasons, and understanding how seasonal patterns influence the availability, diversity and abundance of food is key to implementing strategies to eradicate malnutrition and to improve food security and diets globally.

Recent statistics (FAO et al., 2019) show how malnutrition is still widespread, with 821 million people undernourished and 10% of the world population still severely food insecure, meaning that they lack sufficient accessibility, availability, stability or utilisation of food. The relationship between seasonality and food and nutrition security is no longer an obscure one, as highlighted by a considerable body of literature (Leonard, 1991; Dercon and Krishnan, 2000; Savy et al., 2006). Furthermore, changes in diet quality, both in terms of energy and nutrient intake, as well as dietary diversity, are often due to seasonal variation in food

DOI: 10.4324/9781003044802-16

Seasonal calendars for sustainable diets in Guatemala, Mali and India **175**

availability (Ferguson et al., 1993; Hirvonen et al., 2016; Stelmach-Mardas et al., 2016; Broaddus-Shea et al., 2018; Oduor et al., 2019).

Understanding the role of food availability and its patterns, characteristics and seasonal variation can be an entry point to achieving more diverse, healthy and nutritious diets, and to ultimately improving food security. Suboptimal diets are a major health risk that contributes to the global burden of disease (Hall et al., 2009; Afshin et al., 2019). Furthermore, global diets are becoming less diverse, in a process that has been defined as 'nutrition transition' (Drewnowski and Popkin, 2009), and are also becoming more and more reliant on a handful of economically important animal and plant species. The agricultural intensification required for these changes has had detrimental effects on environmental health and agrobiodiversity, critical for both food system sustainability and climate change resilience (Millennium Ecosystem Assessment, 2005; Mijatović et al., 2013). Despite the estimation that over 5,000 food crops exist worldwide (RGB Kew, 2016), 51% of energy intake relies on rice, wheat and maize (FAOSTAT, 2018). Poor and monotonous diets, characterised by a low consumption of fruits, vegetables and other nutrient-dense food increases the risk of obesity, cardiovascular disease, diabetes, cancer and other non-communicable diseases (He et al., 2006). This biodiversity loss in diets is of severe concern, considering that diet diversity has been associated with nutritional adequacy, health outcomes, food self-sufficiency and food security (Arimond and Ruel, 2004; Kennedy et al., 2007; Sibhatu et al., 2015; KC et al., 2016; Lachat et al., 2018).

Improving dietary diversity and quality is, therefore, a key strategy for alleviating the global burden of malnutrition. Healthy and diverse diets provide a vast and diverse array of macro and micronutrients, such as vitamins, minerals and bioactive compounds, which are fundamental for human health. As a result, understanding seasonal food availability patterns is crucial.

Collecting information

While there are numerous ways to assess local available agrobiodiversity and its fluctuations throughout the year, the best method can be decided based on the scope and purpose of a specific research project. Production and market surveys can enable a more concrete and factual understanding of seasonal availability, but can also be both time and resource intensive, and can often overlook important sources of nutrients in the landscape, such as those stemming from wild resources, or those exchanged within communities. Participatory research approaches, on the other hand, have a strong focus on local perspectives and can capture more efficiently the availability of wild, cultivated and exchanged species, which may be of particular relevance in the local context. Furthermore, participatory approaches enable the creation of a space for mutual learning, where communities can discuss and identify locally tailored action plans (PAR, 2018; Mijatović et al., 2019).

The assessment of seasonal food availability through participatory approaches entails organising focus group discussions (FGDs). Seasonal food availability FGDs are

176 Gaia Lochetti

a very versatile and straightforward participatory data collection method that allows the collection of information on local food species availability and their sourcing, as well as different types of other information (PAR, 2018). A detailed methodology guide on how to collect seasonality data through this participatory approach has been recently published and can be found online (Lochetti et al., 2020).

These FGDs normally last up to a day, depending on the available agro-biodiversity, the scope of the research and the interests of participants. The data collected using this participatory methodology corresponds to *perceived seasonal availability,* and the quality of the results is dependent on participants' knowledge and interest in the foods being discussed. Selecting participants who are familiar with the seasonal dynamics of the foods of interest, such as experienced farmers and merchants, is advisable, in order to gather data of the highest quality. Data collected with this method enables the mapping of seasonal periods of food shortages and gaps in food availability. At the same time, it is also possible to identify periods of high availability of nutrient-dense foods, which are vital entry points to improving year-round availability of healthy and diverse foods.

FGDs can provide key insights to the obstacles and barriers to the production and the consumption of locally available species, and participants should be encouraged and challenged to develop locally relevant solutions to these bottlenecks, with the aim of ultimately leading to the improvement of food and nutrition security, diet diversity and the resilience of smallholder farmers.

Experiences from Guatemala, India and Mali

Within the IFAD-EC NUS project 'Linking agrobiodiversity value chains, climate adaptation and nutrition: Empowering the poor to manage risk', implemented from 2015 to 2020,[1] several FGDs on seasonal availability were organised in the three countries targeted by the project, namely, Guatemala, India and Mali. The project aimed to revitalise the use of local agrobiodiversity, which typically includes a large assortment of nutritious species well adapted to local conditions, but which are underutilized for a variety of reasons, including a lack of awareness and promotion. Overall, 12 villages across the target countries took part in the data collection and contributed significantly to the assessment of local available agrobiodiversity and its seasonal patterns.

The relevance of locally available agrobiodiversity for the livelihoods of the villages involved in the project was immediately apparent. In particular, these FGDs, compared to the production surveys carried out during the baseline assessment, were able to capture the presence of numerous wild-sourced species, which, in the study areas, make important contributions to household diets.

One fundamental legacy of this project was the creation of location-specific seasonal food availability calendars, created from the data collected in the FGDs. In total, four different calendars were created for the three countries, and in each country translations into local languages were also arranged.

The resulting calendars (Figure 14.1) are colourful and graphically pleasing products, easily comprehendible to children and people with a low level of literacy. Furthermore, by drawing on baseline results and closely working with local partners, calendars were enriched with nutritional education and awareness-raising materials in booklets that were distributed back to participating communities (Bioversity International and IER, 2018). These booklets relied on simple graphics and text in the national or local language to present basic nutrition information derived from national guidelines, when available, and included a section with an overview of the state of food and nutrition security specific to each locality, and of local diet quality and consumption levels of fruits and vegetables. A section with easy tips on how to improve diet diversity and year-round consumption of nutritious foods was also included. When available, national food-based dietary guidelines were provided as a starting point to present information on how to achieve a more diversified and healthy diet. Cooperation with national ministries for public health or national institutes of nutrition should be sought to update or improve existing nutrition education material or to develop new guidelines if national ones are not currently available in the country. At the time of the IFAD-EC NUS project, national food-based dietary guidelines were not available in Mali, leading to close cooperation with local nutritionists and the Ministry of Public Health to develop dietary guidelines that were relevant to the local context.

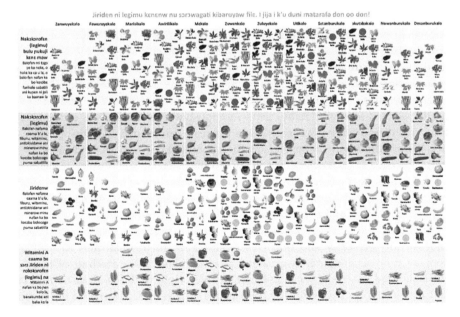

FIGURE 14.1 Seasonal food availability calendar in Bamanankan (one of the local languages) for the region of Segou, Mali.
Source: Bioversity International and IER (2018).

178 Gaia Lochetti

Seasonal calendars for the diversification of diets and ecosystems and the role of traditional knowledge

Building from local experience and knowledge, calendars can thus guide diversification both in terms of food consumption and in terms of production and landscape management. Diversified agroecosystems, leveraging ecosystem services and the interactions between different species are key strategies for strengthening resilience to climate change and other external shocks (Oliver et al., 2015; Rioux et al., 2016). They can optimise yields, provide additional sources of income and contribute to food and nutrition security (Powell et al., 2013, 2015; Boedecker, et al., 2014; Isbell et al., 2017; Mijatović et al., 2019). This is particularly pressing when considering accelerating climate change and widespread ecosystem degradation. The seasonal food availability calendars, and the discussions they build upon, can be very helpful tools for raising awareness of the presence of local food species that may be neglected and underutilized, and can then be promoted through follow-up actions and projects. Finally, the participatory nature of this tool can play a fundamental role in the documentation and preservation of whole sets of traditional knowledge related to local agrobiodiversity, including local names, traditional cultivation practices and uses related to these species.

Conclusion

Seasonal food availability calendars can be used to explore and understand seasonal variations and changes in the availability, abundance and diversity of foods present in both wild and cultivated environments. They can be relevant both at the urban and the rural levels, as consuming a diverse array of foods that are available across different seasons will ultimately lead to positive health gains, a reduced burden of disease and lower impacts on the environment, thereby also eventually improving the overall sustainability of food systems. The understanding of seasonal food availability patterns contributes to the analysis of the local food system and can be fundamental in fostering knowledge and raising awareness of local underutilized species and their potential role in both food and nutrition security, and income generation. When the calendar builds on participatory approaches, it can be used to inform and create awareness of local constraints and barriers to the production and consumption of diverse food, allowing the development of tailored and relevant nutrition education materials.

This tool is an example of a bottom-up strategy that can help improve food and nutrition security and the livelihoods of smallholder farmers as well as people's food sovereignty. A greater knowledge and understanding of locally available food biodiversity can enhance peoples' right to define their own food and agriculture systems, reducing the reliance on external imports and inputs. In fact, this year, the Covid-19 crisis has underscored the world's overdependence on long food supply-chains, and has highlighted the need to diversify locally available options that can ensure communities' food security. These calendars, therefore, represent a valuable tool in increasing food knowledge and access, especially for vulnerable groups. Increasing the awareness in the final consumers of the role of and opportunities in consuming seasonal food, whether they live in rural Guatemala or a major European city, has the potential to lead to positive health outcomes and significantly more sustainable approaches to food production and consumption.

Note

1 http://www.nuscommunity.org/initiatives/ifad-eu-ccafs-nus/

References

Afshin, A., Sur, P.J., Fay, K.A., Cornaby, L., Ferrara, G., Salama, J.S. and Murray, C. J.L. (2019) Health effects of dietary risks in 195 countries, 1990–2017: A systematic analysis for the Global Burden of Disease Study 2017. *The Lancet*, 393(10184), 1958–1972. https://doi.org/10.1016/S0140-6736(19)30041-8

Arimond, M. and Ruel, M.T. (2004) Dietary diversity is associated with child nutritional status: Evidence from 11 demographic and health surveys. *The Journal of Nutrition*, 134(10), 2579–2585. https://doi.org/10.1093/jn/134.10.2579

Bioversity International and Institut d'economic Rural (IER) (2018) Calendrier saisonnier du fruits et légumes pour une alimentation diversifiée (Ségou, Mali). Available online in French and Bamanankan http://www.nuscommunity.org/resources/our-publications/publication/calendrier-saisonnier-des-fruits-et-legumes-pour-une-alimentation-diversifiee-dans-la-region-de-seg/

Boedecker, J., Termote, C., Assogbadjo, A.E., Van Damme, P. and Lachat, C. (2014) Dietary contribution of Wild Edible Plants to women's diets in the buffer zone around the Lama forest, Benin – an underutilized potential. *Food Security*, 6(6), 833–849. https://doi.org/10.1007/s12571-014-0396-7

Broaddus-Shea, E.T., Thorne-Lyman, A.L., Manohar, S., Nonyane, B.A.S., Winch, P.J. and West, K.P. (2018) Seasonality of consumption of nonstaple nutritious foods among young children from Nepal's 3 agroecological zones. *Current Developments in Nutrition*, 2(9). https://doi.org/10.1093/cdn/nzy058

Dercon, S. and Krishnan, P. (2000) Vulnerability, seasonality and poverty in Ethiopia. *Journal of Development Studies*, 36(6), 25–53. https://doi.org/10.1080/00220380008422653

Drewnowski, A. and Popkin, B.M. (2009) The nutrition transition: New trends in the global diet. *Nutrition Reviews*, 55(2), 31–43. https://doi.org/10.1111/j.1753-4887.1997.tb01593.x

FAO, IFAD, UNICEF, WFP and WHO. (2019) *The State of Food Security and Nutrition in the World 2019. Safeguarding Against Economic Slowdowns and Downturns.* FAO. Rome: IGO. Licence: CC BY-NC-SA 3.0

FAOSTAT (2018) Production, Food Balance, and Land Use Data. Retrieved from http://www.fao.org/faostat/en/?#home

Ferguson, E.L., Gibson, R.S., Opareobisaw, C., Osei-Opare, F., Lamba, C. and Ounpuu, S. (1993) Seasonal food consumption patterns and dietary diversity of rural preschool Ghanaian and Malawian children. *Ecology of Food and Nutrition*, 29(3), 219–234. https://doi.org/10.1080/03670244.1993.9991307

Hall, J.N., Moore, S., Harper, S.B. and Lynch, J.W. (2009) Global variability in fruit and vegetable consumption. *American Journal of Preventive Medicine*, 36(5), 402–409.e5. https://doi.org/10.1016/j.amepre.2009.01.029

He, F.J., Nowson, C.A. and MacGregor, G.A. (2006) Fruit and vegetable consumption and stroke: Meta-analysis of cohort studies. *Lancet*, 367(9507), 320–326. https://doi.org/10.1016/S0140-6736(06)68069-0

Hirvonen, K., Taffesse, A.S. and Worku Hassen, I. (2016) Seasonality and household diets in Ethiopia. *Public Health Nutrition*, 19(10), 1723–1730. https://doi.org/10.1017/S1368980015003237

Isbell, F., Adler, P.R., Eisenhauer, N., Fornara, D., Kimmel, K., Kremen, C. and Scherer-Lorenzen, M. (2017) Benefits of increasing plant diversity in sustainable agroecosystems. *Journal of Ecology*. Blackwell Publishing Ltd. https://doi.org/10.1111/1365-2745.12789

KC, K.B., Pant, L.P., Fraser, E.D.G., Shrestha, P.K., Shrestha, D. and Lama, A. (2016) Assessing links between crop diversity and food self-sufficiency in three agroecological regions of Nepal. *Regional Environmental Change*, 16(5), 1239–1251. https://doi.org/10.1007/s10113-015-0851-9

Kennedy, G.L., Pedro, M.R., Seghieri, C., Nantel, G. and Brouwer, I. (2007) Dietary diversity score is a useful indicator of micronutrient intake in non-breast-feeding Filipino children. *The Journal of Nutrition*, 137(2), 472–477. https://doi.org/10.1093/jn/137.2.472

Lachat, C., Raneri, J.E., Smith, K.W., Kolsteren, P., Van Damme, P., Verzelen, K., … Termote, C. (2018) Dietary species richness as a measure of food biodiversity and nutritional quality of diets. *Proceedings of the National Academy of Sciences of the United States of America*, 115(1), 127–132. https://doi.org/10.1073/pnas.1709194115

Leonard, W.R. (1991) Household-level strategies for protecting children from seasonal food scarcity. *Social Science and Medicine*, 33(10), 1127–1133. https://doi.org/10.1016/0277-9536(91)90228-5

Lochetti, G., Meldrum, G., Kennedy, G. and Termote, C. (2020) Seasonal food availability calendar for improved diet quality and nutrition: Methodology guide. Available online: https://www.bioversityinternational.org/e-library/publications/detail/seasonal-food-availability-calendar-for-improved-diet-quality-and-nutrition/

Mijatović, D., Meldrum, G. and Robitaille, R. (2019) *Diversification for Climate Change Resilience: Participatory Assessment of Opportunities for Diversifying Agroecosystems*. Rome: Bioversity International and the Platform for Agrobiodiversity Research

Mijatović, D., Van Oudenhoven, F., Eyzaguirre, P. and Hodgkin, T. (2013) The role of agricultural biodiversity in strengthening resilience to climate change: Towards an analytical framework. *International Journal of Agricultural Sustainability*, 11(2), 95–107. https://doi.org/10.1080/14735903.2012.691221

Millenium Ecosystem Assessment. (2005) *Ecosystems and Human Well-being: Synthesis. Assessment of Climate Change in the Southwest United States: A Report Prepared for the National Climate Assessment*. Washington, DC: Island Press. https://doi.org/10.5822/978-1-61091-484-0_1

Oduor, F.O., Boedecker, J., Kennedy, G., Mituki-Mungiria, D. and Termote, C. (2019) Caregivers' nutritional knowledge and attitudes mediate seasonal shifts in children's diets. *Maternal and Child Nutrition*, 15(1). https://doi.org/10.1111/mcn.12633

Oliver, T.H., Heard, M.S., Isaac, N.J.B., Roy, D.B., Procter, D., Eigenbrod, F. and Bullock, J.M. (2015) Biodiversity and resilience of ecosystem functions. *Trends in Ecology and Evolution*. Elsevier Ltd. https://doi.org/10.1016/j.tree.2015.08.009

PAR (Platform for Agrobiodiversity Research). (2018). *Assessing Agrobiodiversity: A Compendium of Methods*. Rome. Retrieved from http://agrobiodiversityplatform.org/files/2018/10/Assessing-Agrobiodiversity-A-Compendium-of-Methods-highres.pdf

Powell, B., Ickowitz, A., McMullin, S., Jamnadass, R., Padoch, C., Pinedo-Vasquez, M. and Who, F. (2013) The role of forests, trees and wild biodiversity for nutrition-sensitive food systems and landscapes. *FAO and WHO*, (January), 1–25. Retrieved from http://www.fao.org/fileadmin/user_upload/agn/pdf/2pages_Powelletal.pdf

Powell, B., Thilsted, S.H., Ickowitz, A., Termote, C., Sunderland, T. and Herforth, A. (2015) Improving diets with wild and cultivated biodiversity from across the landscape. *Food Security*, 7(3), 535–554. https://doi.org/10.1007/s12571-015-0466-5

RGB Kew (2016) The State of the World's Plants Report. Retrieved from https://stateoftheworldsplants.org/2016/

Rioux, J., Gomez San Juan, M., Neely, C., Seeberg-Elverfeldt, C., Karttunen, Kaisa Rosenstock, T., Kirui, J., Tapio-Bistrom and Marja-Liisa Bernoux, M. (2016) Mitigation of Climate Change in Agriculture Series Planning, Implementing and Evaluating Climate-Smart Agriculture in Smallholder Farming Systems. Rome. Retrieved from www.fao.org/in-action/micca

Savy, M., Martin-Prével, Y., Traissac, P., Eymard-Duvernay, S. and Delpeuch, F. (2006) Dietary diversity scores and nutritional status of women change during the seasonal food shortage in rural Burkina Faso. *The Journal of Nutrition Community and International Nutrition*, 136. Retrieved from https://academic.oup.com/jn/article-abstract/136/10/2625/4746707

Sibhatu, K.T., Krishna, V.V. and Qaim, M. (2015) Production diversity and dietary diversity in smallholder farm households. *Proceedings of the National Academy of Sciences of the United States of America*, 112(34), 10657–10662. https://doi.org/10.1073/pnas.1510982112

Stelmach-Mardas, M., Kleiser, C., Uzhova, I., Penalvo, J.L., La Torre, G., Palys, W. and Boeing, H. (2016) Seasonality of food groups and total energy intake: A systematic review and meta-analysis. *European Journal of Clinical Nutrition*, 70(6), 700–708. https://doi.org/10.1038/ejcn.2015.224

15

ENHANCING THE USE OF UNDERUTILIZED FOOD CROPS

Partnerships in a success story of a pop cereal business in Kenya

Yasuyuki Morimoto, Patrick Maundu, Elizaphan Gichangi Mahinda and Daniel Kirori

Introduction

Many African grains including pearl millet, sorghum, finger millet and some legumes are declining in importance. Crops such as maize, rice and wheat, which enjoy big market shares and attention from researchers, the private sector and government agencies, are replacing these minor crops. Rural women producers of these minor grains are being disenfranchised as their market shrinks. Over the last few decades, agricultural production has received a boost thanks to enhanced cultivation methods, but the majority of agricultural food products in Africa are still sold in local markets as fresh produce or in raw forms. The main challenge experienced by farmers is the lack of appropriate technologies for harvest operations and post-harvest processing before consumption. In particular, research in processing technologies for minor cereals is needed in order to develop novel ways of consuming local grains other than the conventional preparation methods of porridge or "ugali", which are considered highly monotonous foods, especially among younger consumers.

Aim of our intervention

In order to promote the use of local food resources and to enhance the appreciation of neglected and underutilized species (NUS) by people, scientists from the Alliance of Bioversity International and CIAT, in partnership with a multi-disciplinary group, pioneered a community-led intervention that added value to local cereals and legumes through the production and marketing of "pressure popped" snacks (Figure 15.1). Pressure-popping is a technology that uses high temperature and pressure to puff cereals and legumes such as sorghum, finger millet, green gram and a variety of other seemingly "unfashionable" grains (Figure 15.2).

DOI: 10.4324/9781003044802-17

FIGURE 15.1 Local children enjoying a popped snack made from pearl millet. Credit. Y. Morimoto

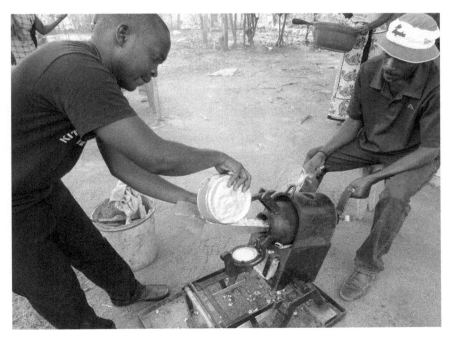

FIGURE 15.2 Production of popped cereals. Credit Y. Morimoto

A multi-stakeholder joint venture

The machine is user friendly and cheap to maintain as it is manually operated and requires little wood fuel to be operated, making it easily accessible in rural areas.

A multi-stakeholder joint venture

Since 2016, through the financial and technical support from the Japanese Ministry of Agriculture, Forestry and Fisheries, and the Japan Association for International Collaboration of Agriculture and Forestry (JAICAF), a prototype machine was tested and fabricated by a Kenyan private company, DK engineering Co. Ltd, using locally available materials with technical support from Japanese engineers. The same financial support enabled IEDA Confectionary Limited, a Japanese private company, to work with three local communities in the Kitui, Embu and Migori counties to initiate the experimental production of popped grain and value addition. The popping properties of a number of local grains were explored and the training of community members on value addition carried out. The groups embarked on marketing the new products, providing cheap, natural and healthy snacks for both rural and urban people, targeting especially children.

Results and outcomes

The intervention motivated local individual entrepreneurs to purchase the machine using their own funds and start local business focused on popping grains. This technology has proved to be a novel way of adding economic value to local food resources through their use as snack, thus providing income and several business opportunities to local groups and individual entrepreneurs.

These efforts have demonstrated great potential to contribute to food security and enhanced livelihoods in Kenya, particularly with regard to:

- The increased use of local food resources (in our case, local grains) and boosting their appreciation as new sources of nutrition and income by local consumers
- Creating new business opportunities, especially for youth and women interested in growing, selling and adding value to local crops
- Enhancing partnerships between farmers and private companies, promoting entrepreneurships and local value chains
- Conserving local agrobiodiversity, production landscapes and the culture associated with local foods.

Because cereal popping is still relatively new to most Kenyans, in order to raise consumers' interest in popped cereal snacks, the project team carried out practical demonstrations in the three target counties and elsewhere in the country about the popping technique and the marketing of its products.

Overall, 25 of these events were held between May 2017 and February 2018. The events, which included agricultural shows, seminars and workshops, were able to be organised also in neighbouring countries. These public demonstrations generated great interest, as evidenced through the high level of public participation and the many enquiries received. People following the events were keen to know more on how to start a business and how to access local manufacturers, besides requesting more information on technical aspects concerning the popping machine and its operation. Over the course of these events, the project team received 37 public awards from 14 agriculture shows attended. The activities were also widely covered by public media (12 events were covered in Kenya and 15 in Japan between March 2019 and January 2020). Annex I lists some relevant media releases covering the events in Kenya.

Worth mentioning as well is that the popping activity helped strengthen the cohesion in communities, bringing success to some groups, while failing to do the same in others (apparently due to some financial mismanagement and disgruntlement among some members, caused by lack of clear role allocation and experience in business activities). Another lesson learnt is related to the case wherein the popping machine was donated to the community by some supporters: here we registered some problems when the machine needed servicing; the repair, instead of being dealt with by the group as a whole, fell upon just a single individual who was not particularly happy with that situation, As a result, the project started to encourage self-financing for the purchase of the machine as a way to assign greater responsibility among all members of the group, including when it came to servicing work.

Some key outcomes and lessons

The following factors were key to the success of this initiative:

1 The economic benefits arising from the initiative were among the strongest factors that led to a shift in the attitude of farmers, traders and consumers towards minor grains, which were no longer seen as crops of the poor with no future.

2 The pop cereal machine could be fully fabricated locally. DK Engineering played a key role in providing the technical backstopping for the machines' servicing, spare parts, training in operating procedures and maintenance. A business relying solely on imported machines tends to progress less due to the high costs of machines and spare parts, the extra time spent obtaining the spare parts and the lack of management information and human resources to offer necessary repairs and follow-ups.

3 Multi-disciplinary partnerships including community groups, entrepreneurs, private companies, development agents, local governments and scientists played a strategic role in providing action-oriented solutions. These

partnerships also created opportunities for individual consultations, providing information and practical trainings to those who were interested in starting the business. They supported, furthermore, the development of unique community products, promoting the use of different varieties of local cereals and seasoning and flavouring through natural products.

4   Steady demonstration sales during public events and business fairs, supported by well-crafted communications messages spread through public and social media. This created opportunities to demonstrate local innovations in their products, i.e., presentations in labelling, packaging and marketing strategies.

Following are two case studies that share the experiences of two private companies who ventured into the popping business. The first case is from Kieru Foods Ltd., owned by Gichangi Mahinda, a new company based in Embu that manufactures popped cereals from local grains such as pearl millet and sorghum. The second is from DK Engineering Co. Ltd. in Nairobi, whose sales manager, Daniel Kirori, offers his perspective on playing a key role in promoting the wider use of NUS.

Case 1: Kieru Foods Ltd., Embu, Kenya

In the year 2008, together with my wife, Lilian Gichangi, I opened a cereal retail shop in the Embu market (Gichangi Cereals & Spices). After earning a living for many years as cereals farmers and traders, we decided to explore new opportunities of value adding technology for our produce. We initially roasted groundnuts and soybean, selling these products in our shop in the Embu market, together with other grains.

In the year 2016, at a farmers' field day organised by Bioversity International, I saw a novel technology for making popped local grains such as pearl millet and sorghum. I was much impressed by this innovation and decided to order a popping machine fabricated locally by DK Engineering and received the necessary training for its operation from Bioversity International and JAICAF. In 2017, JAICAF facilitated my further training at the IEDA Confectionary Ltd. Company in Japan, where I gained additional knowledge on popping techniques and product development. The training in Japan was a real eye-opener for me, as the company I was exposed to was capable of dealing successfully with large volumes of raw material processed and readily sold. Back home in Kenya, I established Kieru Foods Company and spent about 300,000 Ksh (around US$3,000) as initial capital to purchase a popping machine, drying chambers, packaging materials and other items needed for our new business.

Compared with other local businesses, such as the processing of juices from local fruits, popping has several advantages, including easy and low cost operations and management, from production to sales; therefore, starters have a high chance of success.

The work at Kieru Ltd. focused initially on pearl millet, which is often mixed with roasted groundnuts and flavoured with soya flour, baobab, hibiscus, turmeric, ginger and cinnamon powder; sesame oil and honey are also added to churn out a nutritious delicacy. As indicated in Tables 15.1 and 15.2, we found that by implementing this production and sales schedule four times a month, the expenses of the machine would be repaid after 13 weeks, or a little over three months. The company is able to pop about 40 kg of cereals daily, which we package and sell to retailers. One kg of pearl millet, which we buy from farmers at a rate of 70 Ksh, yields more than 144 small-sized bars of snacks called "*kashata*" (Figure 15.3), which we sell at 10 Ksh each, gaining a total of 1,440 Ksh. I regard this income an exponential monetary value appreciation!

Commercialisation

I can say that whatever the reasons are for maintaining these local crops, we have to ensure that these products are able to satisfy the three major needs faced by rural farmers, viz. income generation, food sufficiency and good health. Food processing technologies are important to strengthen the resilience of local farming communities against social and environmental transformations, and, at the same time, to provide opportunities for diversifying diets and generating additional income.

FIGURE 15.3 Value addition to the popped cereals to make snack bars (*kashata*). Credit: Y. Morimoto.

188 Yasuyuki Morimoto

TABLE 15.1 Comparison of profit margins from five types of popped cereals

		Maize	Brown rice	Wheat	Pearl millet	Sorghum
Production	Raw material (kg)	7	7	5	6	6
	Production (kg)	5.06	6.68	4.00	5.48	4.96
Procurement costs	Raw material cost/kg	40	140	45	50	50
	Material cost used	280	980	225	300	300
	Other material cost	126	126	90	108	108
Selection fee	Personnel	217	217	155	186	186
Production cost	Fuel	35	35	25	30	30
	Personnel (3/day)	217	217	155	186	186
Packaging	Wrapping material	149	137	80	110	100
	Personnel (scaling, wrapping)	217	217	155	186	186
Selling	Personnel	289	289	206	248	248
Total cost		1,529	2,217	1,091	1,354	1,343
Amount sold		3,350	4,550	2,650	3,650	3,300
Profit		1,821	2,333	1,559	2,296	1,957
Profit rate (%)		**54.3**	**51.3**	**58.8**	**62.9**	**59.3**

Source: Y. Kanda et al. (2017, p.34). modified.

Kieru Foods Ltd. offers a variety of popped cereal products from local grains that are already familiar to consumers. It should be noted that the diversity of crops used in the snacks is not the only factor behind customers' appreciation. Added flavours (obtained from local fruits, vegetables and aromatic plants) are also very important. To that regard, the company has been investing in innovative ways of flavouring and seasoning popped cereals, as in the case of flour from roasted soybeans and Bambara groundnuts, by-products of popping, that are good for making porridge also. The company is currently marketing its products through bulk selling to retailers and supporting commercialisation with advertisements and public awareness campaigns in crowded places like bus parks (Figure 15.4).

The increased marketing of traditional foods in urban areas will create more demand for the raw materials and, hence, will boost the cultivation of these local crops. At the same time, the traditional knowledge associated with the use of these crops, as well as the genetic diversity of these resources, will be preserved. I also believe that the establishment of local industries revolving around the use of such crops will help tackle the problem of rural–urban migration so common among younger ones. More awareness on the health benefits of these foods will also help popularise their use, which is in great decline due to changing preferences and modernisation of local food cultures. Most of our products in the

TABLE 15.2 Simulated business model in Kenya

| Material | Material (1) Kg | Yield (2) % | Production (3) (1) x (2) kg | Material cost | | | | Personnel | | | | Total |
				Raw material/kg	Sub material	Fuel	Expendables	Production (3/day)	Selection (2/day)	Scaling, wrapping (2/day)	Selling (5 days)	
Maize	4	74	2.96	180			86.9					
Wheat	6	89	5.34	240			107.3					
Brown Rice	4	88	3.52	560			70.5					
Sorghum	5	77	3.85	250			77.3					
Pearl millet	6	94	5.64	300			113.3					
Total cost (4)	25		21.31	1,530	560	125	455.3	960	640	640	1,600	6,510
Total sold (5)												14,225
Profit (5)−(4)=(6)												7,715
Profit rate (%) (6)/(5)												54.2

Source: Y. Kanda et al. (2017, p.38).

FIGURE 15.4 Marketing Kieru products in Embu market, Kenya. Credit: G. Mahinda

market have basic nutritional information printed on the packages. However, a more complete nutrient profile is also needed.

Promotion

Popping cereals represent a new snack to Kenyan people and their promotion is, thus, very important. When we started this business, we had to give free samples of snacks to potential customers in order to popularise the products. According to one of our regular customers, the snack bar is handy food for children, easy to pack and consume at school, and this is certainly helped spread this product among younger kids.

I have been providing training on popped cereal techniques to individuals wanting to start their own popping businesses (Figure 15.5). The trainees also learn tips on how to avoid failures in their business: f.e. we advise trainees on ways to modify product packaging in accordance with the new policy of the Kenyan government, which bans the use of single-use plastic shopping bags.

More needs to be done, however, to support the promotion of popping products in Kenya, and I advocate for greater research support to foster the development of new ideas to help the wider popularisation of these healthy foods.

Networking

In collaboration with some of our trainees, we are creating a formal association of Kenyan poppers. This group will act on behalf of its members, presenting their grievances and requests to relevant authorities and facilitating the acquisition of the Quality Mark from the Kenya Bureau of Standards, which costs entrepreneurs time and money. The association would be tasked with bringing members

FIGURE 15.5 Mr. Mahinda providing business training on pop cereal production techniques. Credit: Y. Morimoto).

together for information sharing and training. The network will provide many innovative ideas for developing new products, as well as approaches for optimising prices, boosting sale volumes, increasing public demand and raising business expectations.

Our venture has now picked up quite well and the good returns so far give us hope for widening our marketing and expanding our business within and beyond Embu, to other towns in Kenya. We envision having a processing factory able to offer a good market for neglected crops, which will be very helpful for struggling smallholder farmers, and we trust that we shall contribute to making better foods of high nutritional value, while also opening up job opportunities to the youth and women.

Effects of COVID-19 on business

Our income has decreased as the number of people in markets decreased, as a result of COVID-19 policies, including a ban on demonstration sales in the open market. Many sectors, such as restaurants and tourism, have been affected by the disruptions caused by the pandemic and many people have lost their jobs. Sectors related to basic food, production and sales, however, tended to be less affected. Many people are eating at home and, as children are staying at home due to restrictions, people are becoming also more conscious of the importance of healthy diets. To that regard, we have registered an increase in consumers' interest in locally made, healthy, traditional foods. In line with the need for greater hygienic and sanitary procedures, our company has also introduced in our factory a new packaging machine to wrap products individually (Figure 15.6).

Additional promising technology

The development of novel marketing strategies is always a continuous need for every entrepreneur. In that regard, we continue to interact with scientists to explore together alternative snack-making machines that cheaper and easier to carry around. For instance, some of the latest technologies developed – the hand-grill and puff-cracker machines – each costs less than a quarter of the price of a popping machine; furthermore, in addition to being affordable, they are easy to maintain since they require little technical expertise and are operated by just one person (Figure 15.7). Whereas the popping machine converts grains to snacks, the cracker utilises flours of sifted cereals like cassava, wheat, rice, pearl millet, maize, amaranth, sorghum and finger millet – among others – to come up with the snacks (Figure 15.8). I strongly believe that these types of innovations can help tackle many challenges facing our society; they can create employment for younger ones and help curb food insecurity, wastage and malnutrition.

FIGURE 15.6 Individually wrapped *"kashata"*. Credit: G. Mahinda

FIGURE 15.7 Locally fabricated puffed-cracker machine. Credit: Y. Morimoto

FIGURE 15.8 Puffed-crackers made using eight different types of cereal and legume flours. Credit: Y. Morimoto

Case 2: DK Engineering Co. Ltd., Nairobi, Kenya

DK Engineering Co. Ltd. was established in 1986 in order to provide engineering technology solutions in food processing and related service support. The company has been a key player in providing machines to industries within and across regions, focusing on bakeries, fruit juices, peanut butter, honey, wet and dry vegetables and institutional/restaurant kitchen equipment.

After inspecting the prototype popping machine brought from Japan, DK Engineering expressed interest in fabricating a facsimile using locally available materials. It studied the design of the Japanese model, investigated the materials used and identified areas requiring redesigns using local materials.

Technical challenges encountered

The major challenges experienced during the development process included the following:

- Developing a cylinder vessel that would hold the required high pressure without leakage.
- Moulding the lid guide and locking the peg that would fit tight on the lid.
- Designing the entire machine with the same precision of the original made in Japan.

FIGURE 15.9 Mr. Yoshimura, on the right, providing technical guidance to DK engineers. Credit: Y. Morimoto)

- In addition, the pressure tight adjustment screw bolt was often bent due to accumulation of pressure, and the locking peg was broken upon hammering to open the lid. The selection and appropriate heat-resistant material was not precise, so we had to engage in trial-and-error methods repeatedly, which consumed time and resources.

To solve all these problems, a request for engineering guidance was made to JAICAF and also to Mr. Yoshimura of Pop Cereal Machine Sales Ltd., Japan, who was the original designer of the popping machine (Figure 15.9). Based on their guidance, we applied a super-alloy containing a large content of sulphur constituent ("En9") to increase the strength and also contracted a steel foundry to cast the lid guide and locking peg.

In terms of challenges encountered in selling the machine, customers expressed some concerns over the loud noise of the machine; others found that although the machine requires the use of relatively little wood fuel, this could be still a limiting factor if the equipment is to be used in urban settings.

Result and prospects

Our knowledge regarding the fabrication of the popping machine has been greatly improved, thanks not only to interactions with the project team, but also

FIGURE 15.10 Comparison of a locally manufactured pressure popping machine in the back and an original, Japanese-made popping machine in the foreground. Credit: Y. Morimoto

to the plentiful feedback received from customers. After several tests, we have now achieved a model that is performing well and we are able to reach large-scale production of this equipment, comparable to what we observed in the factory we visited in Japan (Figure 15.10).

Further improvements to the popping machine are in the making, including replacing the wood stove with a gas stove, adding an insulated panel for reducing the noise and developing mobile equipment that would allow users to reach strategic points of sale.

So far, we have sold about 30 machines in our region; our customers include local farmers groups, individual entrepreneurs, companies, NGOs, local governments and a UN agency interested in promoting the use of local cereals through this technology.

Bibliography

Kanda Y., Morimoto Y. and Nishino, S. (2017). 'Feasibility Survey on Local Production for Local Consumption (Chisan-Chisho) Activity and Extension in Africa. -First year report'. Japan Association for International Collaboration of Agriculture and Forestry (JAICAF) Press. 125p.

Morimoto, Y. (2017). 'Operation and processing manual: Pressure-puffing machine' (Yoshimura-model). JAICAF Press. 32p.

Annex I. Media releases about the popping work carried out in Kenya

Business Daily. (16 September 2019). 'Turning millet into snacks spices up traders' revenues. https://www.businessdailyafrica.com/corporate/enterprise/Turning-millet-into-snacks-spices-up-traders--revenues/4003126-5275714-5434m8z/index.html (Accessed on 10 June 2020).

Daily Nation. (18 March 2017). 'Forget corn, I pop sorghum and rice'. https://www.icrisat.org/forget-corn-i-pop-sorghum-and-rice/ (Accessed on 10 June 2020).

Daily Nation. (3 October 2019). 'Couple pops snacks from good old millet'. https://www.nation.co.ke/kenya/business/seeds-of-gold/couple-pops-snacks-from-good-old-millet-209820 (Accessed on 10 June 2020).

DK engineering (n.d.). 'Popped maize, sorghum, wheat, millet, rice etc'. http://www.dkengineering.co.ke/index.php/2014-07-19-11-37-30/cereal-pop-machine (Accessed on 30 June 2020).

Farm Kenya (21 September 2019). 'Our agricultural future lies in the past'. https://www.farmers.co.ke/article/2001342762/our-agricultural-future-lies-in-the-past (Accessed on 10 June 2020).

Induli, I. (19 March 2019). Panorama solutions for a healthy planet. 'Boosting underutilized nutritious foods as local snack alternatives'. https://panorama.solutions/en/solution/boosting-underutilized-nutritious-foods-local-snack-alternatives (Accessed on 10 June 2020).

Induli, I., Morimoto, Y. and Maundu, P. (1 July 2020). 'Researchers and entrepreneurs bring back forgotten gems: Underutilized crops transformed into healthy snacks'. https://www.bioversityinternational.org/news/detail/researchers-and-entrepreneurs-bring-back-forgotten-gems-underutilized-crops-transformed/ (Accessed on 25 December 2020).

Muriithi, K. (9 October 2019). Standardmedia. 'Trader pops snakes from pearl millet'. https://www.standardmedia.co.ke/mobile/amp/article/2001344940/trader-pops-snacks-from-pearl-millet (Accessed on 30 June 2020).

KBC Channel 1 (16 September 2019). 'Cereals farming counties to benefit from ready market for their crops through value addition venture'. Online video. YouTube. https://www.youtube.com/watch?v=cKNFfE379vI&app=desktop (Accessed on 10 June 2020).

Murigi, M. (19 November 2019). PD online. 'Popping grains to entice growing youth market'. https://www.pd.co.ke/business/agribiz/popping-grains-to-entice-growing-youth-market-13718/ (Accessed on 10 June 2020).

Njata TV (13 September 2019). 'Story aired on primetime news'. Online video. YouTube. https://www.youtube.com/watch?v=UJdtM5ezIDQ&feature=youtu.be (Accessed on 10 June 2020).

NTV Kenya (11 September 2019). 'Cereals for food security'. Story aired on primetime news. Online video. YouTube. https://www.youtube.com/watch?v=p5vLfIJwjqA&-feature=youtu.be (Accessed on 10 June 2020).

Popping Hope (7 January 2020). 'Introducing a new business using popping cereal technology in Kenya'. Online video. YouTube. https://www.youtube.com/watch?v=8oM_D2bxaYQ&t=535s (Accessed on 10 June 2020).

Watsupafrica (12 September 2019). 'Embu farmers learn value add popping cereals for income generation'. http://watsupafrica.com/news/embu-farmers-learn-value-add-popping-cereals-for-income-generation/ (Accessed on 10 June 2020).

Langat, W. (8 January 2020). Thomson Reuters Foundation. 'Traditional crops puff hopes for climate resilience in Kenya'. https://jp.reuters.com/article/us-kenya-climate-change-crops-feature-tr/traditional-crops-puff-hopes-for-climate-resilience-in-kenya-idUSKBN1Z7206 (Accessed on 10 June 2020).

Morimoto, Y. (2018). 'Healthy innovation to local snacks;– Promoting local consumption of locally produced foods to end hunger and malnutrition in Kenya'. Bioversity International Press. 2p. https://panorama.solutions/sites/default/files/puffing_cereal_fact_sheet.pdf (Accessed on 10 June 2020).

16

AGRITOURISM AND CONSERVATION OF NEGLECTED AND UNDERUTILIZED NATIVE ANDEAN CROPS IN SANTIAGO DE OKOLA, BOLIVIA

Stephen R. Taranto, Eliseo Mamani Alvarez and Wilfredo Rojas

Introduction

The Lake Titicaca basin is the center of origin for numerous domesticated crops. Quinoa (*Chenopodium quinoa*), potato (*Solanum tuberosa*), oca (*Oxalis tuberosa*), isaño (*Tropaeolum tuberosum*) and cañahua (*Chenopodium pallidicaule)* are some of the crops domesticated in the region, that have, in some cases, been credited with permitting the development of complex civilizations throughout the Andes (Erickson, 1988; Mann, 2011). Crops such as potatoes serve as globally important staple foods that today are the primary source of nutrition for hundreds of societies (Pearsall, 2008; Bradshaw and Ramsey, 2009).

Over millennia, Indigenous farmers in the Andes not only domesticated single crops from wild plants but, in many cases, developed hundreds and even thousands of varieties of each (Pearsall, 2008). In Bolivia, it is estimated that farmers have developed as many as 2,963 varieties of quinoa and up to 1,944 varieties of potato (INIAF, n.d.). Such high levels of diversity have arisen in part due to the highly variable topography, soil and climate regimes in the region, leading to farmers continually having to select the varieties that are best adapted to local conditions, dietary requirements and constantly shifting cultural practices and preferences (National Academies Press, 1989; FAO, 2009).

Despite its importance for traditional diets and cuisines, alleviating and combating food insecurity and adaptation to climate change, the loss of agrobiodiversity is widespread in the Andes (Padulosi et al., 2011; Zimmerer, 2013). In recent decades, large-scale and rapid migration from rural to urban areas, shifting weather patterns and the globalization of food systems have led to a decline in agrobiodiversity in Bolivia and elsewhere around the world (Padulosi et al., 2011). Furthermore, as rural populations age and older farmers die or cease to farm, knowledge and techniques for maintaining crop diversity are

DOI: 10.4324/9781003044802-18

lost, contributing to the extinction of many varieties (Thrupp, 2000; Zimmerer, 2014).

In response to this genetic erosion, in recent decades Indigenous Aymara and campesino farmers and farming communities in Bolivia and Peru, and their allies from the public and private sectors, have been identifying and experimenting with a variety of practices aimed at halting the loss of agrobiodiversity. This includes *in situ* and *ex situ* conservation of germplasm, disease control, soil management, market development and commercialization of products and agritourism (Gandarillas, 2001; Coca, 2010 cited by Rea, 1995; Córdoba, 2017). This chapter shares the results of an ongoing community-based agrobiodiversity conservation project and assesses the potential of agritourism to contribute to the *in situ* conservation of Andean crops in a traditional farming community on Lake Titicaca.

Background and context

Between 2008 and 2014, Bioversity International and the PROINPA Foundation of Bolivia collaborated on the conservation and commercialization of native Andean crops on the Neglected and Underutilized Species (NUS) project, supported by the International Fund for Agricultural Development. In Latin America, the NUS project was implemented in Bolivia and Ecuador (Rojas et al., 2010, 2014) and worked with Indigenous farming communities, agronomists, government agencies, tour operators, restaurants and other stakeholders to identify and implement sustainable interventions to prevent further loss of Andean agrobiodiversity.

NUS interventions in Bolivia included the design and implementation of a community agritourism pilot project. Community-based tourism (CBT) is an increasingly popular approach to tourism that seeks the greater participation of local communities in tourism activities and services. CBT initiatives are designed to extend a greater share of the benefits of tourism to local communities by reducing the outsourcing of certain services to external actors, such as private tour operators, restaurant owners and transportation providers (Dangi and Jamal, 2016). Since the mid-1990s, dozens of CBT initiatives have been developed in Bolivia and elsewhere, with highly variable results. Challenging to implement and difficult to sustain, CBT projects are often located in remote rural communities associated with high-biodiversity protected areas and under intense pressure from migration (Baldinelli, 2013), climate change (Meave and Lugo-Morín, 2014) and illegal timber extraction and mining, and where food systems tend to decline in diversity (Blundo-Canto et al., 2020).

To identify a pilot site for a CBT agritourism project and assess the feasibility of its sustainable implementation, PROINPA worked with Sendas Altas, a private tour operator in La Paz experienced in supporting and promoting CBT initiatives and familiar with its challenges. As a component of its corporate social responsibility program, Sendas Altas was actively working with several CBT initiatives in protected areas (T. Sivila, personal communication, 20 April 2018).

200 Stephen R. Taranto et al.

For the NUS agritourism pilot project – and based on PROINPA's recommendations – four communities, Cariquina Grande, Cachilaya, Coromata Media, Santiago de Okola and Titijoni, were evaluated as potential pilot project sites under four criteria: agrobiodiversity levels; project acceptance by the community; accessibility of the community and the presence of tourist attractions beyond agrobiodiversity.

Approximately 1.5 hours by ferry from the Island on the Sun in Lake Titicaca, the most important non-urban tourist destination in Bolivia, and four hours from the city of La Paz, the village of Santiago de Okola on Titicaca was selected for the project. Community members expressed interest in developing a tourism project and, in fact, the village had been receiving tourists informally for several years (T. Laruta Hilari, personal comm, 28 February 2008). In addition to the high levels of agrobiodiversity inventoried by PROINPA (2008) and Torresin, who identified 136 useful plant species in the village (2015), the community possesses numerous landscapes, sites and activities of interest, in particular the Sleeping Dragon bluff, an important archaeological site on the shores of the lake (Calla et al., 2013) that until today is used for traditional Aymara rituals (Gil, 2017).

Together with community members and PROINPA, Sendas Altas facilitated the participatory design of the CBT project based on the interests and abilities of participating community members, the identification of complementary attractions and activities, the generation of income and the use of agrobiodiversity. Through group discussions, skill-building workshops, field trips and other participatory methods, tourism skills and services such as interpretation, hygiene practices, food preparation and accommodation were developed. Agrobiodiversity was integrated into the project in various ways, including its inclusion in meals, the design of a cookbook highlighting agrobiodiversity, interpretive walks that include information on agrobiodiversity and traditional production systems and the creation of a community museum highlighting agrobiodiversity.

Since the start of the project in 2008, an estimated 3,500 tourists have visited Santiago de Okola, many crossing Lake Titicaca from the Island of the Sun and others coming by land from La Paz. Furthermore, in parallel with the NUS project and to this day, a sustained series of additional, externally funded projects have been implemented in the community, including the development of a low-impact trail system on the Sleeping Dragon bluff, the organizational strengthening of the Integrated Tourism Association of Santiago de Okola (ASITURSO) – the formal community organization that emerged from the project – and the construction of a community center. Recently, the project has received support from the Italian Cooperation with a particular focus on the remnants of Qhapaq Ñan, the Great Inca Trail that passes directly through the town, which is being developed as a long-distance regional trekking route from Ecuador to Chile (see Figure 16.1).

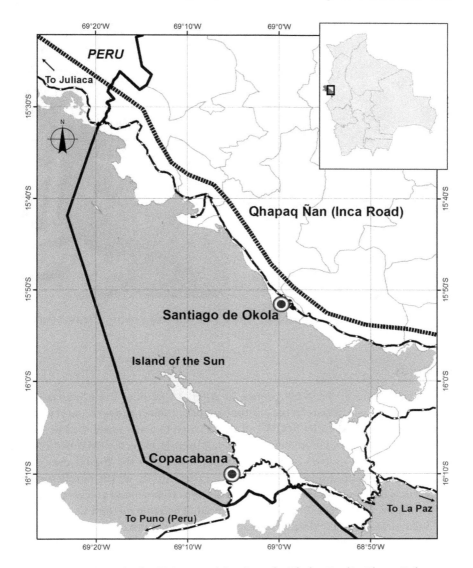

FIGURE 16.1 Map of Lake Titicaca and Santiago de Okola. Credit: Elmer Cuba.

Objectives of the study

The objectives of this study are: a) to evaluate whether the last 12 years of agritourism in Santiago de Okola have contributed to the conservation of the Andean crop varieties inventoried at the beginning of the project; and b) assess the impacts and potential of agritourism to contribute to the *in situ* conservation of Andean agrobiodiversity.

FIGURE 16.2 Participants of the project share a harvest with visitors. Credit: Stephen Taranto.

Methods

At the beginning of the NUS project in 2008, 12 families from Santiago de Okola agreed to participate in the agritourism pilot project and were surveyed in Aymara and Spanish by PROINPA and Sendas Altas to assess established perceptions about tourism and to inventory household agrobiodiversity (Mamani et al., 2008). Researchers visited each family independently after the 2008 harvest and asked respondents to display and name each variety. Once all the families were surveyed, a database of crops and crop varieties was generated and recorded as present (1) or not present (0) in each household. The same method was used for the 2020 surveys and the data were analyzed using descriptive statistics, with the household as the unit of analysis.

To evaluate changes in household agrobiodiversity levels, we compared levels reported in 2008 to those reported in 2020. The 2020 surveys were conducted with the same families and with the same questionnaire. Due to the COVID-19 pandemic, only seven families (58%) were surveyed; however, the remaining families will be surveyed when safe access to the region is secured and the data presented here will be updated.

Results

Today, the agricultural systems of the Lake Titicaca basin reflect five centuries of crop exchange with food systems from around the world and the diets of both rural and urban consumers reflects deep integration with introduced crops such as broad beans, wheat and barley. Farmers in Santiago de Okola cultivate a broad array of native and introduced crops; however, for the purposes of this study, results are presented only for the seven native Andean crops inventoried during both the 2008 and 2020 surveys. Four tubers (potato, oca, isaño and papalisa) and three grains (quinoa, cañahua and maize) were inventoried, while introduced crops such as barley and wheat were excluded, in order to evaluate the impact of the project on native Andean crops rather than all crops present in the systems.

Figures 16.3 and 16.4 summarize the change in the number of varieties of each crop that each household cultivated in 2008 and in 2020. Increases in the number of varieties grown by each household were reported for the seven crops in six of the seven households: oca increased from an average of four to eight varieties, isaño from one to four varieties, papalisa from one to four varieties, maize from three to nine varieties and quinoa from one to six varieties per household. While in 2008 cañahua was not cultivated by any participants, in 2020 one family reported cultivating four varieties of it. The greatest change was registered for potatoes, with an increase in the average number of varieties from 12 to 32 (Figure 16.4). The change in number of varieties of potato is presented separately because while potatoes were not a focus crop of the NUS project, they are central to food security in rural communities and their diversity is crucial to monitor and conserve.

The results also served to evaluate changes in the overall composition of crops and their varieties at the intra-family level between 2008 and 2020 (Figure 16.5). Traditional Andean farming systems are characterized as poly-culture systems and Figure 16.5 summarizes changes in crop variety composition with Families 2 and 3 increasing the total number of crop varieties by 131% and 71% respectively;

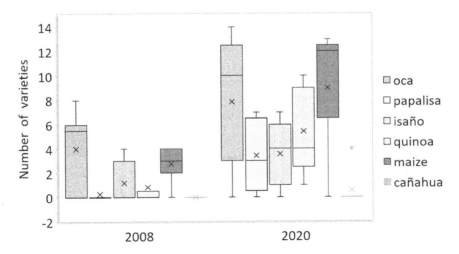

FIGURE 16.3 Change in the number of varieties of oca, papalisa, isaño, quinoa, maize and cañahua between 2008 and 2020.

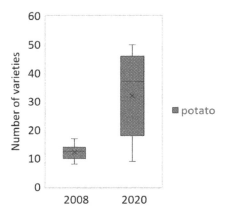

FIGURE 16.4 Change in the number of potato varieties between 2008 and 2020.

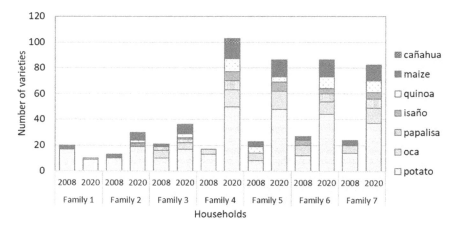

FIGURE 16.5 Overall changes in crop and crop variety composition between 2008 and 2020.

Families 5, 6 and 7 increasing the number of varieties by more than 200% each and Family 4 increasing the number of varieties by more than 500%. Family 1 did not ultimately participate in the project and reported a 50% reduction in number of varieties.

Conclusions

The results of this initial analysis show promise for community-based agritourism projects to contribute to the *in situ* conservation of Andean agrobiodiversity. While five households remain to be interviewed, the consistent and sustained increases both in the number of varieties of each crop cultivated by participating families and in the overall diversity and complexity of household farming systems suggest that varieties at risk of loss are, in fact, being conserved among participating households. While there is little evidence, with the exception of cañahua, of new varieties being introduced from outside the community, the results clearly show that participating families are actively exchanging varieties amongst themselves and, as such, are establishing multiple, highly diverse *in situ* conservation plots within the village.

There are a number of factors that may have contributed to these outcomes. First, it is important to mention that from the early stages of the project there was a steady flow of tourists visiting the community and paying to experience its agrobiodiversity. Santiago de Okola's agricultural landscape was highlighted in promotional materials used by local tour operators to attract visitors, with offers of homestays and traditional meals with families, an agrobiodiversity and medicinal plant walk, cooking workshops and visits to the community museum with exhibits on the food and farming systems of the village and region.

At the same time, project technicians from PROINPA and Sendas Altas were supporting community members by providing skill-building and organizational strengthening, helping participants respond to the needs of visitors, fundraising for new projects and managing the marketing efforts required to sustain the flow of tourists. Together, the internal and external efforts made by project partners and the consistent focus on agrobiodiversity combined to motivate participants to internalize the importance of agrobiodiversity not only for local food security requirements but also as a source of income linked to being the stewards of such a valuable resource.

Qualitative data collected during the 2020 inventories support the notion that the combination of receiving income during and beyond the early stages of the project while also simultaneously developing services and skills associated with tourism focused on agrobiodiversity caused the *revaluing* of crop varieties by both tourists and community members. During the 2008 inventories, for example, many respondents could not remember the names and uses of the varieties surveyed, while in 2020 farmers demonstrated much greater knowledge of varieties and much more interest in conserving them. Further, cultivating a greater number of varieties has become a source of pride and positive competition among project participants as they report the desire to demonstrate their commitment to stewarding cultural patrimony not only for tourists but also for neighbors and nearby communities, ostensibly elevating the village's status among its peers.

Finally, positive developments achieved in Santiago de Okola took place in the context of broader national and international efforts to revalue traditional food and farming systems (M. Taha, personal communication, 12 August 2020). The rise of Indigenous leadership at a national level in Bolivia, for example, greatly influenced increasing pride in traditional cultural practices, including native cuisine. The year 2013, for example, was declared as a Year of Quinoa by the Food and Agricultural Organization. These are just two examples of larger trends that have elevated appreciation for and consumption of native crop varieties and that have influenced tourist interests in the same.

206 Stephen R. Taranto et al.

At the time of writing, and most unfortunately, the 2020 global coronavirus pandemic has brought tourism in Santiago de Okola to a halt and throws into doubt the long term sustainability of agritourism to contribute to agrobiodiversity conservation in the community. The virus' impact on elderly populations is severe, placing Santiago de Okola's aging farmers at risk and, with them, the cultural and agroecological practices associated with the agrobiodiversity they conserve.

References

Baldinelli, M. (2013) "Cuando yo ya no pueda hacerlo, nadie lo hará": La conservación de la agrobiodiversidad en tiempos de migración. [Pdf] Report for Bioversity International. Available at https://www.bioversityinternational.org/e-library/publications/detail/cuando-yo-ya-no-pueda-hacerlo-nadie-lo-haraa/.

Blundo-Canto, G., Cruz-Garcia, G.S., Talsma, E.F., et al. (2020) Changes in food access by mestizo communities associated with deforestation and agrobiodiversity loss in Ucayali, Peruvian Amazon. *Food Security*, Vol. 12, 637–658. Available at https://doi.org/10.1007/s12571-020-01022-1.

Bradshaw, J. and Ramsey, G. (2009) Potato Origin and Production. In Singh, J. and Lovedeep, K. eds. *Advances in Potato Chemistry and Technology*. Academic Press, pp. 1–26.

Calla, S., Vargas, H., Gil, G., Garcia, D., et al. (2013) *Diagnóstico Arqueológico de Santiago de Okola y el Dragón Dormido*. [Pdf] Report for the Asociación de Turismo Integral de Santiago de Okola and Sendas Altas - Operadores en Turismo. Santiago de Okola, La Paz, Bolivia.

Córdoba, D. (2017) Politización, participación e innovación: socializando la investigación agrícola en Bolivia. *Apuntes*, Vol. 44(81), 131–160. ISSN 0252-1865. doi:10.21678/apuntes.81.808.

Consultado el 20 de julio de (2020) Available at http://www.scielo.org.pe/pdf/apuntes/v44n81/a05v44n81.pdf

Dangi, T. and Jamal, T. (2016) An Integrated Approach to "Sustainable Community-Based Tourism." *Sustainability*, 8(5), 475. Available at: http://dx.doi.org/10.3390/su8050475.

Erickson, C. L. (1988) Raised Field Agriculture in the Lake Titicaca Basin: Putting Ancient Agriculture Back to Work. *Expedition*, 30(1), 8–16. Available at: http://repository.upenn.edu/anthro_papers/18.

FAO (2009) Informe Nacional Sobre El Estado de los Recursos Fitogenéticos para la Agricultura y la Alimentación. http://www.fao.org/3/i1500e/Bolivia.pdf

Gil, G. (2017) *Informe de los Sitios Sagrados y Rituales de la Comunidad de Santiago de Okola. [Pdf] Report for the Asociación de Turismo Integral de Santiago de Okola*. Santiago de Okola, La Paz, Bolivia.

Mamani E., Alarcon, J., Paco, V. and Rojas, W. (2008) Inventory of agrobiodiversity in the communities of Santiago de Okola and Coromata Media. In the 2007–2008 Annual Report of the project "Strengthening Income Opportunities and Nutritional Security of the Rural Poor, through the Use and Marketing of Neglected and Underutilized Species." International Fund For Agricultural Development, Bioversity International and Fundación PROINPA. La Paz, Bolivia, pp. 198–222.

Mann, C. (2011) How the Potato Changed the World. *Smithsonian Magazine*, Vol. 11. Available at: https://www.smithsonianmag.com/history/how-the-potato-changed-the-world-108470605/.

Meave, M. and Lugo-Morín, D. (2014) Incidence of Climate Change in the Management of Protected Areas in Bolivia. *International Multilingual Journal of Contemporary Research*, Vol. 2(3), September 2014.

National Institute of Agricultural and Forestry Innovation (INIAF) [n.d.]. National Germplasm Bank of Bolivia. Genetic Resources Database. Retrieved August 23, 2020 from http://germoplasma.iniaf.gob.bo/gringlobal/.

National Research Council (1989) *Lost Crops of the Incas: Little-Known Plants of the Andes with Promise for Worldwide Cultivation*. Washington, DC: The National Academies Press. Available at: https://doi.org/10.17226/1398.

Padulosi, S., Bergamini, N. and Lawrence, T., eds. (2011) On Farm Conservation of Neglected and Underutilized Species: Status, Trends and Novel Approaches to Cope with Climate Change. *Proceedings of an International Conference*, Frankfurt, 14–16 June 2011. Bioversity International, Rome.

Pearsall, D. (2008) Plant domestication and change to agriculture in the Andes. In Silverman H., & Isbell W.H., eds. *The Handbook of South American Archeology*. Springer, New York.

Rea. J. (1995) Manejo y conservación comunitaria de comunitaria de recursos genéticos agrícolas en Bolivia. Revista Biodiversidad. Nro. 21. https://www.grain.org/es/article/entries/802-manejo-y-conservacion-comunitaria-de-recursos-geneticos-agricolas-en-bolivia

Rojas, W., Pinto, M., Bonifacio, A. and Gandarillas, A. (2010) Bancos de germoplasma de granos andinos. En: Rojas, W., Soto, J.L., Pinto, M., Jäger, M. and Padulosi, S. (eds.). *Granos Andinos. Avances, logros y experiencias desarrolladas en quinua, cañahua y amaranto en Bolivia*. Bioversity International, Roma, Italia. https://www.bioversityinternational.org/fileadmin/_migrated/uploads/tx_news/Granos_andinos__avances__logros_y_experiencias_desarrolladas_en_quinua__ca%c3%b1ahua_y_amaranto_en_Bolivia_1413.pdf.

Rojas, W., Pinto, M., Flores, J., Mamani, R. and Padulosi, S. (2014) Los agricultores custodios: Fortalecimiento de la conservación in situ de la agrobiodiversidad en Bolivia. En: INIAF (ed.) *Memoria Primer Congreso Nacional de Recursos Genéticos de la Agrobiodiversidad*. Santa Cruz, Bolivia. http://200.87.120.157/images/BajarINIAF/Memoria-Iniaf-Final.pdf

Thrupp, L.A. (2000) Linking Agricultural Biodiversity and Food Security: The Valuable Role of Agrobiodiversity for Sustainable Agriculture. *International Affairs*, Vol. 76(2), 265–281, https://doi.org/10.1111/1468-2346.00133

Torresin, L. (2015) *Useful Plants of Santiago de Okola*. Master's thesis. University of Padua. Padua, Italy.

Zimmerer, K. (2013) The Compatibility of Agricultural Intensification in a Global Hotspot of Smallholder Agrobiodiversity (Bolivia). *Proceedings of the National Academy of Sciences*, Vol. 110(8), 2769–2774, February 2013. Available at https://doi.org/10.1073/pnas.1216294110.

Zimmerer, K. (2014) Conserving Agrobiodiversity Amid Global Change, Migration, and Non-Traditional Livelihood Networks: The Dynamic Uses of Cultural Landscape Knowledge. *Ecology and Society*, Vol. 19(2). Retrieved on August 23, 2020 from http://www.jstor.org/stable/26269520.

17

MAINSTREAMING AFRICAN VEGETABLES TO IMPROVE DIETS AND LIVELIHOODS

Sognigbe N'Danikou, Maarten van Zonneveld, Fekadu Fufa Dinssa, Roland Schafleitner, Jody Harris, Pepijn Schreinemachers and Srinivasan Ramasamy

Introduction

Diets low in consumption of fruits, vegetables, and other nutrient-dense foods and high in energy, fats, free sugars, and salt are a leading cause of non-communicable diseases in sub-Saharan Africa (SSA), underpinning millions of premature deaths (Afshin et al., 2019). The World Health Organization recommends the consumption of 400 grams or five portions a day of fruits and vegetables for good health. However, healthy diets that include fresh vegetables are unaffordable for more than half the households in SSA (Harris et al., 2019; Hirvonen et al., 2020).

Traditional African vegetables (TAVs) are diverse in taste, flavors, and nutrient content. They present one way to improve people's diets in SSA countries at low cost with food plants suited to local taste and flavors, and which are connected to people's cultural history and adapted to local agro-ecological environments. Yet these vegetables are often neglected and underutilized for multiple reasons, including (i) shifts in food habits and preferences; (ii) agricultural and food policies promoting only a handful of energy-dense staple crops and exotic vegetables; (iii) poor planting materials available to smallholders; and (iv) poor crop management practices and postharvest handling due to inadequate extension and infrastructure.

This chapter gives a brief overview of the World Vegetable Center (WorldVeg) and its partners' efforts to promote TAV production and consumption in SSA and safeguard TAV biodiversity before it is lost. The lessons learned from these efforts will contribute to an African vegetable revolution to improve diets in the continent.

Safeguarding traditional vegetable biodiversity as a primary driver for resilient food systems

TAVs are vegetables native to SSA or vegetables that have been introduced to SSA and adopted by farmers in their traditional production systems. Three

DOI: 10.4324/9781003044802-19

African vegetables, diets and livelihoods **209**

primary regions of crop diversity of TAVs have been identified: (i) Dahomey gap in western lowland tropical Africa; (ii) the Ethiopian highlands; and (iii) South Cameroon (van Zonneveld et al., 2021). Tanzania and Eswatini are two regions of TAV diversity of predominantly wild vegetables; Angola, DR Congo, and South Sudan are areas of potential diversity that require further investigation because of the currently low number of observations (van Zonneveld et al., 2021). Madagascar is another fascinating region of vegetable crop diversity as a historic meeting point of people originating from SSA, Southeast Asia, and the Middle East, who each introduced their own crops and varieties. More than 400 species are reported to be used as traditional vegetables in Africa, with over 60% being leafy vegetables (van Zonneveld et al., 2021). A more detailed study among 126 TAVs indicates that about one-third are domesticated, another third are semi-domesticated, and the rest are predominantly harvested in the wild (van Zonneveld et al., 2021).

Many local varieties of domesticated and semi-domesticated species are lost due to underutilization. For wild species, habitat loss and overexploitation are major conservation concerns. WorldVeg maintains a global key collection of TAVs in the public domain in its genebank in Tanzania, which includes germplasm of amaranth, African nightshade, okra, roselle, and jute mallow, among other crops. This collection comprises about 2,400 accessions of TAV landraces and their wild relatives, collected from 38 African countries. The early collecting missions date back to the 1990s through different grant projects. In partnership with National Plant Genetic Resources Centers in SSA, WorldVeg is implementing a plan to rescue African vegetable biodiversity and support *ex situ* conservation of TAV biodiversity and breeding in Africa. Thanks to these partnerships, new collecting missions have started in western, eastern and southern Africa to fill geographical gaps in the conservation of TAVs.

Access to the germplasm conserved in WorldVeg genebanks to support research and healthy diets

The TAV collection at WorldVeg is held in trust for humanity. It can be accessed for breeding and research in accordance with the guidelines of the International Treaty on Plant Genetic Resources for Food and Agriculture.

In addition to distributing germplasm for research, education, and breeding, WorldVeg distributes vegetable seed kits to international and local NGOs, farmer groups, and local governments, which share the kits with households for the direct cultivation of these crops. Most seed kits aim to improve nutrition through home consumption and to diversify incomes by selling TAVs in local markets. WorldVeg also distributes seed kits to organizations for disaster relief to support smallholders affected by crises such as displacement due to civil unrest or crop failure because of extreme weather events. For example, from 2013 to 2020, the WorldVeg genebank distributed nearly 47,000 seed kits containing about 211,000 seed samples of improved lines and purified promising accessions to smallholder farmers in

Tanzania, Kenya, Uganda, and Madagascar through development projects and programs. Bulk seed distribution in Tanzania, Benin, and Mali was part of the Center's effort to support communities affected by the Covid-19 pandemic. These large numbers reflect increasing recognition among government and development organizations of the relevance of TAVs for income and nutrition.

The lines distributed are open-pollinated, allowing growers to produce and save their own seeds for future seasons. These lines of promising genebank accessions, selections, and WorldVeg improved lines have been tested under local conditions for yield, disease resistance, and consumer preference. Seed kits are distributed along with training in vegetable production and seed saving. Each kit contains a booklet with planting instructions that follow good agricultural practices and information on each crop's nutritional values. Seed kit distributions are intended to be one-time only to avoid dependency or crowding-out local seed enterprises. The provision of healthy seed kits to households coupled with behavior change communication has been shown to raise consumers' nutritional awareness of TAVs (Afari-Sefa et al., 2016). Such actions have resulted in significant improvements in households' dietary diversity, particularly for children under five and women of reproductive age (Ochieng et al., 2018).

In areas where formal and informal seed sectors of TAVs are underdeveloped, vegetable seed kits help introduce good quality seed in local seed systems. Seed kits are also a promising tool for participatory variety evaluation with farmer organizations and local seed enterprises, which, in turn, inform seed enterprises and seed companies about farmer preferences to tailor their supply to local demand. As seed companies and public institutions expand their portfolio with traditional vegetables, the role of the WorldVeg genebank shifts from seed kit distribution to supplying germplasm resources for variety development by seed companies and public institutions. Thus, the WorldVeg genebank optimizes the distribution of vegetable diversity by working with partners in both formal and local seed systems and aims to contribute to integrated seed system development.

Farmers in several SSA countries such as Kenya, Tanzania, and Mali are already adopting varieties developed from WorldVeg genebank materials supplied to seed companies in Africa. For instance, 51% of amaranth areas planted in Kenya and 70% in Tanzania were from WorldVeg improved lines that were released in these countries (Ochieng et al., 2019). Also, 2.9 tons (59%) of amaranth seeds sold by companies in 2016 used WorldVeg-based germplasm. WorldVeg's amaranth lines have reached about 231,000 households in Kenya and Tanzania alone (Ochieng et al., 2019).

Germplasm enhancement for increased productivity and nutrition

The germplasm improvement of TAVs at WorldVeg started in 1992 with a handful of the most popular TAVs, namely amaranth, African eggplant, and Ethiopian mustard. Through repeated selection, purification, and farmers' participatory variety selection, plant breeders quickly identified promising lines for release as improved cultivars of amaranth and African eggplant in Tanzania, Kenya, Ethiopia, and Cameroon (Figure 17.1).

A WorldVeg breeding program for amaranth and African eggplant started in 2015 to develop advanced lines with enhanced yield, nutrients, product color, shape, size, taste, and late and early maturation. These lines have been distributed to national agricultural research systems, private seed companies, and other partners in Africa (Dinssa et al., 2016). They have been adopted at large scales by farmers in several countries such as Tanzania and Kenya (Ochieng et al., 2019). As a result, improved cultivars of amaranth and African eggplant have been released in several countries in the continent. Most seed companies directly use the released varieties and multiply the seed for commercialization. A few advanced companies have started using WorldVeg improved lines in their breeding programs. The increasing commercialization of seed suggests commensurate increasing consumer demand for these vegetables and, consequently, more interest from farmers to invest in amaranth production. Further research will help provide further insights into these dynamics and how to further scale vegetable seed supply in SSA.

To further advance variety release and supply quality vegetable seed, the Africa Vegetable Breeding Consortium (AVBC) was established in 2018. The AVBC is a joint initiative of WorldVeg and the African Seed Trade Association to develop a strong vegetable seed sector in Africa. All WorldVeg improved lines are publicly available and can be requested. AVBC members have limited exclusivity claims to some WorldVeg-developed lines before they become a global public good.

Sustainable production

WorldVeg improved lines include African eggplant accessions resistant to shoot borer *Leucinodes orbonalis*, and sources of resistance in amaranth have been identified for foliar pests and stem weevils (AVRDC, 2003; Othim et al., 2018).

FIGURE 17.1 Two successful amaranth varieties, Madiira 1 (A) and Madiira 2 (B) released in 2011 in Tanzania. Photos: World Vegetable Center.

In combination with appropriate pest and disease management practices, farmers can boost TAV supply to meet growing demand. These practices include removing and destroying infested crop debris, crop rotations with non-host crop species, eradicating weeds, good drainage, and growing healthy seedlings. Grafting African eggplant on suitable resistant rootstocks is an important control measure to overcome bacterial wilt and *Verticillium* wilt diseases. Soil fumigation and soil solarization are important means to control pathogens and root-knot nematodes. Damping-off diseases can be controlled by treating seed with fungicides (including bio-fungicides such as *Trichoderma*) or hot water. Using disease-free seedlings, removing and destroying diseased plants, and protecting fields and nurseries from insect vectors are further recommendations to help manage virus diseases. Yellow sticky traps (for whitefly, leafminer, leafhopper, and aphids) and/or blue sticky traps (for thrips) should be erected on the windward side of fields and nurseries to help trap these pests, some of which also act as vectors of virus diseases. Application of predatory mites and insects and entomopathogenic fungi such as *Neozygites floridana*, *Beauveria bassiana*, and *Metarhizium anisopliae* help reduce infestations of pests such as spider mites. The application of bio-pesticides, including neem formulations, reduces insect pests and also conserves natural enemies such as predatory ladybird beetles and syrphid flies, as well as parasitoids. Extension and training in improved production and postharvest practices support farmers in boosting the TAV supply in SSA, as would adequate infrastructure to get vegetables to appropriate markets.

TAV consumption

Vegetable consumption is generally low in SSA. Only 7% of countries in Africa reach a mean consumption of 240 g/day, as recommended by WHO/FAO guidelines (Kalmpourtzidou et al., 2020), though there is wide variation between regions and countries in SSA (Afshin et al., 2019). Data specifically on TAV intake at individual or population level is not readily available, because national data-collection systems focus predominantly on economically important vegetables (Herforth et al., 2019). Further research is needed to understand the intake of TAVs at household and individual levels in many SSA countries.

Behavior change communication programs have been tested in several countries to promote fruit and vegetable consumption, including some TAVs. Among these approaches are school garden programs used to stimulate interest of children at a young age (Schreinemachers et al., 2019); and integrated home garden programs—coupling garden-based training with nutrition behavior change—to increase the household-level availability of vegetables while simultaneously stimulating demand (Olney et al., 2015). Other approaches include media campaigns, cooking demonstrations and competitions, promoting local food ambassadors/champions, and holding food and seed fairs (Ochieng et al., 2019). Over the last three decades, tens of thousands of households in western (Mali, Burkina Faso), central (Cameroon), eastern (Kenya, Uganda, Tanzania,

African vegetables, diets and livelihoods **213**

Madagascar) and southern Africa (Malawi, Zambia) have been trained on the nutritional benefits of TAVs and how to prepare them by different organizations including WorldVeg, Bioversity International, and their national partners.

Thanks to a series of interventions on both the demand and supply sides, TAVs are successfully being mainstreamed in Kenya and Uganda, especially in the capital cities like Nairobi and Kampala, through a revival of traditional eating patterns among the middle classes. This has been possible because of policy support to create an enabling environment for vegetables in food systems. The recognition of quality declared seeds (QDS) boosted TAV supply to meet increasing urban consumption in these cities (Mabaya et al., 2019). QDS provides farmers with access to quality (though not nationally registered) seeds of less highly bred crops such as TAVs. The QDS process involves local seed enterprises, boosting livelihoods and making diverse vegetables available for healthy diets.

However, in most cases, interventions are fragmented because they address only part of the value chain. There is a need for systems-based actions that encompass the whole value chain and that are enforced by enabling policies to support sustainable production and consumption. Many food and seed policies do not explicitly consider vegetables or desired dietary outcomes for national populations. Therefore, more work is needed to understand how these system-level interventions can connect the production and consumption of nutrient-rich vegetable foods.

Reflection on the way forward for the African vegetable revolution

Despite increased interest from the scientific community in the ongoing work on TAVs, there are still important steps left to achieve an African vegetable revolution to improve diets in the continent. We here identify three significant areas that need a clear commitment from high-level institutions in Africa and worldwide.

a **Enabling policies to mainstream African traditional vegetables into food systems**. Policies to shape the food system to nourish people, not just feed them, must look across the food system from inputs to production and food environments. Multidisciplinary research is key to reveal gaps and opportunities for increasing TAVs in food systems, starting with the desired outcomes—including healthy diets, fair livelihoods, and sustainable environments (McMullin et al., 2021). A critical component of this research is the investigation of seed policies and how to make them more inclusive by (i) recognizing farmer-saved seeds; (ii) supporting the development of nutritious vegetable varieties; and (iii) making these vegetable seeds available to and affordable for farmers. Policies that support farmers in risk management will encourage them to grow TAVs. These include providing subsidies for nutritious foods and insurance programs for TAVs, and connecting growers

214 Sognigbe N'Danikou et al.

with consistent institutional markets such as school feeding programs while demand increases.

b **Promotion and outreach to boost consumption** of TAVs. There is increasing evidence of the impact of promotion campaigns on public awareness of the nutritional value of TAVs and their role in dietary diversity at the household level (Afari-Sefa et al., 2016; Ochieng et al., 2018). Specialized government agencies (e.g., Food and Nutrition Councils and their equivalents) can be tasked with such campaigns.

c **Strengthen regional R&D programs and initiatives** to enhance the use of African vegetable germplasm. This will require governments and funders to invest and stakeholders to organize efforts to rescue the biodiversity of TAVs and strengthen vegetable breeding programs in Africa. An example of such an initiative is the AVBC, which brings WorldVeg and seed companies together to build capacities and exchange knowledge on new breeding techniques. A regional action plan with public, societal, and private actors will be essential to rescue and make efficient use of TAV biodiversity in Africa.

Conclusion

While notable efforts are ongoing to mainstream TAVs in production and food systems in SSA for healthier diets, improved livelihoods, and the protection of genetic diversity, several bottlenecks remain: policies to create an enabling environment are needed to integrate these vegetables into people's diets; awareness campaigns to encourage consumption will enhance the adoption and acceptability of TAVs; consistent investment to safeguard TAV biodiversity is essential for resilient production and supply to meet consumer demand; and supply of high-quality seeds to farmers in combination with extension will help build capacity among farmers for safe and sustainable production practices.

References

Afari-Sefa, V., Rajendran, S., Kessy, R. F., Karanja, D. K., Musebe, R., Samali, S. and Makaranga, M. (2016) Impact of nutritional perceptions of traditional African vegetables on farm household production decisions: a case study of smallholders in Tanzania. *Experimental Agriculture*, Vol 52, no 2, pp300–313.

Afshin, A., Sur, P.J., Fay, K.A., Cornaby, L., Ferrara G., Salama J.S., Mullany E.C., Abate K.H., Abbafati C., Abebe Z., and Afarideh, M. and Murray, C.J.L. (2019) Health effects of dietary risks in 195 countries, 1990–2017: A systematic analysis for the global burden of disease study 2017. *The Lancet*, Vol 393, no 10184, pp1958–1972.

AVRDC (2003) *Identification of sources of genetic resistance to eggplant fruit and shoot borer.* Progress Report 2002. Shanhua, Taiwan: AVRDC-The World Vegetable Center. pp73–74.

Dinssa, F.F., Hanson, P., Dubois, T., Tenkouano, A., Stoilova, T., Hughes, J. d'A. and Keatinge, J.D.H. (2016) AVRDC – The World Vegetable Center's women-oriented improvement and development strategy for traditional African vegetables in sub-Saharan Africa. *European Journal of Horticultural Science*, Vol 81, no 2, pp91–105.

Harris, J., Chisanga, B., Drimie, S. and Kennedy, G. (2019) Nutrition transition in Zambia: Changing food supply, food prices, household consumption, diet and nutrition outcomes. *Food Security*, Vol 11, no 2, pp371–387.

Herforth, A., Masters, W., Bai, Y. and Sarpong, D. (2019) The cost of recommended diets: Development and application a food price index based on food-based dietary guidelines (P10–033-19). *Current Developments in Nutrition*, Vol 3, Supplement 1, p788.

Hirvonen, K., Bai, Y., Headey, D. and Masters, W.A. (2020) Affordability of the EAT–Lancet reference diet: A global analysis. *The Lancet Global Health*, Vol 8, no 1, pp59–66.

Kalmpourtzidou, A., Eilander, A. and Talsma, E.F. (2020) Global vegetable intake and supply compared to recommendations: A systematic review. *Nutrients*, Vol 12, no 6, pp1–14.

Mabaya, E., Mugoya, M., Mubangizi, E. and Ibyisintabyo, C. (2019) Uganda Brief 2018. The African Seed Access Index. 12p.

McMullin S., Stadlmayr B., Mausch K., Revoredo-Giha C., Burnett F., Guarino L., Brouwer I.D., Jamnadass R., Graudal L., Powell W. and Dawson I.K. (2021) Determining appropriate interventions to mainstream nutritious orphan crops into African food systems. *Global Food Security*, Vol 28, no 100465, pp1–15.

Ochieng, J., Afari-Sefa, V., Karanja, D., Kessy, R., Rajendran, S. and Samali, S. (2018) How promoting consumption of traditional African vegetables affects household nutrition security in Tanzania. *Renewable Agriculture and Food Systems*, Vol 33, no 2, pp105–115.

Ochieng, J., Schreinemachers, P., Ogada, M., Dinssa, F. F., Barnos, W. and Mndiga, H. (2019) Adoption of improved amaranth varieties and good agricultural practices in East Africa. *Land Use Policy*, Vol 83, pp187–194.

Olney, D.K., Pedehombga, A., Ruel, M.T. and Dillon, A. (2015) A 2-year integrated agriculture and nutrition and health behavior change communication program targeted to women in Burkina Faso reduces anemia, wasting, and diarrhea in children 3–12.9 months of age at baseline: A cluster-randomized controlled trial. *Journal of Nutrition*, Vol 145, no 6, pp1317–1324.

Othim, S.T.O., Srinivasan, R., Kahuthia-Gathu, R., Dubois, T., Dinssa, F.F., Ekesi, S. and Fiaboe, K.K.M. (2018) Screening for resistance against major lepidopteran and stem weevil pests of amaranth in Tanzania. *Euphytica*, Vol 214, no 182, pp1–21.

Schreinemachers, P., Ouedraogo, M.S., Diagbouga, S., Thiombiano, A., Kouamé, S.R., Sobgui, C.M., Chen, H.-P. and Yang, R.-Y. (2019) Impact of school gardens and complementary nutrition education in Burkina Faso. *Journal of Development Effectiveness*, Vol 11, no 2, pp132–145.

van Zonneveld, M., Kindt, R., Solberg, S.Ø., N'Danikou, S. and Dawson, I.K. (2021) Diversity and conservation of traditional African vegetables: priorities for action. *Diversity and Distributions*, Vol 27, pp216–232.

18

SLOW FOOD AND NUS

Protecting and promoting endangered food products

Charles Barstow, Edie Mukiibi and Dauro Mattia Zocchi

Introduction

Slow Food (SF) is a global grassroots organization and movement founded in 1989 to counter the acceleration of frenetic consumerism by renewing interest in food and the pleasure of eating, protecting local food cultures, and ensuring that everyone has access to food that is good (healthy and delicious), clean (produced and consumed with respect for the environment), and fair (contributing to social justice) (Petrini, 2005). Since its inception, SF has articulated a multifaceted concept of gastronomy that understands food as a web of relationships linking pleasure and wellbeing, people and landscapes, producers and consumers, cultures and ecosystems (Schneider, 2008). SF's activities and initiatives focus on education, traditional knowledge, agroecology, empowerment of women and Indigenous Peoples, participatory management of foodscapes, policy and advocacy, and the creation of short, equitable supply chains. A commitment to defending the biological and cultural diversity on which the resilience of human communities and the global food system depend underpins all of this work.

This is in keeping with the growing political, popular, and scientific interest in the role that traditional food systems can play in combatting and potentially reversing the erosion of biocultural diversity (Maffi, 2001) that has accelerated in recent decades. Particular attention is now paid to neglected and underutilized species (NUS) (Padulosi et al., 2013), which may hold solutions to challenges facing socioecological systems and human health and nutrition. Not just governments, research institutions, and private sector organizations, but also local and international food movements have made important contributions to recovering NUS (Goodman et al., 2012; Counihan and Siniscalchi, 2014). Several grassroots initiatives identify, document, and promote neglected and marginalized products and their associated biocultural and gastronomic value (Nabhan et al.,

DOI: 10.4324/9781003044802-20

2010). In this chapter we discuss aspects of SF's approach to what it calls 'endangered food products'. While this term overlaps in its meaning with NUS, it refers to a broader domain than the latter, as we will illustrate. Our focus is on the systems-based approach and the underlying values of SF's global biodiversity projects.

Defining endangered products

The Ark of Taste, coordinated by the SF Foundation for Biodiversity, is an online catalogue of endangered varieties, breeds, and processed products from cultures and territories around the world (for more information on the Foundation, see www.fondazioneslowfood.com/en/). The criteria that products must satisfy for inclusion on the Ark help define the "ecogastronomic unit of concern" (Nabhan et al., 2010).

First, items must be food products, including domesticated plant varieties and animal breeds, populations, wild species, and processed products. Second, Ark products must be of distinctive organoleptic quality. This criterion points to one of SF's distinguishing features, its concern with pleasure and 'good taste', which implies affording as much importance to what 'tastes good' as to what is healthy. 'Quality' is defined within the relevant local context; understanding the factors that contribute to perceptions of quality requires intercultural dialogue, and pleasure is often what brings people into the conversation. Third, Ark products must be linked to a specific territory and the memory, identity, and traditional knowledge of a particular community. This underscores the importance of landscapes and 'foodscapes', the human-natural systems in which the cultural and ecological values of the ecogastronomic unit (the product) take shape and acquire meaning, and where traditional knowledge and management practices are created, maintained, and transmitted (Stepp et al., 2003; Barthel et al., 2013).

Finally, Ark of Taste products must be produced in limited quantities and at risk or endangered due to various social, economic, and ecological factors. These factors include habitat degradation and land use conversion, forced or voluntary migration, sedentarization of mobile populations, lack of intergenerational knowledge transfer, barriers to market entry, one-size-fits-all hygiene regulations, climate change, environmental and genetic pollution, incentives to adopt improved or foreign varieties and breeds, changing sentiments due to education and the media, and the general mechanization, industrialization, and standardization of the food system.

Looking at these criteria, there are key similarities and differences between SF's concept of endangered food products and the various working definitions of NUS (as elaborated, e.g., in Delêtre et al., 2013). Among the important differences are that, for SF, the local is privileged over the global, the 'what' (i.e., the unit of concern) is more broadly defined, and the 'why' is motivated primarily by questions of culture and identity rather than unrealized market potential. These differences have a number of important implications, as elaborated on below.

218 Charles Barstow et al.

Locating endangered products

SF relies on the wisdom of the crowd—its global network of farmers, fishers, herders, artisans, cooks, educators, academics, activists, Indigenous Peoples, and consumers—to identify endangered products. Ark of Taste products most often result from nominations submitted by people across the globe who may or may not be formally associated with the SF network. If a nominated product meets the criteria described above, it is boarded on the Ark to raise awareness and increase its visibility; individuals and communities can then take action, in their own way or through the implementation of other SF projects. In other cases, Ark products arise from biodiversity mapping undertaken by members of the network with technical support from SF and local experts, or by SF staff during field visits. Mapping is one of the methodologies that appear in the literature on NUS (e.g., Will, 2008), though the order of the steps involved, the elements to be mapped, and the justifications for selecting certain products differ slightly in the SF approach.

Critically, SF does not select products in advance; it works to support and facilitate the efforts of communities in its network, not set an agenda for them. Likewise, SF does not undertake field research or projects in a territory without being invited. This approach seeks to put technical support, networking capacity, and scientific knowledge at the service of local communities, not merely supplement research and development with those communities' traditional knowledge and solicit their input and involvement. In addition to products themselves, SF maps stakeholders within the foodscape. These could be farmers, processors, civil society institutions, tourism and hospitality organizations, and markets. Without understanding this constellation of stakeholders, and particularly how it links rural and urban areas, identifying and promoting endangered products is rarely fruitful. Likewise, it is impossible to promote products according to SF's values without interest and commitment from local communities.

Protecting and promoting endangered products

While 'protecting' and 'promoting' serve and create opportunities for each other, there is potential for opportunities to become threats (Will, 2008), such as when resource use and market attention become too intense in relation to one product at the expense of other endangered products and the environment. Therefore, economic potential alone is insufficient to warrant SF's attention, and when it comes to such potential, emphasis is primarily on local economies and food sovereignty rather than the global market and food security. SF challenges the assertion that global food production must increase 50%–100% by 2050 (Foley et al., 2011; FAO, 2019). Calls to drastically increase production are used to justify a progressist narrative of growth, liberal trade, and technological fixes, the disastrous consequences of which are evident (Ghosh, 2010; Tomlinson, 2013; Hunter et al., 2017). Feeding everybody while protecting biocultural diversity and maintaining production potential in the long term requires multiple local strategies tailored to local complexities (Cunningham et al., 2013). Moreover, though small-scale production alone is insufficient, it is absolutely necessary for achieving these goals—such production already feeds a huge proportion of humanity while simultaneously supporting ecosystem services and high levels of biodiversity (Ricciardi et al., 2018).

With these issues in mind, the SF Foundation for Biodiversity has developed several projects to promote traditional and endangered products—such as those on the Ark of Taste—and their associated biocultural contexts within a broad view of promotion as a means for protection. For example, the Gardens in Africa project is a continent-wide network of school and community gardens designed to protect traditional varieties of plants and promote nutrition and knowledge transfer using gardens as productive and educational spaces. Subsistence is rarely 'mere subsistence'. The importance of subsistence agriculture for biodiversity conservation, nutrition and dietary diversity, and food sovereignty based on access to wild and cultivated traditional foods—particularly among poor and Indigenous communities—is well documented (Kuhnlein and Receveur, 1996; Roche et al., 2008; Galluzzi et al., 2010; Jones et al., 2014; Powell et al., 2015).

The SF Presidium project, a more market-oriented initiative, supports groups of producers committed to reviving or revitalizing endangered products. The SF Foundation provides training and technical assistance and works with producers to create production protocols similar to those for products with Protected Designation of Origin or other geographical indications, but designed to foster

FIGURE 18.1 Women of the Chiapas Milpa SF Presidium (Mexico) processing corn.
Source: Gabriela Sanabria, SF Archive (used with permission).

flexibility, innovation, and subtle differences between producers, all in the spirit of subverting standardization (Friedman and McNair, 2008). Presidia raise the visibility of these products on the market and producers may legally use SF's registered snail logo, which has become a highly regarded seal of approval for good, clean, and fair foods. Furthermore, Presidia use narrative labels to highlight each product's connection to a particular territory and the knowledge of producers, replacing the 'fast' reading of traditional certifications and label texts with the 'slow' reading of biocultural contexts.

One distinguishing feature of SF Presidia is their increasing focus on systems of production rather than single products. An example of this is the recently launched Presidium for the Chiapas Milpa System, which uses agroecology to grow several Ark of Taste products, supports the reproduction and sharing of traditional seeds and knowledge, and empowers women involved in producing and marketing traditional corn products. Another example is the Presidium for dried nettles (*Urtica massaica*) from Kenya's Mau Forest. The Gikuyu and other local Indigenous communities use nettles as food and medicine for themselves and their livestock. Because deforestation has reduced the availability of wild nettles, Gikuyu women have brought them into cultivation in disturbed areas around dwellings and the fertile soils of formerly grazed land. The Presidium assists with production and promotes both fresh and dried powdered nettles in restaurants and in local and regional markets.

Beyond NUS

As some NUS literature points out, value addition can be key for bringing marginalized foods to wider markets (Will, 2008; Padulosi et al., 2013). SF takes this a step further, considering processed products like cheeses, honey, and cured meats. The importance of traditional knowledge, techniques, and technologies associated with these products is clear. What may be less apparent is the biodiversity and cultural value of non-food species behind these products, including

FIGURE 18.2 Women of the Mau Forest Dried Nettle SF Presidium (Kenya) harvesting and processing nettles.
Source: Oliver Migliore, SF Archive (used with permission).

bees, forage plants, and microorganisms responsible for fermentation. Promoting such products can contribute to the protection of these species, their habitats, and their associated ecosystem services. Two examples help clarify this. Honey is vitally important to the gastronomy of the Ogiek, another Indigenous community of the Mau Forest. Many of the melliferous species upon which local bees rely, while inedible to humans, provide materials for tools and beekeeping equipment. One is *Dombeya torrida*, known locally as *silibwet*, a sacred tree that yields the Ogiek's most valued honey (Zocchi et al., 2020). The Ogiek Honey Presidium helps protect this tree and the Mau Forest ecosystem generally.

Another example is the relationship between forage plants and the qualities of dairy products. Sahrawi nomads in Western Sahara know exactly how the plants that camels eat affect the sensory and nutritional qualities of their milk, and believe that this milk retains the medicinal properties of the plants ingested (Volpato and Puri, 2014). Similar observations have been made by Raika camel herders in Rajasthan (Köhler-Rollefson et al., 2013), whose milk is an Ark of Taste product. In Europe, the astonishing diversity of mountain cheeses results directly from biophysical variation across mountain pastures. Promoting these cheeses and their value chains could help reduce agricultural abandonment, which threatens landscape diversity and traditional cultures (MacDonald et al., 2000). SF has recently promoted natural cheeses (those produced with raw milk and natural fermentation) across Europe. Such cheeses have increased organoleptic

FIGURE 18.3 Men of the Ogiek Honey SF Presidium (Kenya) harvesting honey from a traditional log hive.
Source: rootsofafrika.co, SF Archive (used with permission).

FIGURE 18.4 Coazze Cevrin, a cheese from the Alps of Northwest Italy, is made from the raw milk of cows and goats. It is produced only during the months when the animals graze in the mountains. The Coazze Cevrin SF Presidium protects not only the cheese itself, but also the use of natural fermentation starters, the future of the endangered Barà cattle breed (listed on the Ark of Taste), and the Alpine pastures where the animals graze.

Source: Valeria Necchio (image 4a) and Paolo Andrea Montanaro (image 4b), SF Archive (used with permission).

complexity that reflects the specific communities of microorganisms in the pastures and dairy facilities where they're produced, and their producers are not beholden to companies that make and distribute selected fermentation cultures.

These examples refer to landscapes that produce little or no plant-based food for humans, but whose biocultural diversity and contributions to ecological structure and function are worthy of attention. Food can be an effective entry point and rallying cry in this regard.

Conclusions

Over the last three decades, parallel to concerns about the sustainability and equitability of the global food system and its effects on biodiversity, SF has developed a grassroots approach that identifies food systems as both cause and victim of the current ecological crisis, as well as a potential solution. Recognizing that protecting and promoting endangered food products and NUS is insufficient on its own, SF addresses the wider sociocultural and ecological systems of which these products are a part. Social movements, institutional initiatives, and policies must emphasize and rigorously defend the inherent value of these systems. This means using the market in service of these values, not the other way around, and it requires leaving decision-making power in the hands of local communities so that they don't become dependent on market channels over which they have no influence. While global logistics and communications make it increasingly easy to promote and market NUS, endangered foods, dietary diversity, and local, seasonal produce in cities and the Global North—which is desirable—such promotion must never come at the cost of failing to protect these things where they already exist in rural areas and the Global South. Furthermore, financial and other benefits must be returned to and shared with communities. The task, then, is to create a new gastronomy that incorporates biodiversity, cultural diversity, and traditional knowledge into agricultural systems and foodscapes, both rural and urban, and that generates markets and networks of solidarity that incentivize and derive pleasure from these diversities, while recognizing their full value.

References

Barthel, S., Crumley, C. L. and Svedin, U. (2013) 'Biocultural refugia: Combating the erosion of diversity in landscapes of food production', *Ecology and Society*, vol 18, no 4, article no 71.

Counihan, C. and Siniscalchi, V., eds. (2014) *Food Activism: Agency, Democracy and Economy*. Bloomsbury Academic, London.

Cunningham, S. A., et al. (2013) 'To close the yield-gap while saving biodiversity will require multiple locally relevant strategies', *Agriculture, Ecosystems and Environment*, vol 173, pp20–27.

Delêtre, M., Gaisberger, H. and Arnaud, E. (2013) 'Agrobiodiversity in perspective: A review of questions, tools, concepts and methodologies in preparation of SEP2D', revised version, prepared for Bioversity International, Rome. Retrieved from https://www. researchgate.net/profile/Marc-Deletre/publication/281115298_Agrobiodiversity_ in_perspectives/links/58f720300f7e9b67a34bb1cf/Agrobiodiversity-in-perspectives. pdf.

FAO (2019) *The State of the World's Biodiversity for Food and Agriculture*. Commission on Genetic Resources for Food and Agriculture, FAO, Rome.

Foley, J.A., et al. (2011) 'Solutions for a cultivated planet', *Nature*, vol 478, pp337–342.

Friedman, H. and McNair, A. (2008) 'Whose rules rule? Contested projects to certify "Local production for distant consumers"', *Journal of Agrarian Change*, vol 8, nos 2–3, pp408–434.

Galluzzi, G., Eyzaguirre, P. and Negri, V. (2010) 'Home gardens: Neglected hotspots of agrobiodiversity and cultural diversity', *Biodiversity and Conservation*, vol 19, pp3635–3654.

Ghosh, J. (2010) 'The unnatural coupling: Food and global finance', *Journal of Agrarian Change*, vol 10, no 1, pp72–86.

Goodman, D., DuPuis, E.M. and Goodman, M.K. (2012) *Alternative Food Networks: Knowledge, Practice, and Politics*. Routledge, London.

Hunter, M. C., Smith, R. G., Schipanski, M. E., Atwood, L. W. and Mortensen, D. A. (2017) 'Agriculture in 2050: Recalibrating targets for sustainable intensification', *BioScience*, vol 67, no 4, pp386–391.

Jones, A. D., Shrinivas, A. and Bezner-Kerr, R. (2014) 'Farm production diversity is associated with greater household dietary diversity in Malawi: Findings from nationally representative data', *Food Policy*, vol 46, pp1–12.

Köhler-Rollefson, I., Rathore, H. S. and Rollefson, A., edited by Hardy, K. (2013) 'The camels of Kumbhalgarh: A biodiversity treasure', Lokhit Pashu-Palak Sansthan, Sadri. Retrieved from http://www.lpps.org/wp-content/uploads/2013/10/Camels_Of_ Kumbhalgarh_web.pdf.

Kuhnlein, H. V. and Receveur, O. (1996) 'Dietary change and traditional food systems of indigenous peoples', *Annual Review of Nutrition*, vol 16, no 1, pp417–442.

MacDonald, D., Crabtree, J. R., Wiesinger, G., Dax, T., Stamou, N., Fleury, P., Lazpita, J. G. and Gibon, A. (2000) 'Agricultural abandonment in mountain areas of Europe: Environmental consequences and policy response', *Journal of Environmental Management*, vol 59, no 1, pp47–69.

Maffi, L. (2001) *On Biocultural Diversity: Linking Language, Knowledge, and the Environment*. Smithsonian Institution Press, Washington, DC.

Nabhan, G. P., Walker, D. and Moreno, A. M. (2010) 'Biocultural and ecogastronomic restoration: The renewing America's food traditions alliance', *Ecological Restoration*, vol 28, no 3, pp266–279.

Padulosi, S., Thompson, J. and Rudebjer, P. (2013) *Fighting Poverty, Hunger and Malnutrition with Neglected and Underutilized Species: Needs, Challenges and the Way Forward'*, Bioversity International, Rome. Retrieved from https://www.bioversityinternational. org/fileadmin/_migrated/uploads/tx_news/Fighting_poverty__hunger_and_ malnutrition_with_neglected_and_underutilized_species__NUS__1671_03.pdf.

Petrini, C. (2005) *Buono, pulito e giusto: Principî di nuova gastronomia*. Einaudi, Turin.

Powell, B., Thilsted, S. H., Ickowitz, A., Termote, C., Sunderland, T. and Herforth, A. (2015) 'Improving diets with wild and cultivated biodiversity from across the landscape', *Food Security*, vol 7, pp535–554.

Ricciardi, V., Ramankutty, N., Mehrabi, Z., Jarvis, L. and Chookolingo, B. (2018) 'How much of the world's food do smallholders produce?' *Global Food Security*, vol 17, pp64–72.

Roche, M. L., Creed-Kanashiro, H. M., Tuesta, I. and Kuhnlein, H. V. (2008) 'Traditional food diversity predicts dietary quality for the Awajún in the Peruvian Amazon', *Public Health Nutrition*, vol 11, no 5, pp457–465.

Schneider, S. (2008) 'Good, clear, fair: The rhetoric of the Slow Food movement', *College English*, vol 70, no 4, pp384–402.

Stepp, J. R., Jones, E. C., Pavao-Zuckerman, M., Casagrande, D., and Zarger, R. K. (2003) 'Remarkable properties of human ecosystems', *Conservation Ecology*, vol 7, no 3, article no 11.

Tomlinson, I. (2013) 'Doubling food production to feed the 9 billion: A critical perspective on a key discourse of food security in the UK', *Journal of Rural Studies*, vol 29, pp81–90.

Volpato, G. and Puri, R. K. (2014) 'Dormancy and revitalization: The fate of ethnobotanical knowledge of camel forage among Sahrawi nomads and refugees of Western Sahara', *Ethnobotany Research and Applications*, vol 12, pp183–210.

Will, M. (2008) 'Promoting value chains of neglected and underutilized species for pro-poor growth and biodiversity conservation', *Global Facilitation Unit for Underutilized Species*, Rome. Retrieved from https://www.researchgate.net/profile/Marc-Deletre/ publication/281115298_Agrobiodiversity_in_perspectives/links/58f720300f7e 9b67a34bb1cf/Agrobiodiversity-in-perspectives.pdf.

Zocchi, D. M., Volpato, G., Chalo, D., Mutiso, P. and Fontefrancesco, M. F. (2020) 'Expanding the reach: Ethnobotanical knowledge and technological intensification in beekeeping among the Ogiek of Mau Forest, Kenya', *Journal of Ethnobiology and Ethnomedicine*, vol 16, article no 57.

PART III
Integrated conservation and use of minor millets

19

CONSERVATION AND UTILIZATION OF SMALL MILLETS GENETIC RESOURCES

Global and Indian perspectives

Kuldeep Singh, Nikhil Malhotra, Mohar Singh

Introduction

Underutilized crop genetic resources are imperative for sustainable agriculture and small millets belong to this important group of crops (Vetriventhan and Upadhyaya, 2019). Plant genetic resources play a significant role in enhancing the adaptation and resilience of agricultural production systems. With the global population expected to reach 9.8 billion by 2050, there is an urgent need to enhance food production by 60%–70% from the current level. The Green Revolution signified an immense transformation in agriculture in the form of a broad spectrum of species, including locally adapted cultivars and landraces, which are needed to harness livelihood potential to cope with extreme climatic conditions, marginal lands and other biotic and abiotic stresses along with nutritional security (Fita et al., 2015). The Indian subcontinent is bestowed with an immense wealth of agrobiodiversity, which can be utilized to take up this challenge. One such group of vastly promising crops is that of small millets. This group is represented by six species *viz.* finger millet (*Eleusine coracana* L. Gaertner), kodo millet (*Paspalum scrobiculatum* L.), foxtail millet (*Setaria italica* L. Pal), little millet (*Panicum sumatrense* Roth ex Roemer and Schultes), proso millet (*Panicum miliaceum* L.) and barnyard millet (*Echinochloa crusgalli* and *E. colona* L. Link) that have climate-smart features (Saxena et al., 2018).

The cultivation of small millets in India has progressively deteriorated over the last few decades due to their lower economic worth compared to major cereals and, lately, to horticultural crops. This has resulted from a variety of factors including the lack of high-yielding varieties, good quality seed, better cultivation practices, proper food processing technologies and poor value-chains (Upadhyaya and Vetriventhan, 2017). Overcoming these major bottlenecks can allow economically viable opportunities for farmers in areas where the

DOI: 10.4324/9781003044802-22

production of major cereals like rice, wheat and maize may be steadily declining due to climate change. In such scenarios, small millets offer a better alternative, as these are known to possess tremendous nutritional benefits. Overall, time demands immediate thrust for directing more research and development towards these crops.

Present status of conservation of small millets

Globally, the largest ex situ germplasm collections of small millets are maintained by the Consultative Group on International Agricultural Research at the International Crop Research Institute for Semi-arid Tropics (ICRISAT) in Hyderabad, India, for long-term conservation. The ICRISAT gene bank conserves more than 12,000 accessions of six small millets (finger millet, barnyard millet, foxtail millet, proso millet, little millet, kodo millet and proso millet). The majority of these accessions belong to finger millet (5,947 accessions) followed by foxtail millet (1,542 accessions) (http://genebank.icrisat.org/IND/Core?Crop=-Finger%20millet). Other than the ICRISAT gene bank, small millets germplasm is also conserved in national gene banks across the world. Three gene banks *viz.* the National Bureau of Plant Genetic Resources, New Delhi, India, and the national active germplasm site at the All-India Coordinated Small Millet Improvement Project (AICSMIP), Bengaluru; the national gene bank of the Institute of Crop Sciences, Chinese Academy of Agricultural Sciences, Beijing, China; and the gene bank of the United State Department of Agriculture maintain more than 15,000, 22,000 and 4,000 small millets accessions, respectively. Most of these collections are dominated by accessions belonging to finger millet and foxtail millet (Saha et al., 2016). A survey of these collections indicates that although the hot spots of small millets are well represented, gaps exist for wild and weedy relatives belonging to secondary and tertiary gene pools; for example, wild and weedy species constitute less than 10% of finger millet and foxtail millet global germplasm collections (Upadhyaya and Vetriventhan, 2017). Therefore, it is necessary to investigate these gaps and augment these collections with wild and weedy relatives harboring genes for biotic and abiotic tolerance along with agronomic and nutritional traits.

Germplasm characterization and evaluation

Small millets germplasm contains significant variations for agro-morphological, quality and stress tolerance traits, and promising germplasm sources have been reported amongst them. A large-scale phenotypic characterization across the globe has revealed extensive genetic variations for traits of economic importance in germplasm collections. More successful utilization of these collections relies on their wide evaluation for agronomic traits in multiple locations exhibiting high genotype x environment interactions. However, multi-location evaluation of large collections is resource consuming and a constraint to obtain reliable phenotypic data for economic traits to identify trait specific accessions. In addition, for adequate assessment of grain quality and stress tolerance of a wide array of genetically diverse material, it is necessary to reduce the potential number of accessions requiring evaluation. Their use can be enhanced by developing a 'core collection' of reduced sample size, which represents the complete genetic spectrum of the base collection. By virtue of its reduced size, the core collection can be precisely evaluated for economic traits in replications under multiple locations for identification of trait-specific accessions. Germplasm diversity representative subsets (core and mini-core collections) have been developed and evaluated in small millets for agronomic and nutritional traits.

Agronomic traits

Small millets germplasm conserved globally exhibit a wide range of variations in yield and other important agronomic traits. Extensive genetic variations have been reported in core and mini-core collections of small millets developed at ICRISAT. The multi-location evaluation of finger millet core and foxtail millet mini-core collections in India led to the identification of diverse accessions with earlier maturity, higher tillering, larger grain size and higher grain yields than existing control cultivars (Upadhyaya et al., 2011a, b). Likewise, an evaluation of the barnyard millet core collection in the Himalayan region identified promising trait donors for earliness, basal tillering, smut resistance and grain yield (Sood et al., 2015). In proso millet and little millet, a wide variation for morphological traits and trait-specific sources for productivity have been observed (Vetriventhan and Upadhyaya, 2016, 2018).

Nutritional and nutraceutical traits

Core and mini-core collections of small millets have been characterized for grain micronutrients and protein contents. The finger millet mini-core collection has large variations for grain iron (1.71–65.23mg/kg), zinc (16.58–25.33 mg/kg), calcium (1.84–4.89 g/kg) and protein (6.00–11.09%) contents (Upadhyaya et al., 2011b). Likewise, substantial genetic variations for grain calcium (90.3–288.7 mg/kg), iron (24.1–68.2 mg/kg) and zinc (33.6–74.2 mg/kg) have been reported in the mini-core collection of foxtail millet (Upadhyaya et al., 2011a). Wide genetic variations for iron (41–73 mg/kg), zinc (26–47 mg/kg) and calcium (91–241 mg/kg) have been reported and trait-specific donors have been identified in global proso millet germplasm collections (Vetriventhan and Upadhyaya, 2016).

Biotic stresses

Among various biotic stresses, blast in finger millet and foxtail millet, grain smut and sheath blight in little millet and barnyard millet, and sheath blight and bacterial spot in proso millet are important diseases affecting their production

230 Kuldeep Singh et al.

worldwide. These diseases are economically very destructive and are reported to cause over 50% yield losses (Esele and Odelle, 1995). The identification of trait-specific robust resistant sources is the most economic disease-management strategy to overcome such constraints. Mini-core collections have proven to be excellent reservoirs of resistant sources against many of these diseases. In finger millet, a screening of mini-core collection in both field and glasshouse conditions identified 66 accessions with combined resistance to leaf, neck and finger blast (Babu et al., 2013). Similarly, 21 accessions exhibiting resistance to neck and head blast under field evaluation and 11 accessions exhibiting seedling leaf-blast resistance in controlled conditions were identified in the core collection of foxtail millet (Sharma et al., 2014). In barnyard millet, accessions of Japanese *E. esculenta* germplasm were reported to have a strong immune response to grain smut, while a multi-location screening of Indian *E. frumentacea* germplasm did not yield favorable results (Gupta et al., 2010a). However, some landraces of *E. frumentacea* from Karnataka and Bihar are reported to have low grain-smut resistance (Gupta et al., 2010b).

Abiotic stresses

Small millets are suggested crops for drought-hit, marginal and fragile ecosystems. Substantial genetic variations for drought (Krishnamurthy et al., 2016a, b) and salt tolerance (Krishnamurthy et al., 2014a, b) have been reported and trait-specific sources have been identified in mini-core collections of finger millet and foxtail millet. Finger millet accessions (IE5201, IE2871, IE7320, IE2034 and IE 3391) belonging to mini-core collections are reported to exhibit higher root and shoot growth at the seedling stage under low phosphorus stress (Ramakrishnan et al., 2017).

Development of elite cultivars towards germplasm improvement

The ex situ germplasm collections across the world have a huge impact in terms of genetic improvement for yield enhancement and tolerance to various biotic and abiotic stresses. Among the six subtypes, proso millet is the most important species where germplasm augmentation has a major impact on enhancing productivity through cultivar development in the USA (Habiyaremye et al., 2017). Out of the 15 proso millet cultivars, nine were direct selections from the adapted landraces preserved in the USDA gene bank (Rajput and Santra, 2016). Similarly, in Africa, where systematic breeding efforts are limited, elite landraces of finger millet have been released as cultivars in Tanzania (Upadhyaya and Vetriventhan, 2017). Likewise, in India, the active germplasm collection of AICRP-small millets, Bengaluru, has a major impact on promising cultivar development for different agro-climatic zones of the country. So far, a total of 272 varieties in the six small millets have been released in the country and a majority of them are the pure line selections from promising germplasm lines of indigenous and exotic origins. Apart from pure line selection, the utilization of promising germplasm lines in recombination breeding has resulted in many high-yielding cultivars and new plant types in small millets. In a report by Upadhyaya and Vetriventhan (2017), hybridization between trait-specific donors followed by pedigree selection has remained the main breeding method for foxtail improvement in China since 1970s, which has resulted in the development of more than 870 cultivars.

Towards the development of new plant types, isolation of awnless semi-dwarf genotypes of Japanese barnyard millet (*E. esculenta*) is a promising endeavor from the Vivekananda Parvatiya Krishi Anusandhan Sansthan (VPKAS) in Almora, India. These awnless derivatives with reduced plant height and great fodder potential are affirmative initiatives for improvement of barnyard millet (Sood et al., 2015).

Breeding advancements

Yield and yield-contributing parameters are the most targeted traits in the improvement of small millets. Germplasm collections exhibiting substantial variation for various traits, including maturity duration can be subjugated to breed tailored cultivars as per eco-geographical conditions (Upadhyaya et al., 2014; Vetriventhan and Upadhyaya, 2018; Vetriventhan and Upadhyaya, 2019). Hybridization to generate variability followed by assortment in segregating population has been a major breeding method in small millets especially in finger millet, foxtail millet and proso millet. In finger millet, 45% of the cultivars released in India were obtained from hybridization and selection breeding methods, followed by 22% in foxtail millet and 29% in proso millet (AICSMIP, 2014). Emasculation and crossing methods have also been reported in small millets (Gupta et al., 2011). However, the exploitation of hybrid vigor is limited in small millets due to difficulties in hybridization. Thus, developing male sterile lines would be a feasible alternative to use heterosis, which is being effectively implemented in major crops for commercial hybrid seed production.

Genomic resources

Trait-specific germplasm characterization is a primary requirement to identify genotypes contrasting for desirable traits, and genomic resources such as DNA markers, linkage maps and genome sequence are essential for gene tagging, gene mapping and marker-assistant selection for rapid crop-improvement. In small millets, genomes of foxtail millet, finger millet, proso millet, little and barnyard millet have been sequenced till date (Table 19.1). Foxtail millet has the smallest genome size (423 Mb) while finger millet has the largest one (1.5 Gb), followed by barnyard millet (1.27 Gb). However, the draft sequence information is adequately informative for large-scale genotyping applications and gene mining. In a study, the Solexa sequencing technology and the Genome Analyzer II were employed to re-sequence the genome of a foxtail millet landrace 'Shi-Li-Xiang' and to analyze its genetic structures (Bai et al., 2013). The results provided a rich tag library for future genetic studies and molecular breeding of foxtail millet and its related species. In barnyard millet, Wallace et al. (2015) genotyped the core collection using the genotyping-by-sequencing approach to investigate the patterns of population structure and phylogenetic relationships among the accessions. Further, most agronomically important traits are determined by several genes, which is called quantitative trait loci (QTLs). With the rapid development in sequencing technology, marker-assisted crop breeding has resulted in the acceleration of genetic improvement of small millets. Gimode et al. (2016) have identified single nucleotide polymorphism (SNP) and simple sequence repeat (SSR) markers in finger millet by next generation sequencing (NGS) technologies, using both Roche 454 and Illumina platforms. Wang et al. (2017) reported a high-density genetic map and QTL analysis of agronomic traits in foxtail millet using RAD-seq and identified 11 major QTLs for eight agronomic traits along with the development of five co-dominant DNA markers. A finger millet genotype ML-365 was also sequenced using Illumina and SOLiD sequencing technologies (Hittalmani et al., 2017). Later, Kumar et al. (2018) reported high iron and zinc content QTLs using diversity array technology (DArT) and SSRs markers to generate a genetic linkage map using a population of 317 recombinant inbred lines (RILs) in pearl millet. Many other studies have focused on drought tolerance and fungal diseases in pearl millet using marker-assisted selection approaches.

Although genetics and genomics have allowed for noteworthy progress in studying millets, transcriptome-based gene expression profiling has also contributed significantly. In this regard, Rahman et al. (2014) analyzed the salinity receptive leaf transcriptome of distinct finger millet genotypes and identified numerous differentially expressed genes in the tolerant genotype. Similarly, Parvathi et al. (2019) analyzed transcriptome dynamics in leaf samples of finger millet exposed to drought stress conditions. The foxtail millet transcriptome was the first to be released, where Zhang et al. (2012) sequenced the total RNA of root, stem, leaf and spike 'Zhang gu' strain. Recently, Li et al. (2020) sequenced the transcriptome of foxtail millet during *Sclerospora graminicola* infection, and identified many candidate genes for advanced functional characterization.

TABLE 19.1 List of sequenced genomes in small millets

Crop	Genome type	Genome size	No. of genes	Sequence type	Reference(s)
Finger millet	Nuclear	1.5 Gb	62,348	Draft	Hatakeyama et al. (2018)
Foxtail millet	Nuclear	510 Mb	24,000–29,000	Reference	Bennetzen et al. (2012)
	Nuclear	423 Mb	38,801	Draft	Zhang et al. (2012)
Barnyard	Nuclear	1.27 Gb	108,771	Draft	Guo et al. (2017)
millet	Chloroplast	139,851 bp	111	–	Sebastin et al. (2019)
Proso millet	Nuclear	923 Mb	55,930	Draft	Zou et al. (2019)
	Chloroplast	139,929 bp	132	–	Cao et al. (2017)
Little millet	Chloroplast	139,384 bp	125	–	Sebastin et al. (2018)

Further, comprehensive protocols for callus regeneration and transgenics have been reported for all small millet species except little millet (Plaza-Wüthrich and Tadele, 2012). The *Agrobacterium*-mediated system has been predominantly used in small millets transformation as reported in studies by Bayer et al. (2014) and Li et al. (2017). In addition, the availability of standardized protocols for genetic transformation has resulted in the application of a transgene-based approach in a few small millets. The use of clustered, regularly interspaced short palindromic repeat (CRISPR)/Cas mutagenesis was reported in foxtail millet (Lin et al., 2018), where a mutation of the foxtail millet PDS gene by CRISPR/Cas9 system has been achieved through protoplast transfection.

Conclusion and future projections

Small millets have the potential to serve as an alternate to major crops because of their diverse adaptation to adverse conditions and their great nutritional qualities. They fit very well into multiple cropping systems both under irrigated and rainfed conditions. The major bottleneck, however, is inadequate information on their genetic diversity, which limits their effective utilization in crop improvement programs. Therefore, effective germplasm collection is the effective solution to categorize trait-specific resources, which can be utilized in small millets breeding programs. The yield barrier in small millets can be surpassed by a male sterility system and by employing hybrid vigor and genomics-assisted crop improvement. Further, NGS techniques can facilitate the molecular characterization of small millets germplasm along with RNA interference and genome editing tools like CRISPR/Cas, which reduce antinutrients of small millets to make them more useful to feed growing population. Moreover, a coordinated multidisciplinary approach involving farmers and public–private partnerships is needed to speed up research and development programs in small millets. With the ever-changing climate scenario and prevailing conditions of hidden hunger, greater research and developmental focus on small millets is the key to achieving future nutrition security.

References

AICSMIP (2014) 'Report on compendium of released varieties in small millets', Bangalore, India.

Babu, T.K., Thakur, R.P., Upadhyaya, H.D., Reddy, P.N., Sharma, R., Girish, A.G. and Sarma, N.D.R.K. (2013) 'Resistance to blast (Magnaporthe grisea) in a mini-core collection of finger millet germplasm', *Eur J Plant Pathol*, vol 135, pp299–311.

Bai, H., Cao, Y., Quanm, J., Dong, L., Li, Z., Zhu, Y., et al. (2013) 'Identifying the Genome-Wide Sequence Variations and Developing New Molecular Markers for Genetics Research by Re-sequencing a Landrace Cultivar of Foxtail Millet', *PLoS One*, vol 8, ppe73514.

Bayer, G.Y., Yemets, A.I. and Blume, Y.B. (2014) 'Obtaining the transgenic lines of finger millet *Eleusine coracana* (L.) with dinitroaniline resistance', *Cytol Genet*, vol 48, pp139–144.

Bennetzen, J.L., Schmutz, J., Wang, H., Percifield, R., Hawkins, J., Pontaroli, A.C., et al. (2012) 'Reference genome sequence of the model plant *Setaria*', *Nat Biotechnol*, vol 30, pp555–561.

Cao, X., Wang, J., Wang, H., Liu, S., Chen, L., Tian, X., et al. (2017) 'The complete chloroplast genome of Panicum miliaceum', *Mitochondrial DNA Part B*, vol 2, pp43–45.

Esele, J.P.N. and Odelle, S.E. (1995) 'Progress in breeding for resistance to finger millet blast at Serere research station', in *Proceedings of the 8th EARSAM Regional Workshop on Sorghum and Millets* (30 Oct–5 Nov 1992), Wad Medani, Sudan, pp93–99.

Fita, A., Rodríguez-Burruezo, A., Boscaiu, M., Prohens, J. and Vicente, O. (2015) 'Breeding and domesticating crops adapted to drought and salinity: A new paradigm for increasing food production', *Front Plant Sci*, vol 6, pp978.

Gimode, D., Odeny, D.A., de Villiers, E.P., Wanyonyi, S., Dida, M.M., Mneney, E.E., et al. (2016) 'Identification of SNP and SSR markers in finger millet using next generation sequencing technologies', *PLoS One*, vol 11, ppe0159437.

Guo, L., Qiu, J., Ye, C., Jin, G., Mao, L., Zhang, H., et al. (2017) 'Echinochloa crus-galli genome analysis provides insight into its adaptation and invasiveness as a weed', *Nat Commun*, vol 8, pp1031.

Gupta, A., Joshi, D., Mahajan, V. and Gupta, H.S. (2010b) 'Screening barnyard millet germplasm against grain smut (*Ustilago panici-frumentacei* Brefeld)', *Plant Genet Resour*, vol 8, pp52–54.

Gupta, A., Mahajan, V. and Gupta, H.S. (2010a) 'Genetic resources and varietal improvement of small millets for Indian Himalayas', in L.M. Tewari, Y.P.S. Pangtey and G. Tewari (eds) *Biodiversity Potentials of the Himalaya*. Gyanodaya Prakashan, Nainital, India, pp305–316.

Gupta, A., Sood, S., Kumar, P., Jagdish, A. and Bhatt, C. (2011) 'Floral biology and pollination system in small millets floral biology and pollination system in small millets', *Eur J Plant Sci Biotechnol*, vol 6, pp80–88.

Habiyaremye, C., Matanguihan, J.B., Alpoim Guedes, J.D., Ganjyal, G.M., Whiteman, M.R., Kidwell, K.K. and Murphy, K.M. (2017) 'Proso millet (*Panicum miliaceum* L.) and its potential for cultivation in the Pacific North-West, US: A review', *Front Plant Sci*, vol 7, pp1961.

Hatakeyama, M., Aluri, S., Balachadran, M.T., Sivarajan, S.R., Poveda, L., Shimizu-inatsugi, R., et al. (2018) 'Multiple hybrid de novo genome assembly of finger millet, an orphan allotetraploid crop', *DNA Res*, vol 25, pp39–47.

Hittalmani, S., Mahesh, H., Shirke, M.D., Biradar, H., Uday, G., Aruna, Y., et al. (2017) 'Genome and transcriptome sequence of finger millet (*Eleusine coracana* (L.) Gaertn.) provides insights into drought tolerance and nutraceutical properties', *BMC Genom*, vol 18, pp465.

Krishnamurthy, L., Upadhyaya, H.D., Gowda, C.L.L., Kashiwagi, J., Purushothaman, R., Singh, S. and Vadez, V. (2014a) 'Large variation for salinity tolerance in the core collection of foxtail millet (*Setaria italica* (L.) P. Beauv.) germplasm', *Crop Pasture Sci*, vol 65, pp353–361.

Krishnamurthy, L., Upadhyaya, H.D., Purushothaman, R., Gowda, C.L.L., Kashiwagi, J., Dwivedi, S.L., et al. (2014b) 'The extent of variation in salinity tolerance of the mini core collection of finger millet (*Eleusine coracana* L. Gaertn.) germplasm', *Plant Sci*, vol 227, pp51–59.

Krishnamurthy, L., Upadhyaya, H.D., Kashiwagi, J., Purushothaman, R., Dwivedi, S.L. and Vadez, V. (2016a) 'Variation in drought-tolerance components and their interrelationships in the core collection of foxtail millet (*Setaria italica*) germplasm', *Crop Pasture Sci*, vol 67, pp834–846.

Krishnamurthy, L., Upadhyaya, H.D., Kashiwagi, J., Purushothaman, R., Dwivedi, S.L. and Vadez, V. (2016b) 'Variation in drought-tolerance components and their interrelationships in the mini core collection of finger millet germplasm', *Crop Sci.*, vol 56, pp1914–1926.

Kumar, S., Hash, C.T., Nepolean, T., Mahendrakar, M.D., Satyavathi, C.T., et al. (2018) 'Mapping grain iron and zinc content quantitative trait loci in an iniadi-derived immortal population of pearl millet', *Genes*, vol 9, pp248.

Li, J., Dong, Y., Li, C., Pan, Y. and Yu, J. (2017) 'SiASR4, the target gene of SiARDP from *Setaria italica*, improves abiotic stress adaption in plants', *Front Plant Sci*, vol 7, pp2053.

Li, R., Han, Y., Zhang, Q., Chang, G., Han, Y., Li, X., et al. (2020) 'Transcriptome profiling analysis reveals co-regulation of hormone pathways in foxtail millet during *Sclerospora graminicola* infection', *Int J Mol Sci*, vol 21, pp1226.

Lin, C.S., Hsu, C.T., Yang, L.H., Lee, L.Y., Fu, J.Y., Cheng, Q.W., et al. (2018) 'Application of protoplast technology to CRISPR/Cas9 mutagenesis: from single-cell mutation detection to mutant plant regeneration', *Plant Biotechnol J*, vol 16, pp1295–1310.

Parvathi, M.S., Nataraja, K.N., Reddy, Y.A.N., Naika, M.B.N. and Gowda, M.V.C. (2019) 'Transcriptome analysis of finger millet (*Eleusine coracana* (L.) Gaertn.) reveals unique drought responsive genes', *J Genet*, vol 98, pp46.

Plaza-Wüthrich, S. and Tadele, Z. (2012) 'Millet improvement through regeneration and transformation', *Biotechnol Mol Biol Rev*, vol 7, pp48–61.

Rahman, H., Jagadeeshselvam, N., Valarmathi, R., Sachin, B., Sasikala, R., Senthil, N., et al. (2014) 'Transcriptome analysis of salinity responsiveness in contrasting genotypes of finger millet (*Eleusine coracana* L.) through RNA-sequencing', *Plant Mol Biol*, vol 85, pp485–503.

Rajput, S.G. and Santra, D.K. (2016) 'Evaluation of genetic diversity of proso millet germplasm available in the United States using simple sequence repeat markers', *Crop Sci*, vol 56, pp1–9.

Ramakrishnan, M., Ceasar, S.A., Vinod, K.K., Duraipandiyan, V., Ajeesh Krishna, T.P., Upadhyaya, H.D., Al Dhabi, N.A. and Ignacimuthu, S. (2017) 'Identification of putative QTLs for seedling stage phosphorus starvation response in finger millet (*Eleusine coracana* (L.) Gaertn.) by association mapping and cross species synteny analysis', *PLoS One*, vol 12, ppe0183261.

Saha, D., Channabyre Gowda, M.V., Arya, L., Verma, M. and Bansal, K.C. (2016) 'Genetic and genomic resources of small millets', *Crit Rev Plant Sci*, vol 35, pp56–79.

Saxena, R., Vanga, S.K., Wang, J., Orsat, V. and Raghavan, V. (2018) 'Millets for food security in the context of climate change: A review', *Sustainability*, vol 10, pp2228.

Sebastin, R., Lee, G., Lee, K.J., Shin, M., Lee, J., Ma, K., et al. (2018) 'The complete chloroplast genome sequences of little millet (*Panicum sumatrense* Roth ex Roem. and Schult.) (Poaceae)', *Mitochondrial DNA Part B*, vol 3, pp719–720.

Sebastin, R., Lee, K.J., Cho, G., Lee, J., Kim, S., Lee, G., et al. (2019) 'The complete chloroplast genome sequence of Japanese millet *Echinochloa esculenta* (A. Braun) H. scholz (Poaceae)', *Mitochondrial DNA Part B*, vol 4, pp1392–1393.

Sharma, R., Girish, A.G., Upadhyaya, H.D., Humayun, P., Babu, T.K., Rao, V.P. and Thakur, R.P. (2014) 'Identification of blast resistance in a core collection of foxtail millet germplasm', *Plant Dis*, vol 98, pp519–524.

Sood, S., Khulbe, R., Gupta, A.K., Agrawal, P.K., Upadhyaya, H.D. and Jagdish, B.C. (2015) 'Barnyard millet: a potential food and feed crop of future', *Plant Breed*, vol 134, pp135–147.

Vetriventhan, M. and Upadhyaya, H.D. (2016) 'Little millet, *Panicum sumatrense*: an under-utilized multipurpose crop', *1st International Agrobiodiversity Congress* (6–9 November 2016), New Delhi, pp356.

Vetriventhan, M. and Upadhyaya, H.D. (2018) 'Diversity and trait-specific sources for productivity and nutritional traits in the global proso millet (*Panicum miliaceum* L.) germplasm collection', *Crop J*, vol 6, pp451–463.

Vetriventhan, M. and Upadhyaya, H.D. (2019) 'Variability for productivity and nutritional traits in germplasm of kodo millet, an underutilized nutrient-rich climate smart crop', *Crop Sci*, vol 59, pp1095–106.

Upadhyaya, H.D., Dwivedi, S.L., Singh, S.K., Singh, S., Vetriventhan, M. and Sharma, S. (2014) 'Forming core collections in barnyard, kodo, and little millets using morphoagronomic descriptors', *Crop Sci*, vol 54, pp2673–2682.

Upadhyaya, H.D., Ravishankar, C.R., Narasimhudu, Y., Sarma, N.D.R.K., Singh, S.K., Varshney, S.K., et al. (2011a) 'Identification of trait specific germplasm and developing a mini core collection for efficient use of foxtail millet genetic resources in crop improvement', *Field Crop Res*, vol 124, pp459–467.

Upadhyaya, H.D., Ramesh, S., Sharma, S., Singh, S.K., Varshney, S.K., Sarma, N.D.R.K., et al. (2011b) 'Genetic diversity for grain nutrients contents in a core collection of finger millet (*Eleusine coracana* (L.) Gaertn.) germplasm', *Field Crop Res*, vol 121, pp42–52.

Upadhyaya, H.D. and Vetriventhan, M. (2017) 'Underutilized Climate-Smart Nutrient Rich Small Millets for Food and Nutritional Security', in *Regional Expert Consultation on Underutilized Crops for Food and Nutritional Security in Asia and the Pacific*, Asia-Pacific Association of Agricultural Research Institutions (APAARI), Bangkok, Thailand, pp109–120.

Wallace, J.G., Upadhyaya, H.D., Vetriventhan, M., Buckler, E.S., Hash, C.T. and Ramu, P. (2015) 'The genetic makeup of a global barnyard millet germplasm collection', *Plant Genome*, vol 8, pp1–7.

Wang, J., Wang, Z., Du, X., Yang, H., Han, F., Han, Y., et al. (2017) 'A high-density genetic map and QTL analysis of agronomic traits in foxtail millet [*Setaria italica* (L.) P. Beauv.] using RAD-seq', *PLoS One*, vol 12, ppe0179717.

Zhang, G., Liu, X., Quan, Z., Cheng, S., Xu, X., Pan, S., et al. (2012) 'Genome sequence of foxtail millet (Setaria italica) provides insights into grass evolution and biofuel potential', *Nat Biotechnol*, vol 30, pp549–54.

Zou, C., Li, L., Miki, D., Li, D., Tang, Q., Xiao, L., et al. (2019) 'The genome of broomcorn millet', *Nat Commun*, vol 10, pp436.

20
CHALLENGES TO CONSERVING MILLETS IN ECOLOGICALLY SENSITIVE AREAS

Amit Mitra

Introduction

Globally, there is a re-emphasis on the revival and conservation of neglected and underutilized species (NUS), including small millets. Such crops are seen as alternatives to rice and wheat, especially for the poor in the peripheral areas of the Global South (Muthamilarasan and Prasad, 2020). Millets require much less water than rice and wheat and are ideally suited for upland rainfed regions.

Can a species, including small millets, be sustainably conserved without preserving the ecosystem in which it is cultivated? This chapter argues that it is important to consider both the natural and social ecosystems in which these NUS are cultivated as well as the interaction between social and natural systems. It draws on and supplements the findings of an earlier work (Mitra and Rao, 2019) in Koraput district of Odisha.

Koraput had a tribal population of 50.6% in the 2011 census. It is a district where the 'Special Programme for Promotion of Millets in Tribal Areas of Odisha,' also called the Odisha Millet Mission (OMM) has been implemented since 2017 (NCDS, 2019). Small millets are typically cultivated in the rainfed uplands of ecologically sensitive areas of the district. The agrarian landscape, along water flows, is rather complex, but millet cultivation using natural fertilisers like dung and crop residue mulch has been critical to the growing of indigenous rice varieties in the lowlands. These practices were evolved over centuries by tribal women and men, with women playing a critical role in the maintenance and reproduction of these systems. Women decided the species, the area to be planted, household consumption and sale specifics but did not plough or sow. Crop diversity, including mixed cropping on the same plot, was practiced. Despite the existence of the concept of private property in land, the ecosystem

DOI: 10.4324/9781003044802-23

was collectively owned. The overall productivity of the ecosystem, instead of individual crops, was focused on.

Relations between men and women were of asymmetric mutuality. While hierarchies have always existed, today some of these hierarchies have grown and are perhaps behind the near breakdown of the social systems that nurtured NUS.

This chapter argues that with the rapid usurpation of the uplands in the Koraput district for corporate eucalyptus plantations, the ecosystem (natural and social) has transformed drastically. The landraces of paddy and small millets reduced sharply between the 1950s and the 2000s. Only eight small millet landraces were extant in 2000 (Mishra, 2009). The soil quality in the area has deteriorated (Bannerjee, 2015). Long-term field observations reveal a drop in the cattle population.

Small millets are essentially a woman's crop. Eucalyptus, in the domain of men, has skewed gender relations in these areas. The uplands in which small millets were cultivated are now under eucalyptus plantations (Mitra and Rao, 2019). Community relations are giving way to individualisation and increasing wage labour. These transformations are to the detriment of both the NUS and the people belonging to a range of tribes. With rising inequalities, distress migration is the order of the day (Mitra and Rao, 2019).

Small millets

India has the third largest area under small millets cultivation in the world. Cultivated small millets comprise six species – finger millet (*Eleusine coracana*), little millet (*Panicum sumatrense)*, Italian or foxtail millet (*Setaria italica)*, barnyard millet (*Echinochloa crus-galli)*, proso millet (*Panicum miliaceum*) and kodo millet (*Paspalum scrobiculatum*). Taken together, they are grown over approximately two million hectares, mostly in semi-arid, hilly and mountainous regions. Finger millets predominate, comprising 60% of the area and 70% of the production (Bala Ravi et al., 2010). These are hardy crops and are quite resilient to agro-climatic shocks like diminished rainfall and falling soil fertility. They are important in local food cultures, being an important source of food and beverages. The straw is valuable fodder. Nutritionally, they are rich in micronutrients, especially calcium and iron, and are high in dietary fibre. Their grain protein is richer in sulphur and other essential amino acids than all other major cereals (Bala Ravi et al., 2010).

Evolution of the agrarian system

Hills and forests interspersed with numerous streams and rivers are a part of the landscape of the Koraput district, Odisha. Over centuries, local tribes have transformed the landscape into productive systems, both private and public, to meet their food and housing needs. A unique cropping pattern, the *Myda system*, which involves mixing the seeds of long- and short-duration paddy varieties in the same plot, and nurturing them through carefully monitoring the levels of water and vegetative growth, evolved (Mishra, 2009). Land-use patterns developed along water flows, enabling optimal and sustainable use of the environment.

The present day Koraput district was a part of the Jeypore princely state until 1947. Disruptions in local indigenous land-use patterns began in the mid-nineteenth century with the exploitation of forests by the colonial state's forest department. The British extended the Madras Forest Act to Jeypore state in 1891, reservation began in 1900, and protected forests were demarcated in 1916. By 1940, reserved area quantity rose five times, leading to a substantial increase in state revenues (Bell, 1945: 101). Critically, a lot of the village commons and even the uplands were usurped as protected forests and were planted with teak. This impacted adversely the cultivation of NUS and reduced the availability of uncultivated crops (Mitra and Rao, 2019).

Post-Independence, the Forest Conservation Act, 1980, and the Wildlife Protection Act, 1972, further changed land-use patterns by halting shifting cultivation, and also enhanced usurpation of village commons.

The local people, however, classify forests into two types: *bon* and *jangal*. *Bon* corresponds to the protected forests and *jangal* to the reserved. Most subsistence needs like fuel, roots, tubers and leaf litter come from the *bon*. Cattle also graze there. The *bon* are the commons upslope of the habitat. Above the *bon* lies the *jangal*, often consecrated in the name of a deity. There are strict rules about the use of usufructs of these sacred groves but the trees cannot be cut, a norm often violated by the Forest Department for commercial plantations, including of eucalyptus (Mitra and Pal, 1994).

The forest department fells trees from the *bon* too, clearing the land for eucalyptus plantations and changing traditional land-use patterns (Mitra and Rao, 2019). The undulating lands between and along the forests and the settlement has been shaped into four types of croplands: uplands ([*dongar*] on the hill slopes), midlands (*bhettabeda*), lowlands (*khalbeda*), and terraces (*jholas*, that start from the *dongar* and go down to the stream below) (Mishra, 2009). This complex land-use pattern has been developed over centuries through human–nature interactions. It is driven by household consumption needs, maintenance of soil fertility, sustainable water use and safeguarding against the vagaries of nature. Local communities nurture and conserve water by harvesting rainwater, diverting river flows into farm ponds during the monsoons, cultivating varieties – especially of paddy – suited to water availability and not using chemical fertilisers in the *dongar*, to avoid polluting the ground water aquifers (Mitra and Rao, 2019) The leaf litter from the *bon* above the *dongar* fertilises the crops grown there. This underscores the need to consider crop lands and forest lands as parts of an integrated ecosystem essential for the conservation of NUS.

In 2006, the Government of India enacted the Scheduled Tribes and Other Traditional Forest Dwellers (Recognition of Forest Rights) Act that acknowledged that 'traditional forest dwellers' are integral to the very survival and sustainability of the forest ecosystem. By recognising their customary rights, this Act sought to ensure livelihoods and food security, while empowering the people to sustainably use and manage forest biodiversity (Deb et al., 2014). In districts like Koraput, managing forest biodiversity is essential to sustainably managing agrobiodiversity, including that of NUS. At the end of a decade, while close to 78% of individual claims to forestland were met in Koraput district, with about 27,628 households receiving an acre of land each, only 9% of community land claims had been granted. Gaining legal legitimacy for using what they have seen as their land – and, indeed, life – has not been easy (CFR-LA, 2016). The central government's recent enactment of the Compensatory Afforestation Fund Act 2016 appears to be diluting the very intent of the 2006 legislation. Apart from bypassing locally elected bodies and community forest-management groups, and being dominated by the forest bureaucracy, civil society actors fear it will both displace forest dwellers and promote large-scale commercial eucalyptus plantations with little benefit to the local people (Ibid). This reflects a further marginalisation of collective responsibility for social reproduction and care of the environment, and a focus on capital accumulation and individual profits.

While the regulations, as well as state interventions to improve agriculture, did not consider the interconnections and distinctions between different types of land and their contributions to food and consumption needs, cultivation practices and the crops grown remain sensitive to the local landscape. Thus, the *dongar* lands are used to grow a mix of millets, pigeon pea, cowpea, niger, turmeric, sesame and arums, high in nutrient content and not water-intensive, primarily for domestic consumption. The duration of these crops differ. Some are shallow rooted while others are deep rooted. Each crop has an attached ritual. Thus, traditional land-use patterns meet cultural, spiritual, consumption and ecological needs, instead of emphasising the productivity of a single crop (Niyogi, 2020). In the field, it is observed that the nature of the soil and the slope, and whether the plot is on the windward or the leeward side are important determinants of the crop mix and even the landraces of a particular crop. However, not much research has been done in this area.

Men support the processes of land preparation and transportation of the harvest, but the cultivation is mainly managed by women. In fact, *dongar* lands are often considered women's plots, wherein they control decisions around crop choices and use, including income from sales, if any (Mitra and Rao, 2019).

Paddy is cultivated in the *bhettabeda* and *khalbeda* and if water is available, so is a second winter crop of pulses or vegetables. They are considered household plots but fall under male control. Both men and women have to contribute labour to the cultivation of the basic 'household' staple in one season, but can cultivate their own 'individual' crops in the second season to fulfil their shared and separate responsibilities towards household provisioning (Ibid).

The challenges of eucalyptus plantations

Eucalyptus plantations started in Koraput in the 1960s. In 1990, the Jeypore-based JK Paper Mills launched a farm forestry programme under which it added 7,000 hectares annually by distributing over 40 million saplings to farmers (https://jkplantation.wordpress.com/about-jk-paper-ltd/jaykay-paper-mills-ltd-rayagada/). In 2003, the paper industry, through a front NGO under a corporate social responsibility scheme, secured consent from the tribals to plant eucalyptus on their uplands with assured buyback and lump sum payments upon each successive harvest (in the fourth, eighth and twelfth year after planting). Farmers do not have to invest initially as the advances by the company are deducted from the first payment. A study of 2,004 households who planted eucalyptus on 3,360 acres of land showed gross earnings per acre to be Rs 86,000, Rs 150,000 and Rs 82,500 (at the rate of Rs 4,500/metric tonne) for the three successive harvests (Mahana, 2014).

The rapid conversion of land, especially *dongar,* to eucalyptus plantations, has led to a loss of food, especially protein and micronutrients (from millets and vegetables). Soil humus and fertility is likely to decline, as are the water flows, given the displacement of traditional systems of fuel, fodder and water management (NABARD, n.d.; Stanturf et al., 2013). Planting eucalyptus on *dongar* lands impacts fields downstream as the high synthetic fertiliser inputs affect the organic cultivation of aromatic landraces of paddy, including the *kalazeera* rice for which the district is famous.

Apart from education and house construction, a major reason for growing eucalyptus is the desire for large amounts of ready cash. Cropping patterns earlier adhered to community decisions; however, now each household situates the decision to cultivate eucalyptus within the dynamics of their own household structure and its needs, reflecting a transition to a society based on individualism rather than reciprocity.

Yet, some women and men are resisting the commoditisation of their land and labour. While hired wage labour is now widespread in the area, the Parojas and Gadabas still continue the *Palli* (exchange) system. Men have succumbed more easily to both the commodification of their land, seen in the rapid shift to eucalyptus planting, and labour, pushed by the growing need for cash in an increasingly neoliberal, market-driven economy. Migration, still rare amongst the Gadabas, is now rising amongst most tribes, especially in the lean, summer season, prior to the start of paddy cultivation. However, migrant work is effort-intensive, with little rest and restricted diets. The men often return home unwell and are unable to contribute much to household activities. Their absence and subsequent return only enhances the burdens on women (Mitra and Rao, 2019).

Other challenges

Corporate plantations have led to the individualisation of the ecosystem. These plantations on the uplands have consequences for the entire cropping pattern, as well as for social relations. This has been accompanied by the introduction of mono-cropping in contrast to the old mixed-cropping practices led by women. To promote millet cultivation, contemporary state- and NGO-led efforts seek to promote the mono-cropping of finger millets, which has led to a decline of, for instance, vegetable cultivation by women. Thus, in Kadaguda village (name changed) even five years ago, the women grew vegetables and various greens in small plots behind their homestead. Now finger millets are grown in these plots, forcing households to cut down on their vegetable consumption. This is happening in many villages in the area.

Although rice was the major staple consumed by the people, there were many varieties. Millets were a major food supplement. With the distribution of highly subsidised rice and wheat by the state, the consumption of millets is being significantly reduced (Bala Ravi et al., 2010).

Conservation for whom?

There has been a recent spike in urban demand across the country for small millets as health foods (Kulkarni, 2018). It is important to ask whether and to what extent the thrust on conserving and reviving the cultivation of small millets, especially in the mission mode, benefits small and marginal tribal farmers of districts like Koraput. The OMM baseline survey of the Koraput district, in 2017, found that millets were being cultivated in 1,311.9 ha land by 2,605 millet-cultivating households of the seven blocks surveyed. Finger millets comprised 87.5% of this area and 12.5% was under small millets. Interestingly, 95% of these households consumed millets in summer, 90.7% in winter and 80.9% during the monsoons. The dip in consumption during the monsoons is probably due to a decline in household stocks (NCDS, 2019).

The survey does not mention the crop mixes nor the kind of land in which a particular type of millet is cultivated and by whom. No doubt a lot of the millets will be sold in markets by the men, impacting the villagers' already poor dietary diversity (Mitra and Rao, 2017). The OMM's practices are essentially mono-cropping of finger millets with much higher external inputs, without considering the social and natural ecosystem (Sood and Jishnu, 2015; Niyogi, 2020) This is already furthering inequalities, including of gender, and will lead to the loss of agrobiodiversity and Indigenous knowledge.

Conclusion

Small millet cultivation by tribals in districts like Koraput formed an integral part of the Indigenous natural resource management system, involving forests, uplands, midlands, low lands and water flows. Based on exchange labour systems grounded in reciprocity and mutuality between women and men, small millet cultivation involved an element of individual choice that did not adversely affect the welfare of others.

The first blow to this equilibrium came with the introduction of corporate eucalyptus plantations in the uplands. Market penetration individualising the economy and breaking down both the social and natural ecosystems has had deleterious consequences for NUS.

Under these circumstances, efforts to increase the production of small millets and other NUS through a 'kit' approach, of governments giving beneficiary farmers free kits that include fertilisers, bio-fertilisers, micronutrients, seed treatments and plant protection chemicals (Sood and Jishnu, 2015), will have tremendous social and ecological costs. From the policy point of view, it might be more productive to adopt an interdisciplinary, research-based approach that seeks to conserve the ecosystem and the cultures of which these crops are an essential part.

References

Bala Ravi, S., Swain, S., Sengotuvel, D. and Parida, NR. (2010) 'Promoting Nutritious Millets for Enhancing Income and Improved Nutrition: A Case Study from Tamil Nadu and Orissa', in Bhag Mal, Padulosi, S. and Bala Ravi, S. (eds) *Minor Millets in South Asia – Learnings From IFAD-NUS Project in India and Nepal*, Bioversity International and MSSRF, pp19–46. (https://www.bioversityinternational.org/fileadmin/_migrated/uploads/tx_news/Minor_millets_in_South_Asia_1407.pdf).

Bannerjee, K. (2015) '1. Soil Resource Mapping of Koraput District of Odisha 2. LULC of Bhitarkanika Mangrove Ecosystem,' Paper presented to the National Conference in Recent Advancements in Civil and Environmental Engineering, 28–29 November, Conference Proceedings published by International Journal of Engineering and Technology.

Bell, R.C.S. (1945) *Orissa District Gazetteer: Koraput*. Government Press, Cuttack.

CFR-LA (2016). 'Promise and Performance: Ten Years of the Forest Rights Act in India. Citizens', Report on Promise and Performance of the Scheduled Tribes and Other Traditional Forest Dwellers (Recognition of Forest Rights) Act, 2006, after 10 Years of Its Enactment. December 2016. Produced as Part of Community Forest Rights-Learning and Advocacy Process (CFR-LA), India. (www.cfrla.org.in).

Deb, D., Kuruganti, K., Rukmini, V. and Yesudas, S. (2014) *Forests as Food Producing Habitats,* Living Farms, Bhubaneswar.

Kulkarni, V. (2018) 'From Green Revolution to Millet Revolution', *The Hindu Business Line,* March 26 (https://www.thehindubusinessline.com/specials/india-file/from-green-revolution-to-millet-revolution/article23356997.ece#).

Mahana, R. (2014) 'Cultivating *Eucalyptus*: Practising Agro-Forestry in Wastelands for Alleviating Poverty in Koraput, Odisha', Unpublished mimeo.

Mishra, S. (2009) 'Farming System in Jeypore Tract of Orissa, India', *Asian Agri-History,* Vol 13, No 4 pp.271–292.

Mitra, A. and Pal, S. (1994). 'The Spirit of the Sanctuary', *Down to Earth,* January 31

Mitra, A. and Rao, N. (2017) 'Gender Differences in Adolescent Nutrition: Evidence from Two Indian Districts', *LANSA Working Paper Series,* Vol 2017, No 13 pp.1–34.

Mitra, A. and Rao, N. (2019). 'Contract Farming, Ecological Change and the Transformations of Reciprocal Gendered Social Relations in Eastern India', *The Journal of Peasant Studies.* DOI: 10.1080/03066150.2019.1683000.

Muthamilarasan, M. and Prasad, M. (2020). 'Small Millets for Enduring Food Security Amidst Pandemics,' *Trends in Plant Science* (https://www.sciencedirect.com/science/article/pii/S1360138520302557).

NABARD. (no date) 'Eucalyptus Myths and Realities' (https://www.nabard.org/english/forestry_eucalyptus.aspx)

NCDS (2019) *Baseline Survey: Koraput District – 2016–17, Phase 1, Special Programme for Promotion of Millets in Tribal Areas of Odisha, Odisha Millet Mission*, Nabakrushna Choudhury Centre for Development Studies, Bhubaneswar.

Niyogi, D.G. (2020) 'India's Millets Policy: Is It Headed in the Right Direction?' *Mongabay*, July 27 (https://india.mongabay.com/2020/07/indias-millets-policy-is-it-headed-in-the-right-direction/).

Stanturf, J.A. Vance, E.D., Fox, T.R. and Kirst, M. (2013) 'Eucalyptus beyond Its Native Range: Environmental Issues in Exotic Bioenergy Plantations', *International Journal of Forestry Research*. DOI: 10.1155/2013/463030.

Sood, J. and Jishnu, L. (2015) 'Contaminating Millets,' *Down To Earth* (https://www.downtoearth.org.in/coverage/contaminating-millets-38889).

21

COMMUNITY-CENTRED VALUE-CHAIN DEVELOPMENT OF NUTRI-MILLETS

Challenges and best practices in India

E.D. Israel Oliver King, Karthikeyan Muniappan, Prashant K Parida, Sharad Mishra, Kumar Natarajan, Melari Shisha Nongrum, Manjula Chinnadurai, Nirmalakumari, Carl O. Rangad, Somnath Roy, Gennifer Meldrum and Stefano Padulosi

Introduction

Though humans have utilized a vast number of plant species for food (Ulian et al., 2020), today, food security is confined to very few major crops, such as maize, wheat and rice and a few other commodity crops. This has led to the marginalization of many nutritious crops, including 'neglected and underutilized species' (NUS) (see Chapter 1), among which the millet species belonging to the genera *Eleusine, Panicum, Setaria* and *Paspalum,* continue to be cultivated and consumed by smallholder families inhabiting marginal environments.

Landraces of millets are preferred by farm families, but not solely for their nutritional qualities. They are appreciated also because of their inherent resilience to withstand the vagaries of weather as well as pest and disease attacks (Bala Ravi, 2004; Nagarajan and Smale, 2007). These crops are well suited to smallholder agro-ecology and are an important part of local food cultures. Conventional millet farming practices include mixed cropping and crop rotation, and are commonly subsistence oriented. In the past, the selection, preservation and exchange of seeds were popular practices encountered across farming communities. For instance, labour exchange practices within kinship units and communities were conventionally used to minimize cultivation costs, for traditional crops like millets also. The role of custodian families has also been strategic for the conservation and management of landraces and traditional knowledge associated with NUS crops.

In India, small millets are known for their adaptability to marginal and drought-prone areas. Compared to rice, these species contain higher levels of minerals, fibre, vitamins and many essential amino acids (Hulse et al., 1980; Malleshi and Desikachar, 1986). However, in recent decades, developmental

DOI: 10.4324/9781003044802-24

interventions, inadequate processing technologies and market forces have resulted in the decline of on-farm agro biodiversity, aggravating the already fragile conditions of local communities (Gruere et al., 2009). Other constraints include low productivity, lack of good quality seeds and disorganized value chains. Nevertheless, about two million farm households across India still depend primarily on these species for their livelihoods. Furthermore, climate change scenarios and the global food price crisis have brought to the forefront the strategic role that these crops can play in strengthening resilience in agricultural production systems and mitigating the shocks caused by shortages in major cereals that may occur due to various reasons.

The case studies in this chapter highlight some key research and development (R&D) interventions carried out (Plate 21.1.1–21.1.6) in the Indian states of Tamil Nadu, Odisha, Andhra Pradesh, Karnataka, Madhya Pradesh, Meghalaya, Uttarakhand and Jharkhand, since 2000, thanks to the support of various national and international agencies.

Case study 1: integrated value-chain development in Kolli Hills, Tamil Nadu

In the last three decades, Kolli Hills in the Eastern Ghats of Tamil Nadu has witnessed land-use changes due to introduction of cash crops like cassava, coffee and pepper. This fact, accompanied by the expansion of roads and food subsidies, has greatly affected the cultivation and use of millets. Furthermore, the low profitability of millets vis-à-vis cassava, the weakening of traditional seed systems, the drudgery in millet processing and their poor market opportunities have also contributed to relegating these highly nutritious and resilient crops to smaller areas.

Since the early 1990s, the M.S. Swaminathan Research Foundation (MSSRF) has been collaborating with poor smallholder farmers to prevent the erosion of millet diversity by creating an enabling environment for their continued use, thanks to highly participatory R&D efforts. MSSRF has nurtured family farmer groups to conserve and enhance the use of millets and has facilitated the establishment of 'Kolli Hills Agrobiodiversity Conservers Federation' (KHABCoFED, now evolved into the 'Kolli Hills Agro-Bioresource Producer Company Limited' KHABPCOL), to revitalize millet cultivation. Major activities included the scouting of local growers; assessment of traditional knowledge; collection, characterization and evaluation of landraces; multiplication of high-quality seeds; revitalization of seed storage and exchange systems through community seed banks and dissemination of best cultivation practices (including intercropping millets with cassava). Additional measures undertaken included the provisioning of low-cost processing machines accessible to communities, value-addition skill-enhancement, development of novel products, branding, strengthening of community-based institutions and linking local production to peri-urban and local markets.

Community seed banks to secure seeds for survival and resilience

Land-use changes have resulted in the erosion of on-farm millet diversity (Bhag Mal et al., 2010). Since 2001, KHABCoFED has been promoting collection, multiplication and distribution of 21 landraces belonging to five millet species and associated crops, through a network of 15 village seed banks functioning across seven Panchayats (village councils), spread across 45 settlements. The KHABCoFED has developed mechanisms to manage and share germplasm through Community Seed Banks and Seed Exchange Networks, whereby farmers are able to choose varieties apt for changing weather conditions and evolving markets. Custodian farmers are empowered for carrying out effectively their strategic role as conservers of seeds and managers of seed networks.

Village agro-bioresource centres for capacity building

Over the years, the 15 local seed networks have evolved into Village Agro-bioresource Centres, where community members are able to access quality seeds and relevant information regarding millet landraces, conservation, cultivation practices, value-addition and marketing. It also provides training on agricultural best practices, leading to better yields, higher income, more diversified food baskets and, ultimately, strengthened livelihood resilience.

Improved agronomic practices like the usage of farmyard manure, high-quality seeds, line sowing, intercropping with cassava and agroforestry, which reinforce the sustainability of natural resource management, have resulted in an average increase of 24.36% productivity and 37.23% net income. A blend of traditional and modern agro practices are disseminated through capacity-building interventions.

The enhancement of cultivation practices and value-adding technologies has led to 75% to 85% reduction in drudgery and time spent on processing These included the introduction of row seeding (one-meter length with six blades and 20 cm gap between blades), cono-weeder, improved spades for intercultural operations (six inches width and six inches length) and the use of pulverizers (mechanical flourmill pulverizers and de-husking machines – see next section).

Empowering marginal hill-dwellers through knowledge, technologies and value-chain development

Drudgery in processing is a major limitation, as all seeds but finger millet have several layers of hard coats, requiring high abrasive force to break through. Traditional decortication processes using mortar and pestle are physically tedious and are almost exclusively performed by women. The above-discussed interventions facilitated the establishment of improved technology for de-hullers and pulverizers, adaptable to different types of millet species and catering to many households. This has helped the community, especially women, who now are capable of carrying out the processing of millets without suffering fatigue, benefiting them in terms of health and time saved.

The established KHABCoFED greatly facilitates the skill development of women (members of self-help groups) to reap the market potential of millets and other NUS products through training on value-addition, quality standards, packaging, labelling and marketing. Trained women are now able to produce with success – *inter alia* – products like malt, *rava* (semolina), and millets flour mixes, all contributing to raising their income. "*Kolli Hills Natural Foods*" has become a popular brand, successfully marketing 11 products across Tamil Nadu. This is a fully self-sustainable value-chain of local natural products, consisting of seed conservers, farmers and other actors engaged in procurement, value-addition and marketing.

Encouraged by such successes, the District Rural Development Agency (DRDA) of Tamil Nadu provided financial support for establishing small-scale mills along with providing guidance on legal matters to local communities. Individuals and organizations committed to organic farming and natural health foods have also supported the momentum by purchasing and promoting the products of the KHABCoFED. Participation in exhibitions and fairs (*melas*) has provided an additional opportunity to further expand the KHABPCOL, including additional networks, research institutions and Community Based Organisation, and developing new market linkages.

The KHABCoFED has grown steadily, and as of 2020 includes 109 associations consisting of 985 men and 526 women members; its assets include machineries, value-addition units and a procurement centre worth more than US$53,000. Over the years, cultivation, procurement, value-addition, diversification and sales have generated a gross income of approximately US$98,708.

Notably, the KHABCoFED has contributed to strengthening the conservation of millet diversity, making agriculture-based livelihood systems resilient to climate change, improving access to nutritious foods and contributing to an effective interplay of natural, social, economic and infrastructure capitals for the fulfilment of livelihood, empowerment and self-esteem needs of local communities.

Case study 2: alternative seed systems enabling access to quality seeds in Koraput, Odisha

In the tribal areas of the state of Odisha, like in the Koraput district, millets have traditionally occupied a central place in people's diets and in local cropping systems. Most farmers inhabiting these areas are resource-poor and are largely dependent on monsoons. Growing crops in uncertain agro–climatic conditions is challenging because of frequent dryland stresses and limited access to quality inputs, especially seeds and suitable technologies. The seed replacement rate in millets is very low, i.e., 1.62 t in Odisha, which implies that agriculture mostly relies on traditional practices, and seed is either self-saved or acquired through local networks. Saving and exchanging seeds among farmers is certainly a key part

FIGURE 21.1.1 Custodian farmer from Kolli Hills with harvested millet panicles. @ Israel Oliver King/MSSRF.

FIGURE 21.1.2 Bumper harvest of KMR-204 variety of finger millet in alternative seed system – tribal women of Khilloput village interacting with project facilitator. @Jagganth Khillo/MSSRF.

FIGURE 21.1.3 Millet harvest from alternative seed system – tribal women from Machhra village of Koraput processing quality seeds. @ Jeeva/MSSRF.

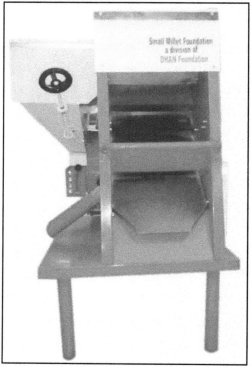

Millet Portable Huller (electric)

Millet Table-top Huller (petrol)

Millet Table-top Huller (electric)

FIGURE 21.1.4 Women-friendly efficient millet hullers developed by the Small Millet Foundation. @ Karthikeyan/DHAN Foundation.

FIGURE 21.1.5 Members of a farmers producer organization from Madhya Pradesh trained on millet value-addition in Kolli Hills. @Israel Oliver King/MSSRF.

FIGURE 21.1.6 Kong Bibiana Ranee, a Khasi custodian farmer and local leader, helped revive millets in Nongtraw and neighbouring villages, East Khasi Hills, Meghalaya, India. @ Raisa Daimary/NESFAS.

of maintaining local diversity, but too often such practices are not being integrated with the introduction of good quality seed. In order to address the diffuse shortage of quality seeds among farmers, an alternate seed system model (Figure 21.2) was tested out. Components of this model included the dissemination to farmers of a package of best practices, as well as reconnaissance surveys to assess seed demand, seed systems and associated constraints faced by local farmers. The model encompassed a plan for improving seed availability and accessibility to improved varieties seeds for resource-poor farmers and capacity building at the community level to enhance productivity. This model involved the Farmer Producer Organization (FPO) the "Kolab Farmers Producer Company Limited" (KFPCL), in order to ensure the timely and sustainable supply of good quality seed at affordable prices and the creation of self-sustainable seed enterprise. Participatory selection of varieties, demonstrations of best practices in farmers' fields, involvement of self-help groups, village-based seed committees; capacity building for seed production, processing, packaging, storage and distribution, business planning and administration were key actions of these interventions.

MSSRF built the capacity of the KFPCL for quality seed production. Improved varieties of seeds were obtained from research institutions and were demonstrated in farmers' fields through participatory variety selection and evaluation trials. The best performing and preferred varieties were selected for large-scale seed production. The KFPCL followed the norms and standards of the Odisha State Seed and Organic Products Certification Agency for seed

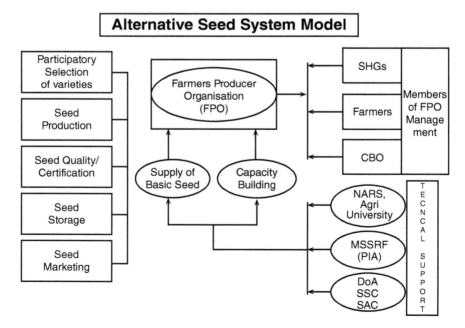

FIGURE 21.2 Alternative seed systems model.

Challenges and best practices of nutri-millets in India   253

purification, roughing, cleaning, drying, seed moisture and germination. The selected varieties were provided to community seed banks for safekeeping, storage and access by farming communities.

A processing unit comprising of an aspirator and grader cum de-stoner was established at Machhra village, Umuri Gram Panchayat to improve seed quality. The KFPCL has an arrangement with the Odisha State Seed Corporation Ltd. to directly procure seed materials at the rates fixed by government. In the very first year, KFPCL was able to sell 132 quintals of certified seeds of finger millet to the Odisha State Seed Corporation Ltd. and around 35 quintals to other farmers, institutions and NGOs. Besides the quality millet seed business, members of the KFPCL are also involved in the preparation of value-added millet products including ragi powder, ragi malt, laddus, chakli and processed grains of little millet, foxtail millet and barnyard millet.

Case study 3: catalysing value-chain development for small millets in Tamil Nadu

Tamil Nadu is one of the few states in India where almost all small millets are cultivated across different agro-ecological zones. However, from 1980 to 2011, the area under small millets fell drastically, with a 49.4% reduction in finger millet areas and 84.8% in other small millets areas. A 2012 study by the DHAN Foundation on the value chains of small millets in the state highlighted, as the reason for the decline, poorly organized supply chains characterized by seasonal uncertain production restricted to few pockets of land, erosion of varietal and crop diversity and lack of assured farm gate prices, leading to uncertainty in supply, quality and prices of raw materials to the small millet processors and food enterprises. Other critical factors included:

- Lack of local processing infrastructure, with few large-scale processors in south Tamil Nadu primarily serving as semi-processors for Maharashtra-based processing units.
- Existence of few millet food enterprises and largely of limited capacities.
- Poor consumer demand due to low awareness of millets, lower culinary skills and low availability, coupled with high prices.
- Inadequate policy support for the production, processing, market development and, moreover, consumption of small millets.

Realizing the importance for the nutritional security and ecological wellbeing, since 2001, the DHAN Foundation has undertaken a set of interrelated interventions, targeting regional value chains in collaboration with the Tamil Nadu Agricultural University (TNAU), Canadian research institutions and several value-chain actors. These efforts resulted in the establishment of the Small Millet Foundation, an exclusive organization dedicated to scaling-up proven interventions across India, which included:

254 E.D. Israel Oliver King et al.

i Streamlining supply chains through (a) capacity building of Farmers Organizations (FOs) in millet clusters (four in Tamil Nadu and two in Odisha) (b) productivity enhancement and (c) collective marketing

ii Promoting decentralized processing through (a) the development and commercialization of improved small-scale millet processing equipment; (b) facilitating their adoption across India and (c) building capacities of eight small-scale equipment manufacturers

iii Developing 56 appealing, small-millet-based food products and their commercialization through training 662 persons and providing onsite incubation support to 66 food enterprises

iv Building capacity of 152 pushcart millet-porridge vendors

v Promoting household consumption through dissemination of promotional materials, organizing promotional events and media campaigns, and building capacity of 85 women and FOs located in Tamil Nadu, Andhra Pradesh and Odisha

vi Linking FOs, millet processors and food enterprises with consumer organizations for shortening the value chain

vii Synthesizing and sharing policy lessons emerging from these efforts with provincial and national governments.

The Overseas Development Institute (ODI), UK, on reviewing the initiatives in 2017 (Keats and Jeyaranjan 2018) and 2019, identified the following positive outcomes:

- Farmers are able to access new varieties and increase cultivation, sell their produce collectively, reduce drudgery and time by using processing equipment and increase consumption (cf. *during 2018–2019, 87 small millet varieties were conserved on-farm, 161.8 tons of produce worth of US$67,030 were supplied to Odisha government and US$60,714 worth of products were marketed*).

- Equipment manufacturers have increased machine sales across India, expand their business and continue the development and launch of new models to meet diverse client needs (cf. *improved processing equipment reached 54 districts in 12 states during 2016–2019 and 174 new village-level and local processors entered the value chain. Small millet processing machines with varied capacities in the market have improved significantly*).

    - Millet-based food entrepreneurs have increased production and sales, developed new millet-based products and improved marketing techniques (cf. *66 existing enterprises improved their millet food business and 69 entrepreneurs and NGOs ventured into the small-scale millet food business. One hundred and fifty two pushcart millet-porridge vendors improved food hygiene and registered with the Food Safety and Standards Authority of India (FSSAI). TNAU has integrated small millet huller and food products developed as part of its ongoing incubation program, thereby sustaining their commercialization*).

    - Low-income consumers enrolled in the women's and farmers' organizations have improved their awareness of the health benefits of millets, have enhanced their skills in cooking healthy recipes and are able to access millets at lower than the market price (cf. *FOs and Kalanjiam Thozhilagam Limited [KTL], an associated enterprise of DHAN Foundation, have supplied 300 tons of small millet rice between 2016 and 2018*).

- With new knowledge and access to millets, women were able to improve their family's diet. Women farmers saved time and energy using processing equipment and women entrepreneurs increased their income and improved their status (cf. *knowledge of 725 women extension staff to promote millet consumption, the skills of 12,993 women on the inclusion of small millets in diets and the entrepreneurial skills of 39 women in millet food businesses were enhanced*).

Such interlinked efforts were key to transforming the value chain of small millets in Tamil Nadu, which was the result of the cooperation of many stakeholders, including the government and media. Large processors in southern Tamil Nadu started focusing on regional consumption and the market network has widened, leading to an increase in the availability of less polished small millets rice and value-added products, even in small towns. The share of the product moving out of Tamil Nadu has decreased with increased inflow from other states. The transformation of small-millets-based food products from the *'elite food'* category towards the *'mass food'* category has been accomplished to an extent.

Case study 4: generating economic benefits through collective marketing in Mandla and Dindori, Madhya Pradesh

Kodo (*Paspalum scrobiculatum*) and Kutki (*Panicum sumatrense*) are central elements in the traditional rainfed farming systems of Gond farmers in eastern Madhya Pradesh, India. Due to their low water requirements and early maturation that helps withstand drought, they are recognized as key assets to supporting farmers' adaptation to climate change. Despite this, the production area of small millets has declined more than 50% in the past two decades. Low productivity, weak market

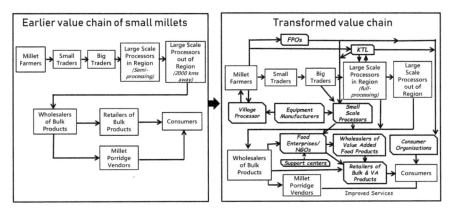

FIGURE 21.3 Transformation of small millet value-chain in the Tamil Nadu region.
Source: DHAN Foundation (used with permission)

256 E.D. Israel Oliver King et al.

channels and difficult processing have contributed to their decline as convenient and profitable crops with better market access and wage labour opportunities have developed. The production of small millets on isolated sloping and rocky lands where other crops are difficult to produce remains important among tribal farmers, yet low yields and poor marketability limit the benefits to these vulnerable groups.

To secure greater benefits from small millets for producers and to encourage their wider use, a holistic approach addressing bottlenecks in their supply and demand was followed in the project "Linking agrobiodiversity value chains, climate adaptation, and nutrition: Empowering the poor to manage risk". Activities sought to connect producers to markets and to enhance and multiply impacts for food security, conservation, profitability and women's empowerment. The stakeholder consultations carried out from 2015 to 2019 guided the pro-poor and gender-sensitive interventions. Farmer producer companies, along with local women shareholders, were a major focal point of interventions to raise productivity and enhance commercial potential (King et al., 2018; Meldrum et al., 2020).

Case study 5: value-chain development of millet among the Indigenous communities of Meghalaya

Of the 104 million Indigenous People inhabiting India, 12% of them, with over 200 tribes and sub-tribes, with their own unique languages, cultures and political structures, reside in the North Eastern region of the country (Xaxa et al., 2014). Meghalaya, a state in this region, is largely inhabited by Indigenous communities (86.1%) like Khasi (includes *Khynriam, Bhoi, War, Pnar* and *Lyngngam* sub-groups) in the east and Garo in the west. Women play a pivotal role in maintaining and preserving the rich biodiversity in the area, as they are closer to nature (Ellena and Nongkynrih, 2017).

North East Slow Food and Agro biodiversity Society (NESFAS), a grassroots organization, works towards the promotion of traditional food for food security and nutrition in the North East. In 2011, it initiated work towards promoting millets in Nongtraw village, on the southern slopes of Meghalaya, in the Khatarshnong Laitkroh Block, in collaboration with the Indigenous Partnership for Agrobiodiversity and Food Sovereignty. In close collaboration with custodian farmers, mostly women, it campaigned the slogan *"No Woman, No Krai"* – millet is known as *Krai* among the Khasis.

For Indigenous communities, millets are not only staple food but also play a central cultural role. Documentary evidence reveals that the Khasis cultivated millets as far back as the late nineteenth century and there are wild species/relatives as well as the edible forms of millets (Singh and Arora, 1972). Millet is usually grown in the *jhum* or shifting cultivation fields, wherein seeds, usually traditional and preserved by custodian farmers, are used (*Nongtraw Village*, 2011). Commonly cultivated millets are finger millets (*Eleusine coracana*) known as *krai-truh* (in *Khasi*) and Job's tears (*Coix lacryma-jobi*), locally known as *adlay*.

Post-harvest processes include sun-drying the grains and storing them in cloth or gunny bags. De-husking is done when needed using a mortar *(Thlong)* and

Challenges and best practices of nutri-millets in India **257**

iron or wooden pestle (*Synrei*), and finally winnowed using an open mat-type container (*Prah*) (Singh and Arora, 1972). This process is quite labour intensive, and both men and women are involved. The Nongtraw community celebrate a harvest festival for millet called "*Bom Krai*", where millet sheaves are laid on a bamboo platform and smoked from underneath with a particular wood to enhance the flavour of the grain. The seeds that escape through the gaps are used for planting the next season. The elders know just when to end the process. Next, the millet is threshed with sticks (*pduh krai*) by men and women, followed by further separation of the seeds by stamping on the millets accompanied by singing, which is most enjoyed by the children (*Nongtraw Village*, 2011; NESFAS, 2016).

Traditionally, millet is cooked and consumed as much as rice or, more often, one part millet is mixed with two parts rice. Previously, a kind of bread was prepared out of millet flour; though it is still made, nowadays it is chiefly consumed as porridge (Singh and Arora, 1972). In the East Khasi hill areas like War Pynursla, millet is used for brewing local liquor (Caritas India, 2018). In winter, millet also provides fodder for cattle. The straw is dumped on cow dung or in pigsties, where pigs trample it and this mixture is then utilized as compost. It has strong roots and acts as a soil binder, reducing soil erosion (Singh and Arora, 1972).

However, the cultivation and consumption of millet reduced after the introduction of the public distribution system, specifically after 1997, with its focus on providing subsidized rice to below-poverty-level families . With rice being more easily available, cultivation of millet became secondary. In addition, rice was easier to cook, while millet required lot of work; thus, given the easy alternative, people reduced the production and consumption of millet. As a result, the market for millet by the late 1990s and early 2000s was very low.

Custodian farmers from Nongtraw shared that by 2010, only two out of the 37 households cultivated millets and, of the four available varieties, only one variety of finger millet was still cultivated. However, with the work of NESFAS and custodian farmers, by 2014, 13 of the 46 villages in Khatarshnong area were cultivating millet, and this revival is spreading also to other areas, including the Mawkynrew Block of East Khasi Hills, Lyngngam area in the West Khasi Hills and Garo Hills. NESFAS has also facilitated the farmers from Nongtraw getting an organic certificate through the Participatory Guarantee Scheme (PGS). In 2019, 12 farmers in two groups received the PGS organic certificate. This has benefitted and boosted marketing, as consumers prefer certified organic products. Around the same period, opportunities opened up with a dealer from Gujarat in west India requiring 200 kgs of organic millet every season.

Traditionally, "millet rice", or even millet mixed with rice, was a staple food, but the youth and younger children do not seem to like it. To encourage their consumption, millet flour is now used to make products like porridge, pancakes, cakes, bread, biscuits, doughnuts, etc. Since 2018, NESFAS has conducted cooking demonstrations monthly to encourage the consumption of millets in different forms at the Meiramew Farmers' Markets. Other steps include sharing the recipe of a simple millet pancake while selling millet flour, and encouraging local bakers to sell millet-based bakery products in the farmers' markets, along

258 E.D. Israel Oliver King et al.

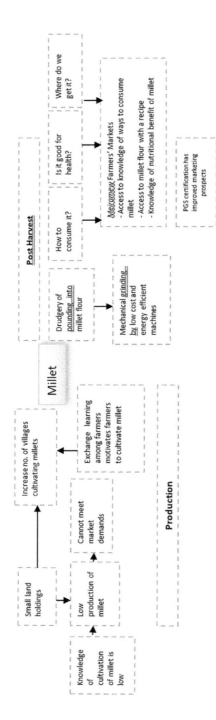

FIGURE 21.4 Enhanced consciousness and motivation among farmers and consumers (used with permission).

with traditional snacks called *pusla*. Farmers' markets are a platform where millet flour is sold, though sales are not limited here, as NESFAS facilitates sales on the basis of orders also, since locally grown millet is not available in the local market.

NESFAS also publishes the nutritional value of millet in its packaging materials and engages with the public through local newspapers. This way, the Meiramew Farmers' Market has been able to enhance consciousness regarding millet being a staple of the Indigenous communities, to educate consumers on the ways to consume millet and its nutritional benefits and to improve access to millet. This has improved the sale of millet; on average, NESFAS sells at least 10–15 kgs of millet flour in every farmers market each time.

There are of course challenges in the value-chain development. Generally, farmers in Meghalaya are marginal farmers, with small landholdings (55.34% own less than one hectare and 27.23% own one to two hectares) (Department of Agriculture, Government of Meghalaya, n.d.). Millet-producing communities also grow other crops and, therefore, the share of millet production is low. Thus, farmers cannot always meet the demand for millet. However, understanding the demand, NESFAS has encouraged other communities who were not growing millet earlier to cultivate it. In early 2020, NESFAS was able to facilitate an exchange of millet seeds through a millet network among four new villages in the East Khasi Hills and Ri Bhoi district.

Exchange visits and learnings among farmers were facilitated locally, to states within North East India such as Nagaland and Arunachal Pradesh and across the region. Such experiences have instilled confidence among farmers and have motivated them to cultivate millet.

Another challenge is in the pounding of millet grains into flour, which is labour intensive. NESFAS has promoted mechanical grinding with a millet machine in Nongtraw, which helped farmers a great deal. This also motivated other farmers to expand their millet production. However, this growth was sustained only for about two years (2015–2017), as the farmers were unable to keep up with expenses of the electricity bill and equipment maintenance, making it unviable for the community. This experience helped NESFAS understand the importance of conducting feasibility studies before installing machinery in a village.

The low participation and interest of youth in promoting millet – and in agriculture, in general – is the greatest challenge. To address this, NESFAS promoted a youth group in Nongtraw and trained them on millet processing, including the packaging, labelling, and costing of millets while following hygienic standards. Throughout the process, they served as a marketing platform for millets and other traditional foods. The group was registered as Nongtraw Multipurpose and Marketing Cooperative Society in 2017.

At present, Farmer Groups that are certified by the PGS are taking forward the work of marketing millets.

With increased consciousness about the nutritional benefits of millet, coupled with the knowledge that it used to be the staple food of our Indigenous communities, there is an interest among farmers and consumers alike in the production and consumption of locally grown millet.

Conclusion

Drought and low soil-moisture conditions are likely to increase in the future, and will affect crop production and further impact negatively farmers inhabiting hilly, rainfed, semi-arid and dry areas, who are most vulnerable to the vagaries of climate change. Millet farming is a climate-resilient system consisting of hardy crops suited to low and erratic rainfall and varied soil nutrient conditions.

Based on the diverse interventions carried out in various Indian states, we would like to recommend that the following pathways and critical issues be taken into consideration for moving the NUS agenda forward:

- The holistic '4C' approach that concurrently addresses conservation, cultivation, consumption and commerce has proven to be an effective pathway for retaining and using millet farming systems in a changing socioeconomic context.
- Village seed banks for revitalizing and promoting farmer-to-farmer seed exchanges, improved agronomic practices, additional income derived through millet-based livelihoods in the supply chain, enhancing on-farm millet conservation, reinforcing seed exchange networks in target communities and creating better linkages with value-chain actors and private companies.
- Collective initiatives harnessed the adaptive capacity of local communities to deal with climate change by leveraging the potential of resilient crops and production systems.
- The availability, consumption and access to diversified diets of micronutrient-rich foods like millets can contribute to better health, enhanced incomes, sustainable agriculture and a more resilient production system.
- Small millets can be a valid instrument of economic and social empowerment for women and vulnerable and marginalized groups. The facilitation of an effective interface between scientific institutions and value-chain actors for participatory research and commercialization of research outputs can aid in value-chain development.
- State-funded medium-term support measures for value-chain actors, including medium and small enterprises (MSEs), are necessary to develop value chains and millet foods, along with identifying and supporting pro-poor stakeholders like vendors, women self-help group federations and farmers' organizations in reaching everyone.
- The active participation of women in Farmer Producer Companies (FPCs) needs special attention, because although their membership is high, cultural values and local norms do not encourage their active engagement.
- More research to validate nutrition and health claims supported by Indigenous knowledge is needed.

Challenges and best practices of nutri-millets in India **261**

- Consistent sensitization of local governments in order to develop supportive policies and projects towards millet production and the use of NUS can contribute at a grassroots level.
- Despite tremendous efforts, still more needs to be done to address the erosion of landraces, lack of improved varieties, knowledge on cultivation practices, post-harvest technologies, skills and processing facilities, disorganized markets, the limited participation of the private sector and poor credit support for value-chain development.

References

Bala Ravi, S. (2004) Neglected millets that save the poor from starvation. *LEISA India*, 6(1), 34–36.

Bhag Mal, Padulosi, S. and Bala Ravi, S. editors. (2010). *Minor Millets in South Asia: Learnings from IFAD-NUS Project in India and Nepal*. Bioversity International, Maccarese, Rome, Italy and the M.S. Swaminathan Research Foundation, Chennai, India. 185 p. ISBN – 978-92-9043-863-2.

Caritas India (2018) Bringing Back Ancestral Millet to Life. Available at: https://www.caritasindia.org/bringing-back-ancestral-millet-to-life/

Department of Agriculture, Government of Meghalaya (no date) Land Use Pattern in Meghalaya. Available at: https://megagriculture.gov.in/PUBLIC/agri_scenario_landuse_pattern.aspx

Ellena, R. and Nongkynrih, K. A. (2017) 'Changing gender roles and relations in food provisioning among matrilineal Khasi and patrilineal Chakhesang Indigenous rural people of North-East India', *Maternal & Child Nutrition*, 13(S3), e12560. doi: 10.1111/mcn.12560.

Gruere, G., Nagarajan, L. and Oliver King, E.D.I. (2009) The role of collective action in the marketing on underutilized plant species: Lessons from case study on minor millets in South India. *Food Policy* 34, 39–45.

Hulse, J.H., Liang, E.M. and Pearson, O.E. (1980). *Sorghum and the Millets: Their Composition and Nutritive Value*. New York Academic Press, London. 1997p.

Keats, S. and Jeyaranjan, J. (2018) India small millets, Country Report, Scaling up small millet post-harvest and nutritious food products (IDRC project #108128), Canadian International Food Security Research Fund (CIFSRF) - Contribution Analysis, Overseas Development Institute and IDA Chennai.

King, I.O., Meldrum, G., Kumar, N., Lauridsen, N., Manjula, C., Padulosi, S., Sivakumar, M.N., Baskar, R. and Madeshwaran, K. (2018). *Research Brief: Value Chain and Market Potential of Minor Millets to Strengthen Climate Resilience, Nutrition Security and Incomes in India*. Bioversity International, Rome, 4 p. Available at: https://cgspace.cgiar.org/bitstream/handle/10568/98367/http://Millet_Israel_2018.pdf

Malleshi, N.G. and Desikachar, H.S.R. (1986) Nutritive value of malted millet flours. *Plant Foods for Human Nutrition* 36, 191–196.

Meldrum, G., Roy, S., Lauridsen, N. and King, O.E.D.I. (2020) *Promoting Kodo and Kutki Millets for Improved Incomes, Climate Resilience and Nutrition in Madhya Pradesh, India*. The Alliance of Bioversity International and CIAT, Rome, 4 p. Available at: https://cgspace.cgiar.org/bitstream/handle/10568/109361/India Impact Brief A4.pdf

Nagarajan, L. and Smale, M. (2007) Village seed systems and the biological diversity of millet crops in marginal environments of India. *Euphytica* 155(1–2), 167–182.

NESFAS (2016) Bringing the Clan Together – Millet | North East Slow Food & Agro-biodiversity Society. Available at: http://www.nesfas.in/2641/ (Accessed: 20 May 2020).

Nongtraw Village (2011) Nongtraw. Available at: https://www.youtube.com/watch?v=ZeHxYpPAo_Y (Accessed: 20 May 2020).

Singh, H. and Arora, R. (1972) 'Raishan (Digitaria sp.)—a minor millet of the Khasi Hills, India | SpringerLink', *Economic Botany*, 26, 376–380. Available at: https://link.springer.com/article/10.1007/BF02860709 (Accessed: 20 May 2020).

Ulian, T., Diazgranados, M., Pironon, S., Padulosi, S., Liu, U., Davies, L., Howes, M.-J.R., Borrell, J., Ondo, I., Pérez-Escobar, O.A., Sharrock, S., Ryan, P., Hunter, D., Lee, M.A., Barstow, C., Łuczaj, Ł., Pieroni, A., Cámara-Leret, R., Noorani, A., Mba, C., Womdim, R.N., Muminjanov, H., Antonelli, A., Pritchard, H.W. and Mattana, E. (2020) Unlocking plant resources to support food security and promote sustainable agriculture. *People Plants Planet* 2(5), 421–445. ISSN: 2572–2611.

Xaxa, V., et al. (2014) Report of the High Level Committee on Socio Economic, Health and Educational Status of Tribal Commuities of India. Ministry of Tribal Affairs, Government of India, pp. 1–431.

22

TAKING MILLETS TO THE MILLIONS

Experiences from government-driven value chains

Dinesh Balam, Saurabh Garg, Srijit Mishra, Bhagyalaxmi, Mallo Indra, Jayshree Kiyawat

Introduction

Millets are C4 crops with high nutrition and climate-resilience qualities and are native to the rainfed areas of India (NAAS, 2013). Among them, finger millet and other small millets were largely grown, historically, and were part of the diet of the tribal communities of Odisha.

Among small millets, little millets are grown in the districts of Bolangir, Koraput, Malkangiri, Mayurbhanj, Sundergarh and Rayagada (DMD, GoI [2014]). In the case of finger millets (*Eleusine coracana*), the area of cultivation declined from 0.25 million hectares in 1990–1991 to 0.11 million hectares in 2017–2018 and the yield per hectare in the same period declined from 10.23 quintals per hectare to 8.80 quintals per hectare in 2017–2018 (Status of Agriculture in Odisha). The decline is further escalating due to a lack of support for cultivation, lack of appropriate processing technology and poor market development (Bhag Mal, et al., 2010). However, since 2010 onwards, there has been a growing interest in promotion of millets both by the Union and state governments. The Initiative for Nutritional Security through Intensive Millets Promotion, Rainfed Area Development Programme and Integrated Cereals Development Programmes in Coarse Cereals-based Cropping Systems Areas (ICDP-CC) under the Macro Management of Agriculture are some of the initiatives that were launched by the Union government. But these have not been successful in bringing about changes in the millet landscape (RESMISA, 2012).

Multi-stakeholder consultation

It is in this context of the decline of millets that the Planning & Convergence Department, Government of Odisha (GoO) and Nabakrushna Choudhury

DOI: 10.4324/9781003044802-25

Centre for Development Studies (NCDS), in partnership with the Revitalising Rainfed Agriculture Network and Alliance for Sustainable and Holistic Agriculture (ASHA) Network, jointly organised a multi-stakeholder consultation under the chairmanship of R Balakrishnan IAS, Development Commissioner cum Additional Chief Secretary (DC-cum-ACS) to develop a strategy for revival of millets in Odisha. Based on the experiences of experts, civil society and government officials, the Department of Agriculture & Farmers' Empowerment (DA&FE) launched the *"Special Programme for Promotion of Millets in Tribal Areas (Odisha Millets Mission)"*. This programme is also known as the Odisha Millets Mission (OMM) (NCDS, 2016).

The key objectives of the OMM were to increase household consumption of millets by about 25% to enhance household nutrition security and to create demand for millets with a focus on women and children; promote millet-processing enterprises at the Gram Panchayat (GP) and block levels to ease processing in households and for value-added markets; improve productivity of millets crop systems and make them profitable; develop millet enterprises and establish market linkages to rural/urban markets, with a focus on women entrepreneurs, and include millets in state nutrition programmes and the public distribution system.

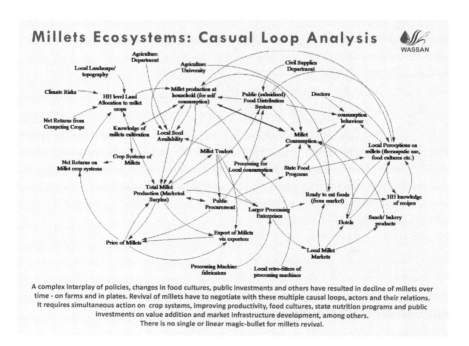

FIGURE 22.1 The causal loop analysis of the different stakeholders in millets value-chain (WASSAN 2016).

Scale

The Block was chosen as a unit of the implementation and the OMM targets were extended from 1,000 to 2,500 acres within a Block in five years. The OMM started in 30 blocks across seven districts and later extended to 76 blocks covering 14 districts in a phased manner. About 11 blocks are funded through the District Mineral Fund (DMF). The GoO has increased the funding for the OMM from US$8.96 million in 2017 to US$73.38 million. Out of which US$30.60 million is for project implementation and US$42.78 million is for the procurement and distribution of finger millet in PDS and ICDS.

Stakeholders

While the DA&FE is the nodal department for the OMM, the Directorate of Agriculture and Food Production is a implementing directorate and also hosts project man agreement unit of the OMM at the state level. The Agriculture Technology Management Agency (ATMA) is the nodal agency at the district level. The programme is implemented by Farmer Producer Organisations (FPOs) with support from facilitating agencies (FAs) or non-governmental organisations (NGOs) at the block level. Community resource persons support local NGOs and FPOs. A state secretariat (SS) anchored by NCDS was formed for the OMM. The SS consists of a programme secretariat (PS) and a research secretariat (RS). The Watershed Support Services and Activities Network (WASSAN) was selected as PS. NCDS was chosen as the research secretariat in addition to SS.

Components of the OMM

The annual reports of OMM (Garg and Muthukumar 2017, 2018, 2019) elaborates on the design and initiation of the project, progress and achievements.

Promoting household level consumption

Awareness on millets were built up on a massive scale through food festivals, cooking competitions, awareness campaigns, district-level *melas* and millet recipe training events with active participation from different district-line departments.

Setting up decentralised processing facilities

To reduce drudgery during the post-harvest and processing of millets, the OMM aimed to set up post-harvest and processing machinery at the GP and block levels. It was proposed that about 10 post-harvest/processing units (threshers, destoners, graders, dehullers and pulverisors) be set up in each block through WSHGs/FPOs. The technical specifications of the machines are finalised through a state-level committee on processing consisting of different experts. WSHGs/FPOs are selected through a expression of Interest process at the district level through another committee consisting of members from ATMA, District Social Welfare Officer, Odisha Livelihood Mission and FAs. Against the target of 444 threshers, 299 are placed in the district.

266 Dinesh Balam et al.

Improved agronomic practices

Encouraging farmers through the provision of incentives for a period of three years for adoption of improved agronomic practices such as System of Millet Intensification, line transplanting, line sowing and intercropping. Incentives to farmers are designed to be gradually reduced over a period of three years. No chemical inputs are supplied to farmers under the OMM, with the farmers encouraged to adopt agro-ecological practice (Table 22.1).

Establishing diverse seed centres

Farmer-preferred and locally appropriate seeds were promoted under the OMM. This was done through participatory varietal trials with varieties released by Indian Council of Agricultural Research (ICAR), State Agricultural Universities (SAUs), and the government. The best-performing varieties are then selected and multiplied through seed farmers. A working group has been formed for developing alternate seed system for indigenous landraces under the OMM. For the first time, four finger millet landraces – namely, Bati Mandia, Kalia Mandia, Bhadi Mandia and Mami Mandia – are being considered for release by DA&FE and the GoO (Table 22.2).

Custom hiring centres

Locally appropriate implements are supplied on a for-hire basis through custom hiring centres to reduce drudgery in various intercultural operations. Eight sub-centres within block-level Custom hiring centres are anchored by SHGs/FPOs/CBOs per block under the OMM.

TABLE 22.1 Summary of year-wise outreach under the improved agronomic practices of the OMM

Sl. no.	Year	No. of districts	No. of blocks	Area (ha) (target)	No. of farmers covered
1	2017–2028	7	30	3,333.5	8,030
2	2018–2019	11	55	12,758.2	29,056
3	2019–2020	14	72	21,552.6	51,045
4	2020–2021 (Target)	14	76	56,118	In progress

TABLE 22.2 Summary of year-wise information on PVTs and landraces tried out in the OMM

Sl. no	Year	No. of blocks	Total no. of landraces/ varieties in PVT	No. of local landraces	No. of check/ govt varieties
1	2018–2019	15	147	125	17
2	2019–2020	39	226	185	41
3	2020–2021	13	86	79	7

Farmer producer organisations

The comprehensive revival of millets in a Block requires service delivery. The community/farmer-level institutional base varies from Block to Block. FPOs will be formed at the block level. These FPOs will also anchor the different enterprises by linking with SHGs and other smaller collectives (Table 22.3).

Procurement of finger millet

The GoO approved the initiation of ragi procurement in 2018. Accordingly, a three-year proposal for US$42.78 million was also approved. The Tribal Development Cooperative Corporation Odisha Limited was selected as the nodal procurement agency. The Department of Agriculture & Farmers Empowerment issued the finger millet procurement guidelines, with the concurrence of the Food Supplies & Consumer Welfare Department, to the concerned district administration. The District-Level Procurement Committee, under the chairpersonship of the Collector and District Magistrate, finalise the timelines for farmer registration, the points and dates for procurement, procurement infrastructure and effective awareness campaigns in their respective districts.

Implementation in KMS 2018–2019 (year 1)

In KMS 2018–2019, the first year of the procurement programme, unforeseen challenges restricted procurement to 17,986 quintals against the target of 75,000 quintals. Finger millet was distributed at the rate of 1 kg per ration card through the PDS scheme in six districts. In Malkangiri District, it was distributed at the rate of 2 kg per ration card. Millet procured in Sundergarh District was provisioned for inclusion in ICDS in Keonjhar District.

Implementation in KMS 2019–2020 (year 2)

The learnings from KMS 2018–2019 were used to lay the groundwork for KMS 2019–2020. A procurement target of 100,000 quintals was approved for 2019–2020. The Government of India declared the Minimum support price for the KMS 2019–2020 as US$43.04 per quintal. By making changes based on past experiences, the procurement increased more than five times, totalling 94,745 quintals in KMS 2019–2020 (Table 22.4).

TABLE 22.3 Status of different community institutions of the OMM

Sl. no	Institution type	Total numbers	Currently operational
1	Seed centres	76	72
2	Custom Hiring Centre	76	72
3	Farmer Producer Organisations (FPOs)	76	65

268 Dinesh Balam et al.

TABLE 22.4 District-wise information on procurement of finger millet for FY 2018–2019 & FY 2019–2020

Sl. no.	District	Quantity procured (in qtls.) in KMS 2018–2019 @ US$36.85 per qntl	Quantity procured (in qtls.) in KMS 2019–2020 @ US$43.04 per qntl
1	Bargarh	NA	718.35
2	Bolangir	NA	2,056.53
3	Gajapati	671.77	4,345.6
4	Ganjam	NA	603.61
5	Kalahandi	920.29	5,377.57
6	Kandhamal	1,365.02	4,952.93
7	Keonjhar	NA	46.91
8	Koraput	8,904.69	36,164.59
9	Malkangiri	1,793.54	12,673.37
10	Mayurbhanj	NA	1,011.6
11	Nawarangpur	NA	1,983.01
12	Nuapada	247.25	1,288.74
13	Rayagada	3,589.78	21,448.32
14	Sundergarh	493.44	2,031.4
Total		**17,895.78**	**94,702.530**

Implementation in KMS 2020–2021 (year 3)

In the KMS 2020–2021, a procurement target of 1,65,000 quintals has been finalised by DA&FE and GoO. For the first time, the procurement of finger millet shall also be done by 16 FPOs in different blocks. Finger millet procured by these FPOs is proposed to be utilised in the ICDS and PDS.

Millet inclusion in Integerated Child Development Scheme (ICDS), Midday Meal Scheme (MDM)

Following the successful procurement witnessed in year 2, finger-millet-based entitlements were included in the PDS and ICDS schemes. Finger millet *laddu* was piloted in the anganwadi menu in Keonjhar District on July 2, 2020 and in Sundergarh District on August 15, 2020. Distribution of finger millet in PDS is planned in the 14 procurement districts in lieu of rice. The High Powered Committee of the OMM has approved the inclusion of millets in ICDS and MDM through DMF.

Recognition and impact

The NCDS baseline report has shown that among the beneficiaries of the OMM, yield has doubled and income has tripled. Based on the successful outcomes, the Government of India has asked all states to adopt the OMM model for the promotion of millets, pulses and oilseeds. Niti Aayog has chosen Odisha and

Karnataka as two progressive models and is willing to facilitate learnings for the same. The Governor of Maharashtra has requested the GoO to extend support to initiate a similar project on millets in the state of Maharashtra. The Government of India has set up a task force to understand the framework of the OMM and include the learnings of the OMM into new guidelines of national sub-mission on nutri-cereals.

Box 22.1 Piloting the inclusion of sorghum and foxtail millet recipes in ICDS in Vikarabad District, Telangana

Keeping in view the crop failures, farmer indebtedness and poor nutrition profiles, the District Administration of Vikarabad, Government of Telangana, initiated a pilot project to include sorghum in ICDS with the support of WASSAN. A series of three millet food festivals were organised to build consensus around the inclusion of millets in ICDS. Children and mothers were given millet-based foods at the festivals. To understand the public perception on including millets in diets, feedback was taken from the community, mothers, people's representatives, anganwadi teachers, helpers and children. Resolutions passed in the mothers' committees were also taken to understand the view on these millet-based efforts. Through such participatory approaches, diverse menu options were selected, which also included inputs from the community and from scientists at the National Institute of Nutrition. Such innovative, nutritious and appealing preparations included a preliminary menu of foxtail millet *kitchidi* and sorghum *upma*, which were given to the children in 45 Anganwadi Centres (AWCs). The initiative was launched on April 14, 2017.

The cost of one normal rice-based meal per child per day is US$0.08. As millets were not subsidised under PDS, cost of the korra (foxtail millet) *kitchdi* and jonna (sorghum) *upma* is about US$0.11 and US$0.14 respectively. Additional funds were provided from the District Collector through the Flexi-Funds Scheme. For the pilot of three months, covering 45 AWCs and 1,000 children, an additional expenditure of Rs. 1.73 lakhs was incurred. The Civil Supplies (CS) Department procured millets through farmer cooperatives as per approved specifications.

Based on the success of the Vikarabad initiative, the Women Development & Child Welfare Department, Government of Telangana, is now planning to champion the inclusion of millets in ICDS on a large scale. A proposal for decentralised pilots for three aspirational districts was also submitted to Niti Aayog. The proposal was approved in the 13th Empowered Committee meeting of Niti Aayog on July 10, 2020. In addition, the state of Telangana has also received an outlay of 355 Metric Tonnes of jowar and 607 MT of bajra under ICDS for piloting millets in ICDS (Deverajan, D. [2020] "Note on inclusion of millets in Telangana").

Box 22.2 Integrated value-chain development of finger millet (*Eleusine coracana*), Khed, District Ratnagiri, Maharashtra

Mahila Arthik Vikas Mahamandal (MAVM), Department of Women & Child Development, Government of Maharashtra, encouraged WSHGs to take up the value addition and marketing of millet products. As a result, 404 WSHG members started cultivating finger millets in 78 acres of land and harvested 300 quintals of the crop; of this, 200 quintals were kept aside for household consumption, while the rest was purchased by the Community Managed Resource Centre (CMRC) for the preparation of different finger-millet-based products such as malt, biscuits, laddoo, etc. CMRC also opted for FSSAI certification and proper branding and packaging was tackled.

The revenue generated from this was US$19,677.51, against a total cost of US$14,129.54. The CMRC with the support of MAVIM tied up with Dr. Balasaheb Konkan Agriculture University, Dapoli for technical support. CMRC also linked with Shri Vivekanand Research and Training Institute, at Lote, and the corporate social responsibility wing of Excel Industries Pvt. Ltd to purchase small machinery.

The MAVM initiative is a good example of how proper micro-planning combined with strong grassroot institutions and convergence can make a substantial difference in the lives of millet-growing farmers.

Box 22.3 Integrated value-chain development for millets - The Kodo - Kutki way

The Mahila Vitt Evam Vikas Nigam (MVEVN), Government of Madhya Pradesh, through the International Fund for Agriculture Development (IFAD), supported the Tejaswini Project, which initiated the promotion of kodo (kodo millet) and kutki (little millet) in the Mehandwani area in Dindori District in 41 tribal villages in 2013. Initially, 1,497 women farmers from SHG Federations were trained on improved agricultural practices to take up millet cultivation in at least 0.5 acres of land. By 2017, the area under millet cultivation had increased from 748.5 to 3,750 acres, with an associated growth in total production from 2,245.5 quintals to 15,000 quintals. The total income increased from US$122,738.45 to US$819,896.15, with a net profit increase from US$53,416.23 to US$562,995.35. Apart from the increase in area, productivity, income and profit, the number of women farmers who took on the goal of millet cultivation rose to over 7,500 farmers. Inspired by this success, farmers from other nearby blocks like Sambalpura and Shahpura are also adopting these processes and are reaping benefits through millet cultivation.

Further, with support of MVEVN, the SHG Federations shouldered the responsibility for processing, marketing, FSSAI certification and branding. For a wider reach, marketing and promotion was undertaken in various avenues like women's clubs, exhibitions, *melas* etc., covering cities like Nagpur, Nasik, Jabalpur and Bhopal. A pilot on the inclusion of Kodo-Kutki Bar in 226 Aanganwadi Centres for about 5,000 children was taken up. The efforts of Tejaswini in spreading the use of millets in everyday diets has been recognised at multiple forums and has also won the Sitaram Rao Livelihood India Case Study Award and SKOCH Order of Merit Award.

Conclusion

All these interventions show that a multi-stakeholder approach involving different departments, civil society and farmers'/women's collectives and that focuses on the value chain within and outside government programmes leads to success. Transforming food systems for meeting the emerging needs of climate change and malnutrition will need a multi-stakeholder approach.

References

Bhag Mal, Padulosi, S. and Bala Ravi, S., editors. (2010). *Minor Millets in South Asia: Learnings from IFAD-NUS Project in India and Nepal.* Bioversity International, Maccarese, Rome, Italy and the M.S. Swaminathan Research Foundation, Chennai, India. 185 p.

Deverajan, D. (2020) "Note on inclusion of millets in Telangana" to Director, Women and Child Development, Niti Aayog".

Garg, S. and Muthukumar, M. (2017) "Annual Report 2017–18-Odisha Millet Mission" DA&FP, Odisha, http://www.milletsodisha.com/resources/publications

Garg, S. and Muthukumar, M. (2018) "Annual Report 2018–19-Odisha Millets Mission", DA&FP, Odisha, http://www.milletsodisha.com/resources/publications

Garg, S. and Muthukumar, M. (2019) "Taking Millets to Millions - Odisha Millets Mission, Directorate of Agriculture and food Production, Odisha, http://www.milletsodisha.com/resources/publications

Government of India - GoI (2014) 'Status paper on Coarse Cereals", Directorate of Millet Development, Ministry of Agriculture and Farmers Welfare.

Kiyawat, J. (2018) Presentation on Integrated Value Chain Development For Millets, The Kodo - Kutki Way in Dindori - Millets, Monsoon and Markets Conference of M S Swaminath Research Foundation.

Mallo, I. (2018) Presentation on Integrated Value Chain Development of Finger Millet (*Eleusine coracana*), Khed, District Ratnagiri - Millets, Monsoon and Markets Conference of M S Swaminath Research Foundation.

NAAS (2013) *Role of Millets in Nutritional Security of India.* Policy Paper No. 66, National Academy of Agricultural Sciences, New Delhi, 16 p.

NCDS Summer Internship Report (2016) Special Programme for Promotion of Millets in Tribal Areas.

RESMISA (2012) Supporting Millets in India; Policy Review and Suggestions for Action. DHAN Foundation and WASSAN.

"Status of Agriculture in Odisha, Department of Agriculture & Farmers' Empowerment https://agriodisha.nic.in/Home/StatusofAgriculture

23

MILLET-BASED INTERCROPPING SYSTEMS FACILITATED BY BENEFICIAL MICROBES FOR CLIMATE-RESILIENT, SUSTAINABLE FARMING IN TROPICS

Natarajan Mathimaran, Devesh Singh, Rengalakshmi Raj, Thimmegowda Matadadoddi Nanjundegowda, Prabavathy Vaiyapuri Ramalingam, Jegan Sekar, Yuvaraj Periyasamy, E. D. Israel Oliver King, Bagyaraj Davis Joseph, Thomas Boller, Ansgar Kahmen and Paul Mäder

Introduction

According to the FAO, an increasing number of people are going to bed hungry, and more than a billion people are known to be nutritionally poor (Swaminathan 2010). Agroecology-based intercropping – growing more than one crop at the same time on a given piece of land – is now regarded as a promising approach for addressing food security in an environmentally and socially sustainable way (Brooker et al. 2015; Duchene et al. 2017). Many studies have shown that intercropping provides greater resource use-efficiency, reduced soil erosion and nutrient losses, and improved soil moisture (Maitra 2020; Triveni et al. 2017). Water is arguably the single most important factor that limits crop production in agriculture, particularly in rainfed or dryland ecosystems, and a consideration of plant hydraulic lift of soil water may help in designing a sustainable intercropping system (Liste and White 2008).

Soil microbes such as arbuscular mycorrhizal fungi (AMF) and plant-growth-promoting rhizobacteria (PGPR) may be beneficial as well, particularly the former with their capability to form a common mycorrhizal network (CMN) that extends beyond plant root systems, facilitating long-distance nutrient mobilization and water transfer (Aroca and Ruiz-Lozano 2009). Sustainable intensification through intercropping may be achieved with many different crops but requires the integration of various sciences such as agronomy, soil, microbial and social sciences (Brooker et al. 2015). This chapter deals with millet-based intercropping systems as an example. Specifically, we show results from our recent studies, revealing the contribution of CMN in mediating "bioirrigation", a process where the hydraulically lifted water by a deep-rooted legume such as

DOI: 10.4324/9781003044802-26

Millet-based intercropping systems

Millets can be intercropped with many crop species, particularly with oil seeds such as sesame, or with legumes such as black gram, cowpea and pigeon pea. The choice of companion crop with the millets is usually decided by the farmers based on the economic value/benefit-cost ratio, land equivalent ratio (LER) and social factors like labour availability. However, legumes are particularly interesting because of their potential to fix atmospheric nitrogen in symbiosis with rhizobia and thereby to increase soil fertility, as shown, for example, in grain legume-pearl millet intercropping systems (McDonagh and Hillyer 2003).

In addition, studies have shown facilitation in intercropping systems, where one species promotes the growth of the other (Callaway 2007; Li et al. 2009). Studies by Li et al. (2016) clearly show that in a maize and faba bean intercropping system, the root exudates of maize increase nodulation and enhance nitrogen fixation of faba bean. A trial conducted by Dass and Sudhishri (2010) with a combination of finger millet with pulses revealed enhanced productivity and increased resource conservation by a significant reduction of soil runoff. Several studies have reported the yield of millet to increase under legume intercropping systems, particularly in low fertile soils and low-input systems (Runkulatile et al. 2015; Triveni et al. 2017; Bitew et al. 2020). Significantly increased LER, more efficient water and better nutrient utilization were observed in intercropping systems than in monocropping (Yu et al. 2016; Daryanto et al. 2020). Among different legume–millet intercropping systems, the pigeon pea–finger millet system is a widely adapted practice by the farmers, particularly in the rainfed regions of south India. We discuss this example in more detail below.

Bioirrigation-based intercropping systems facilitated by microbes

Bioirrigation – definition and concept

Many herbs, grasses, shrubs and trees are known to lift water via roots from subsoil and redistribute it to topsoil in a process known as hydraulic redistribution. The lifting of water by roots is passive and driven by leaf transpiration and stomatal closure. The redistributed water into topsoil might provide sufficient moisture for shallow-rooted plants growing adjacent to a deep-rooted plant (Caldwell et al. 1998). Using a stable isotope deuterium tracer, Sekiya and Yano (2004) found that pigeon pea could redistribute hydraulically lifted water to neighboring maize plants. Furthermore, the hydraulically redistributed water into dry topsoil is known to trigger microbial activities, which, in turn, could initiate various biogeochemical cycles, eventually leading to plant nutrient availability. This important ecophysiological process is an inherent part of many intercropping

Millet-based intercropping systems in the tropics **275**

systems, particularly where a deep-rooted legume such as pigeon pea is grown with shallow-rooted cereals such as finger millet.

Bioirrigation facilitated by beneficial microbes

Numerous studies have shown that below-ground niche complementarity and resource sharing, combined with improved plant water relations, forms the basis for the overall productivity of the system. Measurements using various techniques, including stable isotope tools, have shown improved plant water relations, particularly under drought conditions (Zegada-Lizarazu et al. 2006). A classic work by Querejeta et al. (2012) showed the facilitative role of mycorrhizal hyphae in redistributing the water hydraulically lifted by oak. Catabolic response profiling of the microbial community obtained from the native shrubs grown in drylands at Sahel showed greater microbial diversity and a more active microbial community in the rhizosphere compared with a monocropping system (Diakhaté et al. 2016). Microbiome analysis in millet rhizosphere showed that an intercropping system can harbor potentially huge number of beneficial microbes (Debenport et al. 2015). Although there is ample information about rhizosphere microbial diversity, only limited knowledge is available with regard to ecophysiological and/or functional ecology of microbes involved in complex below-ground interaction. In particular, our current knowledge on the role of microbes in the facilitation of nutrients and water (Duchene et al. 2017) is limited.

Microbial-facilitated, bioirrigation-based finger millet-pigeon pea intercropping system

Despite some studies showing role of beneficial microbes in facilitating hydraulic lifts, mostly in tree species, currently there are only few studies showing the role of CMN in combination with PGPR in facilitating "bioirrigation" in crop species, particularly in cereals and legumes. In this regard, we conducted a series of research, first starting to unravel facilitative role of CMN in combination with PGPR in pigeon pea-finger millet intercropping systems (Figure 23.1).

In one of our earliest studies in this context, we tested whether finger millet, a shallow-rooted cereal, can profit from neighboring pigeon pea, a deep-rooted legume, in the presence of "biofertilization" with AMF and PGPR, under drought conditions. Our results showed that "biofertilization" with AMF alleviates the negative effects of drought conditions on finger millet, indicating that CMN connecting pigeon pea and finger millet exert clearly a positive influence in this simulated intercropping system (Saharan et al. 2018). In our latest work in this regard (Singh et al. 2020), we used stable isotope and classical physiological measurements such as stomatal conductance to better understand the ecophysiology of bioirrigation. We found that pigeon pea can indeed promote the water relations of finger millet during a drought event. The observed facilitative effects of pigeon pea on finger millet were partially enhanced by the presence of a

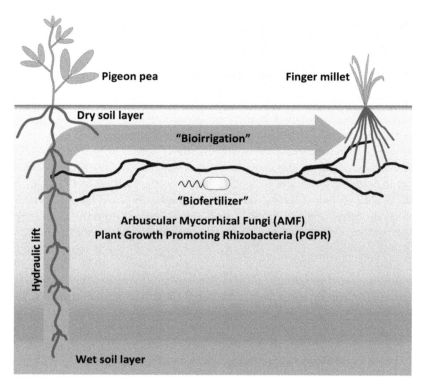

FIGURE 23.1 Scheme depicting a bioirrigation- and biofertilizer-based pigeon pea-finger millet intercropping system in a dryland agroecosystem.

CMN. In contrast to the facilitative effects under drought, pigeon pea exerted strong competitive effects on finger millet before the onset of drought. This hindered the growth and biomass production of finger millet when intercropped with pigeon pea, an effect that was even enhanced in the presence of a CMN. The results from our study, thus, indicate that in intercropping, deep-rooted plants may act as "bioirrigators" for shallow-rooted crops, and that a CMN can promote these facilitative effects. In more general terms, our study shows that the extent to which the antagonistic effects of facilitation and competition are expressed in an intercropping system strongly depends on the availability of resources (Singh et al. 2020).

Furthermore, we conducted field trials over three consecutive seasons at two sites in India, an intercropping and biofertilization scheme to boost their yields under low-input conditions. Our major findings are (i) the effects of the biofertilizers were particularly pronounced at the site of low fertility; (ii) the dual inoculation of AMF+PGPR to finger millet and pigeon pea crops showed increased grain yields more effectively than single inoculation; (iii) the combined grain yields of finger millet and pigeon pea in intercropping increased up to

Millet-based intercropping systems in the tropics **277**

128% due to the biofertilizer application and (iv) compared to direct sowing, the transplanting system of pigeon pea increased their average grain yield up to 267% across site, and the yield gains due to biofertilization and the transplanting system were additive (Mathimaran et al. 2020).

Technology adoption and the upscaling potential of bioirrigation-based intercropping systems

Finally, as millets are considered neglected crops, at least in the recent past, the diversity of millets is yet to be brought back the mainstream farming for nutritional security. King et al. (2009) have reviewed the importance of agrobiodiversity and conservation of millets for tribal populations in south India. The adoption of millets–legume intercropping systems by smallholders are governed by several socioeconomic factors. In one of our studies in south India we found that access to market, the choice of companion intercrops and the cost of cultivation are major influencing factors for adopting a legume–millet intercropping system by smallholder farmers. Similar scenarios have been seen in several African countries, corroborating our study in south India. For example, a study by Ortega et al. (2016) in Malawi found that labour constraints and market access are the key factors governing farmers' adoption of maize–legume intercropping practices despite their ecological benefits. Another important variable that is considered in the decision-making process is the availability of family labour while adopting intercropping systems in Uganda (Ekepu and Tirivanhu 2016). Last, but not least, the value-chain models for the intercropping system have yet to be developed for different combinations of crops and biological inputs, including the biofertilizers, so as to make millet-based a model for sustainable farming, both in socioeconomic and environmental dimensions. A study by Magrini et al. (2016) has pointed out the need for combined actions both in the research and development of catalytic technologies, institutional innovations and market support to upscale the adoption of millet–legume intercropping systems. Similarly, access to quality seeds and markets, as well as policy support, were highlighted as key factors essential to upscaling millet and legume intercropping systems in sub-Saharan Africa (Mugendi et al. 2011).

Conclusions

Agroecological farming methods, such as intercropping, have been in use by farmers for centuries. However, practicing intercropping on a larger scale has been always a challenge for various reasons that include limited understanding of complex interactions, particularly below-ground microbial interactions at the rhizosphere, involving various biotic and abiotic factors. Thanks to recent advancements in molecular and ecophysiological tools, there has been an increasing amount of research on the role of microbes in the intercropping system, eventually allowing us to explore possibilities for adapting microbial-facilitated

278 Natarajan Mathimaran et al.

intercropping to a larger scale. Yet mainstreaming intercropping requires further basic and applied research to optimize productivity and economic sustainability, as shown by our studies of the millet-based intercropping systems.

References

Aroca, R. and Ruiz-Lozano, J. M. (2009) Induction of plant tolerance to semi-arid environments by beneficial soil microorganisms – a review. in Lichtfouse, E. (ed.) *Climate Change, Intercropping, Pest Control and Beneficial Microorganisms.* Springer, Dordrecht, The Netherlands, pp. 121–135.

Bitew, Y., Alemayehu, G., Adgo, E. and Assefa, A. (2020) Competition, production efficiency and yield stability of finger millet and legume additive design intercropping. *Renewable Agriculture and Food Systems, 36*(1), pp. 1–12.

Brooker, R. W., Bennett, A. E., Cong, W. F., Daniell, T. J., George, T. S., Hallett, P. D., Hawes, C., Iannetta, P. P. M., Jones, H. G., Karley, A. J., Li, L., McKenzie, B. M., Pakeman, R. J., Paterson, E., Schob, C., Shen, J. B., Squire, G., Watson, C. A., Zhang, C. C., Zhang, F. S., Zhang, J. L. and White, P. J. (2015) Improving intercropping: A synthesis of research in agronomy, plant physiology and ecology. *New Phytologist, 206*(1), pp. 107–117.

Caldwell, M. M., Dawson, T. E. and Richards, J. H. (1998) Hydraulic lift: Consequences of water efflux from the roots of plants. *Oecologia, 113*(2), pp. 151–161.

Callaway, R. M. (2007) Indirect mechanisms for facilitation. In: *Positive Interactions and Interdependence in Plant Communities.* Springer, Dordrecht. The Nehtherlands, pp. 117–177.

Daryanto, S., Fu, B., Zhao, W., Wang, S., Jacinthe, P. A. and Wang, L. (2020) Ecosystem service provision of grain legume and cereal intercropping in Africa. *Agricultural Systems, 178,* (Article 102761).

Dass, A. and Sudhishri, S. (2010) Intercropping in fingermillet (*Eleusine coracana*) with pulses for enhanced productivity, resource conservation and soil fertility in uplands of Southern Orissa. *Indian Journal of Agronomy, 55*(2), pp. 89–94.

Debenport, S. J., Assigbetse, K., Bayala, R., Chapuis-Lardy, L., Dick, R. P. and Gardener, B. B. M. (2015) Association of shifting populations in the root zone microbiome of millet with enhanced crop productivity in the Sahel region (Africa). *Applied and Environmental Microbiology, 81*(8), pp. 2841–2851.

Diakhaté, S., Gueye, M., Chevallier, T., Diallo, N. H., Assigbetse, K., Abadie, J., Diouf, M., Masse, D., Sembène, M., Ndour, Y. B., Dick, R. P. and Chapuis-Lardy, L. (2016) Soil microbial functional capacity and diversity in a millet-shrub intercropping system of semi-arid Senegal. *Journal of Arid Environments, 129*, pp. 71–79.

Duchene, O., Vian, J. F. and Celette, F. (2017) Intercropping with legume for agroecological cropping systems: Complementarity and facilitation processes and the importance of soil microorganisms. A review. *Agriculture Ecosystems & Environment, 240*, pp. 148–161.

Ekepu, D. and Tirivanhu, P. (2016) Assessing socio–economic factors influencing adoption of legume-based multiple cropping systems among smallholder sorghum farmers in Soroti, Uganda. *South African Journal of Agricultural Extension, 44*(2), pp. 195–215.

King, E. D. I. O., Nambi, V. A. and Nagarajan, L. (2009) Integrated approaches in small millets conservation: A case from Kolli Hills, India. in Jaenicke, H., Ganry, J., Hoeschle Zeledon, I. and Kahane, R., (eds.) *International Symposium on Underutilized Plants for Food Security, Nutrition, Income and Sustainable Development.* International Society of Horticultural Science, Leuven, Belgium, pp. 79–84.

Li, B., Li, Y. Y., Wu, H. M., Zhang, F. F., Li, C. J., Li, X. X., Lambers, H. and Li, L. (2016) Root exudates drive interspecific facilitation by enhancing nodulation and N2 fixation. *Proceedings of the National Academy of Science USA, 113*(23), pp. 6496–501.

Li, Y.-Y., Yu, C.-B., Cheng, X., Li, C.-J., Sun, J.-H., Zhang, F.-S., Lambers, H. and Li, L. (2009) Intercropping alleviates the inhibitory effect of N fertilization on nodulation and symbiotic N 2 fixation of faba bean. *Plant and Soil, 323*(1–2), pp. 295–308.

Liste, H.-H. and White, J. C. (2008) Plant hydraulic lift of soil water - implications for crop production and land restoration. *Plant and Soil, 313*(1–2), pp. 1–17.

Magrini, M.-B., Anton, M., Cholez, C., Corre-Hellou, G., Duc, G., Jeuffroy, M.-H., Meynard, J.-M., Pelzer, E., Voisin, A.-S. and Walrand, S. (2016) Why are grain-legumes rarely present in cropping systems despite their environmental and nutritional benefits? Analyzing lock-in in the French agrifood system. *Ecological Economics, 126*, pp. 152–162.

Maitra, S. (2020) Intercropping of small millets for agricultural sustainability in drylands : A review. *Crop Research, 55*(4), pp. 162–171.

Mathimaran, N., Jegan, S., Thimmegowda, M. N., Prabavathy, V. R., Yuvaraj, P., Kathiravan, R., Sivakumar, M. N., Manjunatha, B. N., Bhavitha, N. C., Sathish, A., Shashidhar, G. C., Bagyaraj, D. J., Ashok, E. G., Singh, D., Kahmen, A., Boller, T. and Mäder, P. (2020) Intercropping transplanted pigeon pea with finger millet: Arbuscular mycorrhizal fungi and plant growth promoting rhizobacteria boost yield while reducing fertilizer input. *Frontiers in Sustainable Food Systems, 4*, p. 88.

McDonagh, J. E. and Hillyer, A. E. M. (2003) Grain legumes in pearl millet systems in northern Namibia: An assessment of potential nitrogen contributions. *Experimental Agriculture, 39*(4), pp. 349–362.

Mugendi, D. N., Waswa, B. S., Mucheru-Muna, M. W., Kimetu, J. M. and Palm, C. (2011) Comparative analysis of the current and potential role of legumes in integrated soil fertility management in East Africa. in *Fighting Poverty in Sub-Saharan Africa: The Multiple Roles of Legumes in Integrated Soil Fertility Management.* Bationo, A., Waswa, B., Okeyo, J.M., Maina, F., Kihara, J., Mokwunye, U. (eds.) Springer, Dordrecht, The Nehtherlands. pp. 151–173.

Ortega, D. L., Waldman, K. B., Richardson, R. B., Clay, D. C. and Snapp, S. (2016) Sustainable intensification and farmer preferences for crop system attributes: Evidence from Malawi's central and southern regions. *World Development, 87*, pp. 139–151.

Querejeta, J. I., Egerton-Warburton, L. M., Prieto, I., Vargas, R. and Allen, M. F. (2012) Changes in soil hyphal abundance and viability can alter the patterns of hydraulic redistribution by plant roots. *Plant and Soil, 355*(1–2), pp. 63–73.

Runkulatile, H., Horama, K., Horie, T., Kurusu, T. and Inamura, T. (2015) Land equivalent ratio of groundnut-fingermillet intercrops as affected by plant combination ratio, and nitrogen and water availability. *Plant Production Science, 1*(1), pp. 39–46.

Saharan, K., Schütz, L., Kahmen, A., Wiemken, A., Boller, T. and Mathimaran, N. (2018) Finger Millet Growth and Nutrient Uptake Is Improved in Intercropping With Pigeon Pea Through "Biofertilization" and "Bioirrigation" Mediated by Arbuscular Mycorrhizal Fungi and Plant Growth Promoting Rhizobacteria. *Frontiers in Environmental Science, 6*(46), pp. 1–11.

Sekiya, N. and Yano, K. (2004) Do pigeon pea and sesbania supply groundwater to intercropped maize through hydraulic lift? Hydrogen stable isotope investigation of xylem waters. *Field Crops Research, 86*(2–3), pp. 167–173.

Singh, D., Mathimaran, N., Boller, T. and Kahmen, A. (2020) Deep-rooted pigeon pea promotes the water relations and survival of shallow-rooted finger millet during drought-Despite strong competitive interactions at ambient water availability. *PLoS One, 15*(2), pp. e0228993.

Swaminathan, M. S. (2010) Achieving food security in times of crisis. *New Biotechnology,* 27(5), pp. 453–460.

Triveni, U., Sandhya Rani, Y., Patro, T., Anuradha, N. and Divya, M. (2017) Evaluation of different finger millet based intercropping systems in the north coastal zone of Andhra Pradesh. *IJCS,* 5(5), pp. 828–831.

Yu, Y., Stomph, T. J., Makowski, D., Zhang, L. and van der Werf, W. (2016) A meta-analysis of relative crop yields in cereal/legume mixtures suggests options for management. *Field Crops Research, 198,* pp. 269–279.

Zegada-Lizarazu, W., Izumi, Y. and Iijima, M. (2006) Water competition of intercropped pearl millet with cowpea under drought and soil compaction stresses. *Plant Production Science, 9*(2), pp. 123–132.

24

STATUS OF MINOR MILLETS PROCESSING TECHNOLOGIES IN INDIA

An overview

B. Dayakar Rao and Vilas A Tonapi

Introduction

Minor-grained cereal grasses are collectively described as 'millets' and are one of the oldest cultivated foods known to humans. The major group of millets includes crops like sorghum and pearl millet, while minor millets consist of finger millet, foxtail, kodo, barnyard, proso and little millet.

Millets are a traditional staple food of the dryland regions of the world. In India, they are grown in about 15 million ha with an annual production of 16 million tonnes and contribute slightly less than 10% to the country's food-grain basket. Nutri-cereals are known to have high nutrient content, including protein, essential fatty acids, dietary fiber, B-vitamins, and minerals such as calcium, iron, zinc, potassium and magnesium. Although millets are nutritionally rich like other fine cereals, their consumption has significantly declined over the last three decades. The decline is mainly due to the rather laborious and time-consuming process involved in the preparation of millet and government policy to supply fine cereals at subsidized prices. Therefore, it has become necessary to reorient efforts on millets to generate demand through value addition of processed foods. Value addition in food processing has a high degree of interdependence with forwarding and backward linkages, which can play an essential role in accelerating economic development. Through value addition, the shelf life and storage quality of food can be improved.

Limitations of minor millet processing in India and world

- Lack of appropriate processing technologies that yield stable shelf products constitutes a significant limitation in the utilization of millet grain to develop value-added products.

DOI: 10.4324/9781003044802-27

- Consumer-level preferences include food that is convenient, tasty and attractive, with good texture, which may not always be the case with millet products.
- The size of grain is minor so processing is a challenge.
- The outer seed coat of most minor millets has about 1–7 layers, making it further difficult to process with available means.
- Polishing minor millets as done for major food cereals may lead to nutrition losses, defeating the purpose of consuming it.
- Lack of proper mechanism to separate unhulled grains from de-hulled grains.
- No serious processing interventions are attempted because of lack of demand for these millets.
- The market is limited as primarily poor farmers from tribal regions and typical dryland regions produce and consume millets.
- Absence of grades and standards.
- Lack of awareness with regard to grain and product quality standards.
- Regular supply of grains is inadequate for demand from large-scale manufacturing lines.
- Non-availability of local millet processing units.
- Lack of grain storage options.

Importance of processing

Processing grains enhances its consumer acceptability and adds to its convenience. Though millet is known for its nutritional qualities, its consumption has declined due to the unavailability of ready-to-eat (RTE) and ready-to-cook (RTC) millet products. It is also regarded in many countries as an inferior food. A decrease in millet consumption is found to be proportional to an increase in expenditure. Besides, increased income is accompanied by increased consumption of wheat and rice, as products from these fine cereals are easy to prepare and have better 'keeping' quality. At the same time, people now tend to eat a greater variety of foods. Technological change could perhaps shift the status of millets, improving its production and utilization. Processing also improves the food value in terms of increasing the variety of products and improving carbohydrate and protein digestibility.

Primary processing

The primary processing of millets revolves around the grading of millets (which depends on the size of the grain), de-stoning (removal of stones and other impurities), de-hulling (removal of outer indigestible husk layer), and polishing. Its essential aim is the removal of waste, stones, and glumes from the grain, which are necessary for proper storage and consumer acceptance. The majority of primary processing is still performed with inefficient machinery and demands skilled operators, making it uneconomical for farmers and minor-scale entrepreneurs. The primary processing of millets is a vital step to converting the grain into an edible form and thereby enhancing its quality. Although processing millets without husk (naked grains) – i.e., sorghum, pearl, and finger millets – are easy, processing millets that do have husk – i.e., little, proso, kodo, barnyard, and foxtail millets – are more complicated, as they have an inedible husk that needs to be removed through primary processing (Figures 24.1, 24.2).

Constraints identified in the existing primary processing machinery

1 As of now, no machinery is available that can effectively perform the process of cleaning, grading, de-hulling, and separation. The development of

FIGURE 24.1 Cleaning, grader, and aspirator. A: Single-stage De-huller, B: Double-stage De-huller, C: Millet De-huller, D: Minor De-huller.

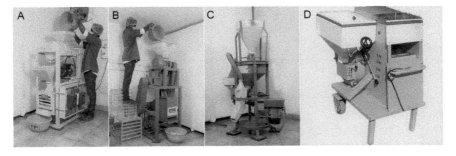

FIGURE 24.2 Different de-hullers available for primary processing of millets.

284 B. Dayakar Rao and Vilas A Tonapi

such machinery will prove to be a helping aid for various entrepreneurs and minor-scale industries.

2   The de-hulling efficiency of millets is influenced by impeller speed, hence a provision to control the working speed of machines should be incorporated.

3   Separation of de-hulled and raw grains from the de-huller output is due to the slight difference in specific gravity. Hence, because of improper sieve sizes and the design of reciprocating sieves, nearly 1.0–1.5 kg of material falls through the sieve holes and accumulates deep inside the machine, leading to wastage.

4   The collection of husk is burdensome, and causes spillages all over the working station, sometimes getting mixed with de-hulled output.

5   After the de-hulling activity, the next stage is to separate the de-hulled grains and raw grains. The separation stage is imperative and dramatically affects overall grain recovery; nearly 1 kg of de-hulled grains (for every cycle of separation) is prone to wastage due to inefficient separation methodology.

ICAR-IIMR studies on primary processing to increase efficiency

Identified suitable machines for different millet grains

Not all minor millets can be de-hulled by one machine due to the different size and shape of each millet. Specific and suitable machine is required for proper recovery.

A study was conducted on minor millets by using different existing de-hulling machines, and the recovery was observed (Table 24.1). The study indicated that the minor de-huller was best suited for barnyard, foxtail, and kodo millet, while the single-stage de-huller best fit proso and the double-stage was suitable for little millet.

Secondary processing

Secondary processing involves using the primary processed raw material for different RTE and RTC millet products such as flour, millets-based multigrain flour, semolina (fine and medium), millet cookies, vermicelli, pasta, etc., which minimize the cooking time needed and make them convenient foods. It also increases the shelf life of products. Busy lifestyles and the lack of inspiration to

TABLE 24.1 De-hulling efficiency of different millet grain by using different de-hullers

S. No	Type of De-huller	Barnyard millet	Foxtail millet	Kodo millet	Little millet	Proso millet
01	Single-stage	50.23	61.5	22.07	77.43	**85.82**
02	Double-stage	34.13	52.2	51.46	**85.36**	52.87
03	Millet mill	53.63	38.49	54.03	72.7	19.52
04	Minor de-huller	**73.88**	**88.08**	**57.17**	78.7	84.85

cook make convenient nutritious RTC and RTE products more appealing and accessible (Alavi et al., 2019). These products have a multi-billion dollar market worldwide. ICAR-IIMR has developed and standardized millet-based products such as *atta*, semolina (fine and medium), flakes, cookies, cold extruded products (pasta and vermicelli), RTE snacks, etc. using different processing technologies.

Flaking technology

Flaking is the process where moistened and roasted grain is pressed into flattened flakes and dried. It can be applied to any millet grain, irrespective of variety, size, and shape. However, the pericarp color of the grain impacts the color of the flakes.

The machines used for different flaking technology are:

1. Edge runner – produces thick flakes like rice flakes
2. Roller flaker – produces thin flakes like corn flakes

This technology comprises many sub-processes like soaking the grains in excess water to hydrate their equilibrium moisture content. After soaking, the grains are washed appropriately by changing the water thrice; otherwise, the grain's sliminess will affect the storage life of the flakes. This soaking method is similar for both edge-runner and roller-flakes processing (Figures 24.3, 24.7, 24.8).

Cold extrusion technology

Cold extruded products such as vermicelli, pasta, and noodles are generally made from durum wheat or refined wheat flour. These products can also be manufactured with millets by the same machine, called a "cold extruder." In millet-based cold extruded products, the raw material should be fine-sized millet semolina (355 µ); otherwise, the binding nature will be very low, which would increase the cooking losses. Currently, every millet-based product is blended with wheat semolina/*suji* (500 µ), as the millet grains are devoid of gluten, and so lack the binding properties and strength of the texture. Extruded millet products are mixed in a blender with different ingredients and desired moisture levels to make a wet homogenous mass. Subsequently, it is forced through an opening in a perforated plate or dies with a design specific (it varies for vermicelli, pasta, and

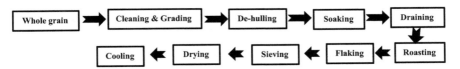

FIGURE 24.3 Processing steps of flaking line. **Outputs**: Flakes (thick, thin, and extruded), broken and bran.

noodles), and is cut to a specified size by blades. Now this moist product goes through the drying process (Figures 24.4, 24.9).

Puffing technology (grain expansion technology)

The popular and world-famous snack popcorn is commonly made from maize. Likewise, expanded rice or *murmura* is also a very popular and simple traditional snack. However, similar products from other cereals are rare. "Gun popping" is another traditional method of cereal popping, widely used in East Asian countries, but not popular in India. Recent R&D has shown that millet-expanded snacks can also be processed using the gun popping method. Mainly this technology requires decorticated and polished grains to deliver puffs/pops with high functional and highly acceptable sensory qualities. The puffing occurs due to the quick transmission of heat from the barrel to the grain and converts its moisture levels into steam form. It gelatinizes the starch and produces high pressure in the core part of the grain, resulting in the expansion of the grain, forming its puffy nature. These products are milky white and have a pleasant aroma, with an attractive appearance and soggy texture. The roasting process can be applied to improve its crispiness (Figures 24.5, 24.10).

Baking technology

Baked goods are RTE and the most popular snack foods across the world. They include cookies, biscuits, bread, buns, rusks, doughnuts, etc., with cookies being the most consumed by the population. Currently, the ICAR-IIMR has developed 100% pure millet-based cookies and cakes with improved textures by applying different emulsifiers, which are closely comparable to famous wheat-based cookies in the market. The machine used for cookie and cake processing

FIGURE 24.4 Processing steps of cold extrusion line. **Outputs**: Vermicelli, pasta, noodles, etc.

FIGURE 24.5 Processing steps of gun puffing line. **Outputs**: Puffed and unpuffed grains

are the cookie/biscuit-cutting machine, rotary rack oven, cake-filling machine, planetary mixer, etc. In the cookie process, the millet in the form of fine flour (150 μ and 180μ particle size) is used. Millet cookies are manufactured by the creaming process (vegetable fat/vegetable oil and sugar) and the addition of leavening agents, flavors, and millet flour, along with desired moisture levels, and making it into soft dough, cutting and baking at 180 °C for 18–20 minutes. Its attractiveness can be improved by incorporating different ingredients like cashew nuts, almonds, chocolate chips, etc. These products lack the sponginess of wheat bread and pizzas because they do not contain gluten protein (Collar et al., 2019) (Figures 24.6, 24.11).

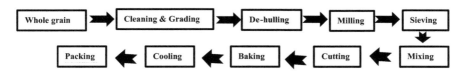

FIGURE 24.6 Processing steps of cookies line. **Outputs**: Biscuits and cookies.

FIGURE 24.7 Edge runner machine.

FIGURE 24.8 Roller flaking machine.

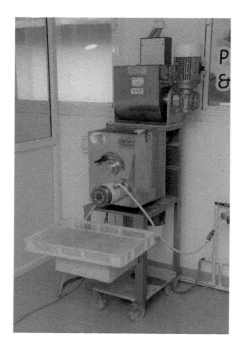

FIGURE 24.9 Cold extruder.

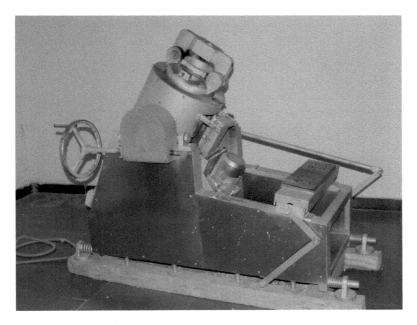

FIGURE 24.10 Gun puff machine.

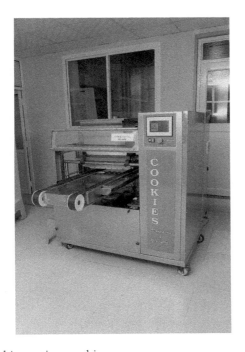

FIGURE 24.11 Cookie-cutting machine.

Inclusion of minor millets in public-funded schemes – ICDS, mid-day meal program

The government programs spearheaded by the Ministry of Agriculture were primarily focused on improving the supply of millets. Minor millets are naturally nutrient-dense cereals (Malleshi et al., 1982), and so making them available through public-funded schemes can help deal with the problem of micronutrient deficiency or hidden hunger among the less-privileged sections of the population. Within the Indian context, as a healthy food, minor millets have great potential to address this challenge due to their nutritional qualities (Durairaj et al., 2019). The effective delivery of millets under public-funded schemes could have far-reaching implications for tackling the problem of malnutrition. Most public food programs do not include millets, except for the inclusion of finger millet in Integrated Child Development Scheme (ICDS) in a few states of the country. There are no exclusive government schemes/projects/programs for minor millets. However, many Indian states are gradually starting to support the consumption of millets through government-sponsored schemes; for instance, Karnataka was the first state to distribute millets through the PDS; Odisha has started the Millets Mission; Andhra Pradesh supplies finger millets and jowar through PDS, and Tamil Nadu has included millets in its MGR Nutritious Meal Programme 2011-12 and in ICDS. It is necessary to include millets in public-funded schemes in all states to minimize malnutrition and mortality rates throughout India.

Policy support

There are limited policies and schemes that explicitly include millets, with not many exclusive government schemes/projects/programs for minor millets. Of the extant schemes, the most important ones are Initiative for Nutritional Security through Intensive Millets Promotion as part of the Rashtriya Krishi Vikas Yojana (RKVY); Rainfed Area Development Programme as part of Rashtriya Kristi Visas Yolanda (RKVY); and Integrated Cereals Development Programmed in Coarse Cereals-based Cropping Systems Areas (ICDP-CC) under the Macro Management of Agriculture. There is a lot of variation across the states on how they utilize these opportunities for promoting millets.

Conclusion

The processing of minor millets has come a long way; however, one grey area is the lack of availability of efficient de-hullers and separators in primary processing that have a greater than 80% efficiency. This is an important determinant for secondary processing, as quality raw material will lead to greater scope of diversified millet products through secondary processing. Secondary processing technologies like flaking, puffing, milling, and baking are standardized, but current raw material costs do not favor viable products except as niche markets. The greater the efficiency in primary processing, the higher the share of farmer producers in the consumer's rupee. Further bioavailability studies of millet-based products need to be conducted to develop standards and grades, to identify suitable minor millet cultivars for specific end-products for large-scale production, and to strengthen the use of minor millets, which are storehouses of nutrition. There should be an in-depth policy shift from the government towards accommodating the inclusion of minor millets in public-funded schemes to address malnutrition and lifestyle diseases.

References

Alavi, S., Mazumdar, SD., and Taylor, J.R. (2019). Chapter 10 - Modern Convenient Sorghum and Millet Food, Beverage and Animal Feed Products, and Their Technologies. In John R.N. Taylor, Kwaku G. Duodu, Sorghum and Millets (eds.), *Sorghum and Millets Book* (Second Edition), pp. 293--329. AACC International Press. doi: 10.1016/B978-0-12-811527-5.00010-1

Collar, C. (2019). Gluten-Free Dough-Based Foods and Technologies. In Sorghum and Millets Book (Second Edition), pp. 331--354. AACC International Press.

Durairaj, M., Gurumurthy, G., Nachimuthu, V., Muniappan, K., and Balasubramanian, S. (2019). Dehulled minor millets: The promising nutricereals for improving the nutrition of children. *Maternal & Child Nutrition*, 15, p.e12791.

Malleshi, N.G., and Desikachar, HSR. (1982). Formulation of a weaning food with low hot paste viscosity based on malted ragi (*Eleusine coracana*) and green gram (*Phaseolus radiatus*). *Journal of Food Science and Technology*, 19(5), pp.193–197.

PART IV

Nutritional and food security roles of minor millets

25
MODELLING THE FOOD SECURITY ROLE OF MILLETS UNDER CLIMATE CHANGE IN EASTERN MADHYA PRADESH

Gennifer Meldrum, Victoria Rose, Somnath Roy and Ashis Mondal

Introduction

Small millets are well adapted to poor quality soils and rainfed, low moisture conditions, while also producing grains that have superior nutrient values compared to paddy and wheat (Saleh et al. 2013; Behera 2017, Figure 25.1). For these reasons, small millets are recognized as key assets for ensuring food security under climate change (Padulosi et al. 2015; Davis et al. 2019). To inform strategies to leverage their role in climate change adaptation, the precise mechanisms by which small millets support food security should be clarified in consideration of the many parameters and constraints of farm and livelihood systems.

Millets may contribute to climate change adaptation through several mechanisms. Because of their drought tolerance, including millets in a diversified

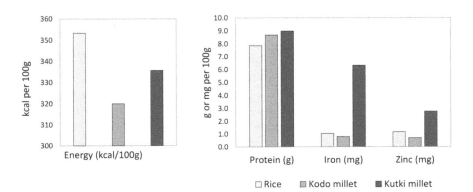

FIGURE 25.1 Nutrient content of minor millets.
Source: Saleh et al. (2013), Shobana et al. (2013), Longvah et al. (2017).

DOI: 10.4324/9781003044802-29

farm portfolio can support harvest security in drought years (e.g., Prieto et al. 2015; Renard and Tilman 2019). If drought stress becomes very severe, minor millets could replace more drought-sensitive staples (e.g., Kurukulasuriya and Mendelsohn 2008). Because of their capacity to produce a harvest on marginal soils, minor millets could enable more areas in the landscape to be brought into production, which can help mitigate the effect of yield losses resulting from climate change. The ability of millets to be stored for long periods can also provide a fallback source of nutrition and income under variable weather conditions. The contribution of millets to climate change adaptation may derive from one or a combination of these mechanisms, among other possibilities.

To sharpen understanding of the mechanisms by which minor millets support food security under climate change, a scenario modelling exercise was carried out. The study focused on the Mandla and Dindori districts of eastern Madhya Pradesh, India, where kodo (*Paspalum scrobiculatum*) and kutki (*Panicum sumatrense*) millets are cultivated by tribal farmers in marginal lands. Kodo and kutki are grown especially on rocky and sloping lands, whereas flatter richer soils tend to be used for paddy cultivation. Millets have received limited attention from the research and development perspective in India. Agricultural schemes in Madhya Pradesh have even pushed to replace these "low-value crops" with higher-value ones like oil seeds and pulses in recent decades (Gupta 2003; Catalyst Management Services Ltd. 2009). Improved rice varieties have been promoted in the region, while improved varieties of millet have only been promoted more recently and their adoption is still limited (Bioversity International and ASA 2016).

Methods

A simple model for household production and consumption was devised as summarized in Figure 25.2. The model considered household production of rice and minor millets, storage of rice and millet, and consumption of millets and rice in fulfilling household energy requirements. The production levels of millet and rice varied depending on whether or not drought was faced in a given year. Once family requirements were met, surplus grain from the current year's production would be placed in storage. The model assumed that the household would preferentially consume rice over millet and fresh grains over stored grain.

The model was built as a series of linked equations as presented in Table 25.1 using the R statistical programing language (R Core Team 2018). The parameters for the model are summarized in Table 25.2. A single iteration of the model provided data for an individual household over a 20-year span. For a given set of parameters, the model was applied over the full range of drought risk values (0–1) and repeated for 1,000 iterations. The initial amount of rice in storage was set as half of the household's annual energy requirement, as was done for the initial store of millets. Three distinct investigations were completed with the model in which the yield values or the area planted with rice and millet were varied, while

Millets and climate change in Madhya Pradesh **297**

the remaining parameters were left constant. The questions explored in these investigations were as follows:

- Investigation 1: How do different values for crop yield in drought and good years affect household food security under increasing levels of drought risk?
- Investigation 2: How does the area planted with small millets and rice affect household food security under increasing levels of drought risk?
- Investigation 3: How would increasing yield values for small millets affect household food security under increasing levels of drought risk?

The first investigation explored the effect of different yield values in drought and non-drought years. The values for millet and rice yields under drought and non-drought conditions were defined with reference to the government's yield and weather data for a 35-year period (1997–2014). General linear models of millet and paddy yields obtained for total rainfall by district were developed and the predicted values were used in defining the parameters for the scenario model. For drought years, we used two alternative yield values, which were the predicted yields under the minimum level of precipitation experienced over the previous 20 years (Min) and the predicted yields for below-average precipitation levels (Q1). For non-drought years, we used four alternative yield values, which were the predicted yields for mean precipitation levels (Avg); the predicted yields for above average precipitation (Q3); the predicted yields under the maximum level of precipitation (Max); and the absolute highest yields for rice and millet in the dataset (AbsMax). The combinations of yield values that were simulated are shown in Table 25 A1.

The area planted with millet and rice was held constant in the first investigation. The area values were defined based on a household survey in the focal area with 297 households across 30 villages (Bioversity International and ASA 2016). The survey results showed that the mean area households assigned to paddy was 0.6 Ha and 0.6 Ha to small millets. These values were used as the parameters for crop area in the first investigation. Local consultations suggested that farmers use their best quality soil with black color for rice and their rocky red soils not suitable for paddy to grow millet. These discussions suggested that farmers could grow millet in areas where rice is currently grown but that rice may not be easily cultivated in areas where farmers currently grow millet. The second investigation explored this nuance, in considering that intensifying drought conditions may mean that planting millet on black soils – effectively replacing some area of rice cultivation – could bring greater harvest and food security. The combinations of areas planted to rice and millet are shown in Table 25A2, which assumes millets could be grown on land currently used by rice but not vice versa. The simulations in the second investigation used the most optimistic yield values (Q1 in drought years, AbsMax in good years).

The third investigation assessed the role of improved millet varieties in supporting climate change adaptation. Starting with more optimistic yield values

TABLE 25.1 Scenario model as a series of linked equations.

No	step	Description
1	**Ereq**=5,325,350 **RiceE**=3,533,200 **MilE**=3,278,500 **RiceYLow**=variable **RiceYHigh**=variable **MilYLow**=variable **MilYHigh**=variable **DroughtProb**=variable **RiceEStore0**=variable **MilEStore0**=variable	The first step is to define the initial parameters for the model: • The family energy requirements (**Ereq**) was set to 5,325,350 kcal per annum, which corresponds to the energy requirement for a five-person family (one man + one woman + one avg of man and woman two children [average for all child values]) with a heavy work requirement, using values in NIN (2009). • The energy content of rice (**RiceE**) was set to 3,533,200 kcal/ton (Longvah et al., 2017; Saleh et al., 2013; Shobana et al., 2013) • The energy content of millet (**MilE**) was defined as 3,278,500, which is the mean value for kodo and kutki (combined: 327.85 kcal/100g; kodo: 320 kcal/100g; kutki: 335.69 kcal/100g; Longvah et al., 2017; Saleh et al., 2013; Shobana et al., 2013) • Yield values for drought years (**RiceYLow, MilYLow**, value greater than 0) and non-drought years (**RiceYHigh, MilYHigh**, value greater than 0) were defined with reference to government yield data (see Tables 25A1 and 25A2) • The drought probability (**Drought Prob**) was set to a value ranging from 0 to 1 • The initial quantity of rice and millet in storage (**RiceEStore0** and **MilEStore0)** was set to half the energy requirement of the household in the first year and as the amount remaining from the previous year for subsequent iterations (**RiceEStore1**: defined in Step 7; **MilEStore1:** defined in Step 11).
2	**Drought**=sample(c(0:100),size=1,replace=T) <(**DroughtProb**★100)	Whether a drought occurs in given year (**Drought,** TRUE/FALSE) depends on the probability of drought (**DroughtProb:** defined in Step 1).

3	RiceYield= if(**Drought**==TRUE) **RiceYLow** else **RiceYHigh** **MilYield**= if(**Drought**==TRUE) **MilYLow** else **MilYHigh**	Rice and millet yield in a given year (**RiceYield** and **MilYield,** Ha) depend on whether drought occurs (**Drought:** defined in Step 2). In drought years, yields are lower than in non-drought years (**RiceYLow**<**RiceYHigh** and **MilYLow**<**MilYHigh**: defined in Step 1).
4	**ProdRiceE**=**RiceYield*AreaRice*RiceE** **ProdMilE** =**MilYield*AreaMil*MilE**	The total energy secured from the rice and millet harvest in a given year (**ProdRiceE** and **ProdMilE,** kcal) is a product of the area of the crop planted (**AreaRice** and **AreaMil**: defined in Step 1), the yield (**RiceYield** and **MilYield**: defined in Step 3) and the energy content of the crop (**RiceE** and **MilE:** defined in Step 1).
5	**RiceESurProd**=if(**ProdRiceE-Ereq**>0) (**ProdRiceE-Ereq**) else 0	If the energy from the rice harvest (**ProdRiceE**: defined in Step 4) is greater than family requirements (**Ereq**: defined in Step 1), then the surplus (**RiceESurProd**, kcal) is the difference between production (**ProdRiceE:** defined in Step 4) and use (**Ereq**: defined in Step 1). Otherwise, the surplus (**RiceESurProd**, kcal) is 0 because all the rice produced will be consumed.
6	**RiceEConSto**=if(**ProdRiceE-Ereq**>=0) 0 else if (**RiceEStore0-(Ereq-ProdRiceE)**>=0) **Ereq-ProdRiceE** else **RiceEStore0**	It was assumed that the household would consume fresh production over stored grains. Under these conditions, if the production of rice (**ProdRiceE**: defined in Step 4) is greater than family needs (**Ereq**: defined in Step 1), the consumption from storage (**RiceEConSto**, kcal) will be 0. It was also assumed that if rice production from the current year does not meet family needs, then the household will try to make up their energy needs from stored rice before consuming millet. Under these conditions, if the production of rice (**ProdRiceE**: defined in Step 4) and availability of rice in storage (**RiceEStore0:** defined in Step 1) meet family energy needs (**Ereq**: defined in Step 1), the amount of rice consumed from storage (**RiceEConSto**, kcal) will be the difference between family energy requirements (**Ereq**: defined in Step 1) and the amount of rice energy produced (**ProdRiceE**: defined in Step 4). If the availability of rice from production (**ProdRiceE**: defined in Step 4) and storage (**RiceEStore0:** defined in Step 1) does not meet family needs, then the household will consume all the rice available in storage (**RiceEConSto**, kcal).

(Continued)

No	step	Description
7	**RiceEStore1=RiceEStore0-RiceEConSto+RiceESurProd**	The amount of rice energy in storage at the end of a given year (**RiceEStore1**, kcal) is calculated as the amount remaining in storage from the previous year (**RiceEStore0**: defined in Step 1) minus the amount consumed (**RiceEConSto**: defined in Step 6) or plus the amount added from surplus production (**RiceESurProd**: defined in Step 5)
8	**RiceECons=if((ProdRiceE+RiceEStore0)>=Ereq) Ereq else (ProdRiceE+RiceEStore0)**	If production (**ProdRiceE**: defined in Step 4) and storage availability of rice (**RiceEStore0**: defined in Step 1) meet family requirements (**Ereq**: defined in Step 1), then the amount of rice consumed (**RiceECons**, kcal) will be equal to the energy requirement of the family (**Ereq**: defined in Step 1). Otherwise, the family will consume all of their available rice from the current year's production (**ProdRiceE**: defined in Step 4) and their stored grains (**RiceEStore0**).
9	**MilESurProd=if((ProdRiceE+RiceEStore0)>=Ereq) ProdMilE else if (ProdMilE >= (Ereq - (ProdRiceE + RiceEStore0))) (Ereq - (ProdRiceE + RiceEStore0)) else 0**	If rice production (**ProdRiceE**: defined in Step 4) and storage (**RiceEStore0**: defined in Step 1) meet the family needs (**Ereq**: defined in Step 1), then all millet production (**ProdMilE**: defined in Step 4) is surplus (**MilESurProd,** kcal) and will go to storage (see Step 11). Otherwise, if millet production from the current year (**ProdMilE**: defined in Step 4) is able to make up the gap, then the surplus millet (**MilESurProd**, kcal) will be the amount remaining after complementing the shortfall in rice. If millet production cannot fully make up the gap in energy required by the household, the surplus millet (**MilESurProd**, kcal) will be 0 because all the millet produced will be consumed.
10	**MilEConSto=if((ProdRiceE+RiceEStore0)>=Ereq) 0 else if (ProdMilE >= (Ereq - (ProdRiceE + RiceEStore0))) 0 else if ((ProdMilE+MilEStore0+ProdRiceE + RiceEStore0)>=Ereq) (Ereq - (ProdMilE + ProdRiceE + RiceEStore0)) else MilEStore0**	If rice from production (**ProdRiceE**: defined in Step 4) and storage (**RiceEStore0**: defined in Step 1) has met the subsistence need of the family (**Ereq**: defined in Step 1), then no millet will be consumed (**MilEConSto**, kcal). If rice does not meet the needs of the family and millet production from the current year (**ProdMilE**: defined in Step 4) is able to fill the gap, then the amount of millet consumed from storage (**MilEConSto**, kcal) will be 0. If the current year of millet production is unable to meet the gap and the millet available in storage (**MilEStore0**: defined in Step 1) can meet the family needs, then the amount of millet consumed (**MilEConSto**, kcal) will be the gap after rice production (**ProdRiceE**: defined in Step 4), rice in storage (**RiceEStore0**: defined in Step 1) and current millet production (**ProdMilE**: defined in Step 4) are used. If the total energy available from production and storage does not meet subsistence needs, then all millet in storage will be consumed (**MilEConSto**, kcal).

11	**MilEStore1=MilEStore0-MilEConSto+ MilESurProd**	The amount of millet energy in storage at the end of a given year (**MilEStore1,** kcal) is calculated as the amount remaining in storage from the previous year (**MilEStore0:** defined in Step 1) minus the amount consumed (**MilEConSto:** defined in Step 10) or plus the amount added from surplus production (**MilESurProd:** defined in Step 9).
12	**MilECons=if((ProdRiceE+RiceEStore0)> =Ereq) 0 else if ((ProdMilE + MilEStore0)>= (Ereq - (ProdRiceE + RiceEStore0))) (Ereq - (ProdRiceE + RiceEStore0)) else (ProdMilE + MilEStore0)**	If production (**ProdRiceE:** defined in Step 4) and storage (**RiceEStore0:** defined in Step 1) availability of rice meet the family's needs, then no millet will be consumed (**MilECons,** kcal). Otherwise, the family will attempt to make up their energy requirements with millets, consuming (**MilECons,** kcal) just enough to fill the gap if quantities are available, or consuming the full amount in storage (**MilEStore0:** defined in Step 1) if there is a shortfall.
13	**Econs=RiceECons + MilECons**	The total energy (**Econs**, kcal) consumed by the family in a given year is assumed to be the total of rice consumed (**RiceECons:** defined in Step 8) and millet consumed (**MilECons:** defined in Step 12)
14	**FoodSecure=(Econs>=Ereq)**	A family is food secure in a given year (**FoodSecure**, TRUE/FALSE) if their consumption (**Econs:** defined in Step 13) is greater or equal to their energy requirements (**Ereq:** defined in Step 1).

in a normal year and a drought year (Q1:AbsMax), the effect of progressive yield increases from 125% to 350% over the base values were modelled, assuming equal percent yield increases in drought and normal years.

Results

Regression of yields by precipitation

Yields of paddy recorded in the region ranged from 0.364 to 2.640 ton/ha with a mean of 0.947 ton/ha. The yields of small millets (kodo-kutki) ranged from 0.089 to 0.903 ton/ha with a mean of 0.266 ton/ha. Precipitation ranged from 480.2 mm to 1595.0 mm with a mean of 1090.0 mm. Above average precipitation (Q3) was 1163.0 mm and below-average precipitation (Q1) was 965.9 mm. The regression analysis showed a significantly increasing trend for yields of both paddy (slope 0.00104±0.00045, t=2.330, p=0.0267) and small millets (slope 0.00038±0.00014, t=2.634, p=0.0132) with higher monsoon precipitation. The slope was steeper for paddy, meaning its yields were more impacted by poor rainfall than the yields of millets, but the yields of small millets were always considerably lower than those for paddy (Figure 25.2).

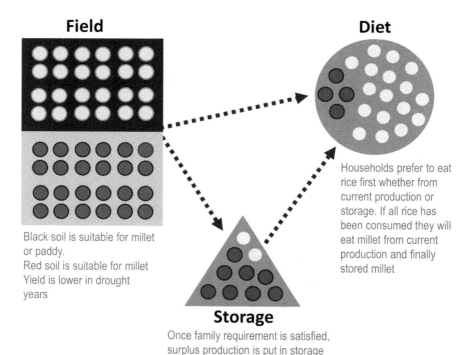

FIGURE 25.2 Overview of the scenario model.

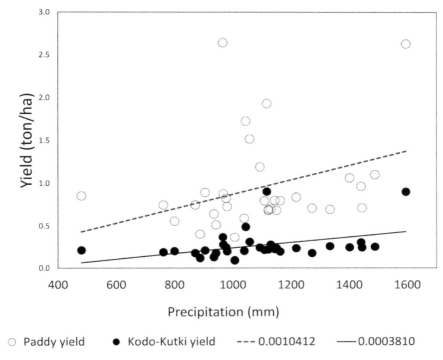

FIGURE 25.3 Regression of paddy and millet yields against precipitation over past 20 years in the Mandla district.

Series 1: varying yields for good and bad years

The scenario modelling revealed that households only achieved their annual energy requirements under the most favourable climate conditions (Figure 25.3 top-left panel). The role of millets for food security was most significant when yields were high in good years, which allowed households to store millets for use in drought years.

Series 2: area planted to small millets

Optimal food security was achieved under the current 'average' situation, with rice grown to the maximal area available with adequate soil quality. The worst outcomes for household food security were seen under the scenario where millet replaced or nearly replaced rice. This was because rice is higher yielding than millets, so producing it on the largest area possible maximized the harvest and the household energy consumption. Millets showed an important food security role when surpluses could be stored in good years.

Series 3: percentage yield increases

The scenario models investigating the effect of yield-enhancements revealed that yield increases of millet as low 150% notably increased food security when drought risk was high (0.5 to 0.8).

Discussion and conclusions

This modelling exercise provided useful insights into the mechanisms by which millets may contribute to climate change resilience and adaptation. A key result was that the low yields of millet limit their role in food security in the face of increasing drought risk. Modest yield increases resulted in food security benefits in our modelling, and the values would be realistic to achieve with existing

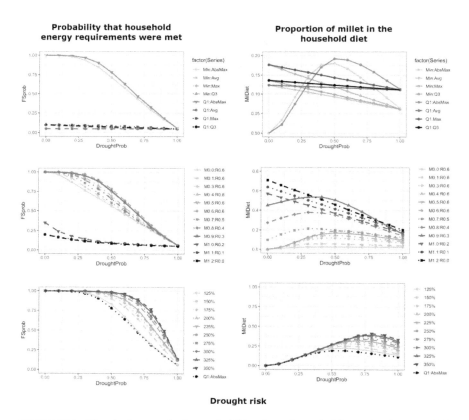

FIGURE 25.4 Simulation results for probability that a household was food secure (left panels) and the proportion of household energy requirements met by millets (right panels). Top row: Results of investigation 1 in which the values for yields in drought and good years were varied. Middle row: Results of investigation 2 in which the area planted with millet and rice was varied. Bottom row: Results of investigation 3 exploring the effect of different percentage yield increases for millets.

varieties. For instance, yields as low as 0.350 tonne/ha in drought years and 1.355 tonne/ha in good years showed notable food security benefits in the model results. Kodo variety DPS 9-1 has a yield ranging from 2.7 to 3.0 tonnes/ha, while the kutki variety DHL M36-3 has a range of 1.4–1.6 tonnes/ha (ICAR nd). The yields of these varieties obtained in farmer fields may be lower than these values achieved under more controlled experimental conditions, but the numbers reveal good potential for these improved varieties (and others that are already available in the region) to contribute to improved food security outcomes.

The major role of millets for household food and climate security was seen in terms of their storability. Millets provide a source of food from marginal soils with little effort that provides a fallback for food security. Discussions with farmers in the project area have highlighted the storability of millets as a cherished quality and our results point to this characteristic as one that deserves more attention in discourses and research regarding the role of millets in climate change adaptation.

It is acknowledged that the model oversimplifies the farm and livelihood system, such as by excluding any links to the market and the diversity of other crops and livestock, as well as wild harvesting, which is quite prevalent. We note that this simplicity was a conscious choice to focus on the tradeoffs in cultivating and consuming the most important cereal crops of the region. The fact that the households generally did not achieve their energy requirements in our models may reveal that there is an important role for other crops, the market, or the public distribution system in supplementing food security, which is not captured by the model. Otherwise, this result may reflect upon the extent of the poor nutrition that exists in Madhya Pradesh, where 60% of children under five years of age are underweight as compared to 43% at the national level (IIPS and Macro International 2007, p 269 and p 273). Scheduled Castes and Tribes, which are the predominant population in the focal area, are disproportionately affected by malnutrition (WFP and IHD 2008; Das and Bose 2015; Jain et al. 2015; Kapoor and Dhall 2016).

It is noted that this was a quick exercise and that the model could benefit from further validation and verification. More sophisticated modelling approaches for cropping systems and household nutrition exist and could provide more reliable results. This simple approach was nevertheless useful because it was possible to devise using open source software. It has clarified expectations and understanding within a short project timeline. The model can easily be modified to consider other nutrients aside from energy and to vary additional parameters that have not been explored in this paper.

Acknowledgments

This paper was prepared in the context of the project 'Linking Agrobiodiversity Value Chains Climate Adaptation and Nutrition: Empowering the Poor to Manage Risk', funded by the European Commission and the International Fund for Agricultural Development (Grant 2000000978) and implemented as part of the CGIAR Research Programmes on Climate Change, Agriculture and Food Security and Agriculture for Nutrition and Health. Discussions with Shambhavi Priyam,

306 Gennifer Meldrum et al.

E.D.I. Oliver King, and Stefano Padulosi were fundamental in conceptualizing the model. Thank you for sharing your insights on the role of millets in improving the livelihoods of farming communities in the focal area and in other regions of India.

Annex

TABLE 25 A1 Yield values derived from government yield and weather data, tonne/ha

Parameter	Conception	Name	Yield value for rice	Yield value for millet
Drought year	Predicted yield for below-average (1st quartile) precipitation in Mandla District (965.9 mm)	Q1	0.878	0.233
	Predicted yield for minimum precipitation experienced over last 20 years in Mandla District (480.2 mm)	Min	0.568	0.096
Non-drought year	Predicted yield for mean precipitation in Mandla District (1090.0 mm)	Avg	0.957	0.268
	Predicted yield for above average (3rd quartile) precipitation in Mandla District (1163.0 mm)	Q3	1.004	0.289
	Predicted yield for maximum precipitation experienced over last 20 years in Mandla District (1595.0 mm)	Max	1.280	0.410
	Absolute maximum yield documented overall (regardless of precipitation or district)	AbsMax	2.640	0.903

TABLE 25 A2 Parameters for millet and rice area simulated in series 2

Parameter combination	Rice area (Ha)	Millet area (Ha)
M0.0:R0.6	0.6	0.0
M0.1:R0.6	0.6	0.1
M0.2:R0.6	0.6	0.2
M0.3:R0.6	0.6	0.3
M0.4:R0.6	0.6	0.4
M0.5:R0.6	0.6	0.5
M0.6:R0.6	0.6	0.6
M0.7:R0.5	0.5	0.7
M0.8:R0.4	0.4	0.8
M0.9:R0.3	0.3	0.9
M1.0:R0.2	0.2	1.0
M1.1:R0.1	0.1	1.1
M1.2:R0.0	0.0	1.2

TABLE 25 A3 Yield increases modelled with base values from Q1:AbsMax (tonne/ha)

Parameters	Non-drought year	Drought year
Q1:AbsMax	0.903	0.233
125%	1.129	0.291
150%	1.355	0.350
175%	1.580	0.408
200%	1.806	0.466
225%	2.032	0.524
250%	2.258	0.583
275%	2.483	0.641
300%	2.709	0.699
325%	2.935	0.757

References

Behera, M.K. (2017) 'Assessment of the state of millets farming in India' *MOJ Ecology & Environmental Science*, vol 2, no 1, pp 16–20.

Bioversity International and Action for Social Advancement (ASA) (2016) Underutilized crops in the livelihoods, diets, and adaptation practices of Gond farmers in Eastern Madhya Pradesh, India: Baseline results from the programme "Linking agrobiodiversity value chains, climate adaptation and nutrition: Empowering the poor to manage risk" Rome, Italy.

Catalyst Management Services Ltd. (2009) Impact assessment of agriculture interventions in Tribal areas in Madhya Pradesh Bhopal, India.

Davis, K.F., Chhatre, A., Rao, N.D., Singh, D., Ghosh-Jerath, S., Mridul, A., Poblete-Cazenave, M., Pradhan, N. and DeFries, R. (2019) 'Assessing the sustainability of post-green revolution cereals in India' *PNAS*, vol 116, no 50, pp 25034–25041.

Indian Council of Agricultural Research (ICAR) (nd) 'Recently released varieties of small millets (2005 to 2018)' ICAR-All India coordinated research project on small millets, http://www.aicrpsm.res.in/Research/Released%20varieties/Small%20Millet%20varieties-2005-2018.pdf

Gupta, S.K. (2003) Agricultural policy and Madhya Pradesh: A policy matrix in a Federal Structure J.N.K.V.V. Agro-Economic Research Centre for Madhya Pradesh & Chhattisgarh, Jabalpur, Madhya Pradesh, India, http://jnkvv.org/PDF/AERC/Study-89.pdf

IIPS and Macro International (2007) National Family Health Survey (NFHS-3), 2005-06: India: Volume I. Mumbai: International Institute for Population Sciences. pp 1–531. https://dhsprogram.com/pubs/pdf/frind/frind3-vol1andvol2.pdf

Jain, Y., Kataria, R., Patil, S., Kadam, S., Kataria, A., Jain, R., Kurbude, R. and Shinde, S. (2015). 'Burden & pattern of illnesses among the tribal communities in central India: A report from a community health programme' *The Indian Journal of Medical Research*, vol 141, no 5, pp 663–672.

Kapoor, A. K. and Dhall, M. (2016) 'Poverty, malnutrition and biological dynamics among tribes of India' *Health Science Journal*, vol 10, no 3, pp 1–5.

Kurukulasuriya, P. and Mendelsohn, R. (2008) 'Crop switching as a strategy for adapting to climate change' *African Journal of Agricultural and Resource Economics*, vol 2, no 1, pp 1–22.

Longvah, T., Ananthan, R., Bhaskarachary, K. and Venkaiah, K. (2017) *Indian Food Composition Tables*. National Institute of Nutrition, Hyderabad, India.

National Institute of Nutrition (NIN) (2009) *Nutrient Requirements and Recommended Dietary Allowances for Indians*. A Report of the Expert Group of the Indian Council of Medical Research. Hyderabad, India.

Padulosi, S., Mal, B., King, O.I. and Gotor, E., (2015) 'Minor millets as a central element for sustainably enhanced incomes, empowerment, and nutrition in rural India' *Sustainability*, vol 7, no 7, pp 8904-8933.

Prieto, I., Violle, C., Barre, P., Durand, J.L., Ghesquiere, M. and Litrico, I. (2015) 'Complementary effects of species and genetic diversity on productivity and stability of sown grasslands' *Nature Plants*, vol 1, no 1, pp 1–5.

R Core Team (2018) R: *A Language and Environment for Statistical Computing*. R Foundation for Statistical Computing, Vienna.

Renard, D. and Tilman, D. (2019) National food production stabilized by crop diversity. *Nature*, vol 571, pp 257–260.

Saleh, A.S.M., Zhang, Q., Chen, J. and Shen, Q. (2013) 'Millet grains: Nutritional quality, processing, and potential health benefits' *Comprehensive Reviews in Food Science and Food Safety*, vol 12, no 3, pp 281–295.

Shobana, S., Krishnaswamy, K., Sudha, V., Malleshi, N., Anjana, R., Palaniappan, L. and Mohan, V. (2013) 'Finger millet (Ragi, *Eleusine coracana* L.): A review of its nutritional properties, processing, and plausible health benefits' *Advances in Food and Nutrition Research*, vol 69, pp 1–39.

WFP and IHD (2008) Food Security Atlas of Rural Madhya Pradesh. New Delhi The UN World Food Programme and Institute for Human Development. The UN World Food Programme (WFP) and the Institute for Human Development (IHD). pp1–142. http://www.ihdindia.org/pdf/FSA_Rural-Madhya-Pradesh.pdf

26

GERMPLASM CHARACTERIZATION AND NOVEL TECHNOLOGIES TO UNLEASH THE NUTRITIONAL POTENTIAL OF SMALL MILLETS

Vasudevan Sudha, Nagappa Gurusiddappa Malleshi, Chamarthy Venkata Ratnavathi, Shanmugam Shobana, Mani Vetriventhan, Krishna Hariprasanna, Bakshi Priyanka, Viswanathan Mohan and Kamala Krishnaswamy

Introduction

The main challenges of the 21st century are population growth, climate change, water scarcity and soaring food prices, all of which could trigger a great threat to agriculture and food security worldwide (Kulkarni et al., 2018). Globally, the burden of malnutrition, in all its forms, remains a challenge. Nearly 8.9% (~690 million) of the world population is undernourished according to 2019 estimates; and 21.3% of children under five years of age are stunted, 6.9% wasted and 5.6% overweight (FAO et al., 2020). Nutrition insecurity is a major threat to the world's population, which is highly reliant on cereal-based diets (seen in both developing and under-developed countries), particularly refined cereals that are deficient in micronutrients. Today the bigger threat comes from the triple burden of malnutrition – under-nutrition co-existing with micronutrient deficiencies and over-nutrition. Economic losses equal to 5%–6% of Gross National Product (GNP) were estimated due to deficiencies of iron, zinc and vitamin A in South Asia, leading to illness and poor performance (Shivran, 2016). Both micronutrient deficiencies and economic loss could be averted with improvements in nutritional status through enhanced quality, agriculture productivity (De Benoist et al., 2004) and cereal staples consumption. Agricultural research needs to engage in novel technologies and strike a balance between boosted food production with better crop genotypes and improved nutritional quality of food crops (Saha et al., 2016). The process of biofortification is one means to improve the micronutrient content (vitamins and minerals) of food grains is improved, through plant breeding and agronomic practices. Staple crops, when biofortified and consumed routinely, could have a greater positive impact on the nutritional status of vulnerable populations.

DOI: 10.4324/9781003044802-30

Millets are a highly promising alternative option to strengthening rice- and wheat-based diets, as they can survive in arid conditions, leave a smaller carbon footprint and still offer higher nutritional value (Bergamini et al., 2013). These ancient grains have been used as cereal staples by millions of people in arid zones across sub-Saharan Africa and Asia, and for fodder, feed and industrial purposes in developed economies (Pasha et al., 2018). Despite India being among the largest producers of small millets (SM), consumption of these crops is currently low (Kulkarni et al., 2018), as they have been replaced largely by rice and wheat since the Green Revolution. One of the factors hindering the wider use of SM is their cumbersome traditional processing; but today, this can be effectively replaced with novel techniques capable of creating more value-added and readily acceptable products. Such innovations represent, indeed, a robust strategy to increase the consumption of these (Kulkarni et al., 2018).

SM are a group of small-seeded cereals of the grass family *Poaceae*. They include finger millet (FM) (*Eleusine coracana* (L.) Gaertn.), kodo millet (KM) (*Paspalum scrobiculatum* L.), little millet (LM) (*Panicum sumatrense* Roth), foxtail millet (FXM) (*Setaria italica* (L.) P. Beauv.), barnyard millet (BM) (*Echinochloa spp.*), proso millet (PM) (*Panicum miliaceum* L.) and brown top millet (BTM) (*Brachiaria ramosa* (L.) Stapf) (Figure 26.1). FM, FXM and PM are important crops cultivated globally, while the other millets are region or country specific. For example, KM, LM and BTM are cultivated as cereals in India (Vetriventhan et al., 2020).

The climate-resilient and nutrient-dense characteristics of SM make them highly relevant crops for food and feed, even in agriculturally marginalized areas

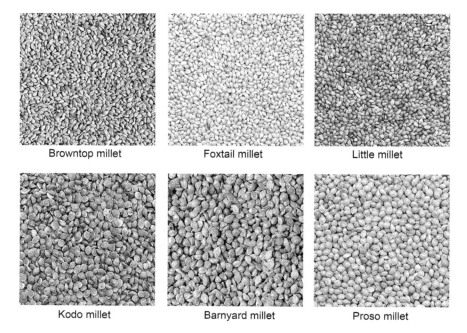

FIGURE 26.1 Small millet grains.

that are characterized by scarce water and poor soil. SM are excellent sources of carbohydrates, micronutrients and phytochemicals with nutraceutical properties. They are also excellent sources of B-group vitamins. Although their proteins are poor in lysine and tryptophan, their consumption can be easily complemented with lysine-rich (leguminous) vegetables and animal proteins. Because of their high nutritional value, millets have been recently renamed 'nutri-cereals' by the Government of India through a Gazette Notification (Ministry of Agriculture and Farmers Welfare, 2018).

In recent times, there has been noticeable progress in the development of millet processing machinery, enabling a more effective preparation of value-added products from SM. Each subtype of SM has unique physical, nutritional and processing qualities. Varietal selection within each SM with desired nutritional and processing traits is a prerequisite for making value-added products or novel food items and recipes. Today, there are many new varieties of SM, including biofortified ones, which could be utilized to boost nutritional security.

Germplasm characterization for nutritional quality

Germplasm characterization is the process allowing a detailed description of accessions at phenotypic, biochemical and molecular levels. Owing to their unique genetic diversity, combined with great nutritional properties, SM offer a tremendous opportunity for diet diversification, which should be better capitalized on. Over 133,000 germplasm of SM have been conserved globally (Vetriventhan et al., 2020). The International Crops Research Institute for the Semi-Arid Tropics (ICRISAT) conserves over 12,000 accessions of six SM, and germplasm diversity sub-sets evaluated for grain nutrients composition have revealed significant variability; numerous trait-specific germplasm have been identified so far in support of crop improvement (Table 26.1).

Some of the millets, in general, contain lower amylose compared to wheat and rice, which may be a factor that results in high glycaemic index. There is, thus, a need to identify varieties of SM with higher amylose for slower sugar assimilation that would, thus, be better suited for diabetics.

Small millets processing technologies

SM are neither ready-to-eat (RTE) nor ready-to-cook (RTC) grains and need to be processed (Saleh et al., 2013; Jaybhaye et al., 2014). Commonly followed, conventional, contemporary and novel processing techniques to exploit the nutritional and value-addition potential of SM (Figure 26.2) are discussed below.

Milling

The term 'milling' in the context of SM refers to de-husking and de-branning operations, similar to those done in rice, and to size reduction and gradation as

TABLE 26.1 Variation in the grain nutrients content of global collections of small millets conserved at ICRISAT, India

Crop	Fe (mg/kg)	Zn (mg/kg)	Ca (mg/kg)	Protein (%)	Reference
Finger millet	21.7–65.2	16.58–25.3	1,840–4,890	6.0–11.19	Upadhyaya et al. (2011a)
Foxtail millet	24.1–68.0	33.6–74.2	90.0–288.7	10.7–18.5	Upadhyaya et al. (2011b)
Kodo millet	14.4–56.4	17.0–31.5	121–321	5.6–11.3	Vetriventhan and Upadhyaya (2019)
Proso millet	41.0–73.0	26.0–47.0	91–241	11.0–19.0	Vetriventhan and Upadhyaya (2018)
Little millet	17.6–58.0	19.4–39.5	92.1–390	6.0–15.6	Vetriventhan et al. (2021)

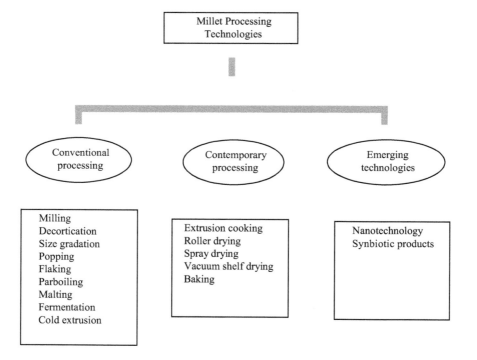

FIGURE 26.2 Millet processing.

done in wheat. In the case of SM except FM, the grains can be de-husked in a centrifugal sheller, rubber roll sheller or even in the traditional disc mills and then de-branned to a desirable degree of polish. Decorticated as well as de-branned millets can be cooked as rice or may be flaked or used to prepare expanded grains or pulverized and size-graded to prepare flour or semolina for conventional food preparations. Hardening the endosperm through hydrothermal treatment improves dehulling efficiency. Milled millets have improved bioavailability of

Unleashing the nutrition potential of small millets **313**

nutrients and higher consumer acceptability. The bran fraction from SM has higher oil and nutraceutical content whereas FM seed coat, rich in phenolics, may serve as functional food additive (Devi et al., 2014). Large-scale SM milling systems involve high capital investments. However, today, small-scale, low-cost millet dehullers and integrated millet mills equipped with pre-cleaner, destoner, dehuller and gravity separator are also available.

Popping and puffing

Grain popping is a popular dry-heat (high temperature short time– HTST) treatment, which is simple and least energy-intensive processing method followed to prepare RTE products. Wholegrain SM are used to prepare popped/ puffed grains. Normally, the expansion volume of popped SM ranges from 7 to 12 ml/g (the highest being for PM and the lowest for BM). A few good popping varieties of SM have been identified (Malleshi, 1989; Delost-Lewis et al., 1992, Mirza et al., 2015). During popping, due to the heat, the lipases get denatured and this enhances the shelf life of the product. Now-a-days, expanded SM are also emerging, which are prepared using dehulled/debranned grains following gun-popping process or processingparboiled millets similar to rice (*murmura*). Expanded millets have high potential for confectioneries, low-fat breakfast cereals, cereal bars and also as a thickener in soup mixes.

Parboiling or hydrothermal treatment

Parboiling or hydrothermal treatment hardens SM and improves their decortication or milling, as well as cooking, characteristics. The process also facilitates resistant starch formation, minimizes the loss of vitamins and minerals during milling and improves the storage qualities of millets (Shobana and Malleshi, 2007). Normally, SM with denser endosperm and high amylose content yield better quality parboiled millets. Parboiling is not practiced extensively but it has very high potential for extended uses of SM.

Flaking

Flakes from millets are prepared by traditional edge runner, roller flaker and extrusion cooking methods. RTE flakes result from the process of starch gelatinization; they are used for a wide range of snacks, roasted flakes find applications in granola bars, breakfast cereals and muesli formulations. Generally, bold grains with well-filled starchy endosperm yield better quality flakes.

Malting

FM has unique potential for malting and its malt is a storehouse of active hydrolytic enzymes, andtermed as"Amylase Rich Food". Thus, it finds a place in the preparation of infant, weaning and enteral food formulations. Germination lowers anti-nutrients and enhances mineral bioavailability. Varietal variations of malting characteristics in FM have been observed and varieties 'Indaf' 1, 3 and 9, and Annapurna have been identified as good malting strains (Malleshi and Desikachar, 1979). Further screening of germplasm for identifying or breeding malt-millet accessions deserve attention.

Fermentation

Popular fermented FM-based alcoholic beverages in India are *'Kodo ko jaanr'*, *'Koozh'* and *'Chhang'*, while lactic fermented beer is popular in sub-Saharan Africa. Fermented products have higher protein and starch digestibility and lower antinutrients. *Ambali*, a mild, auto-fermented thin porridge, is a popular product made from FM. Currently *idli, dosa* and other traditional Indian fermented products are prepared from all types of SM.

Extrusion (cold and hot)

Various food products from SM are produced by adapting both cold and hot extrusion technologies. SM are non-glutinous cereals and, hence, need special pre-treatments for the preparation of cold extruded products such as vermicelli and pasta (Malleshi, 2007; Ushakumar and Malleshi, 2007). Shobana et al. (2018) successfully prepared medium GI gluten-free vermicelli from SM by cold extrusion. RTE snacks, breakfast cereals and specialty foods such as 'Smoothix' are produced from SM (Ushakumari et al., 2004). Normally, millets with higher amylose content are more suitable for preparing cold extruded products with a firm and chewy texture, whereas hot extruded products are generally not varietal specific. Information on the suitability of SM varieties for such products would be valuable so as to enhance the production of these high-value products for both the national and international markets.

Baking

Due to an absence of gluten, SM are a challenging ingredient to use for bakery and confectionary products similar to those obtained from wheat. However, composite flours consisting of wheat and SM flours (up to 30%) can be used for the preparation of pizza, cookies, cakes, muffins, flatbread, etc. FM varieties GPHCPB-149 and VL 324 are more suitable for biscuits (Kaur et al., 2020). According to Desai et al. (2010), 50% of malted FM flour can be utilized in cake formulations. Based on these findings, there is a need for greater screening of SM germplasm in order to identify varieties suitable for each of these products.

Miscellaneous products

There is ample scope for the preparation of SM-based RTE and RTC products such as *papad, sandige, murukku, chakkuli* and ready-to-warm *rotis* (similar to Mexican tortillas) on industrial scale, as such products are in great demand. *Papads* from FM are very popular whereas multi-millet baked as well as deep-fried snacks are also gaining popularity. The use of SM for making ice cream, *halwa, idli, dosa* and traditional sweet and savoury products is now possible thanks to the latest findings in food technology.

Starch, dextrins and ethanol

SM are starchy grains and can be used for the industrial-level production of starches and starch derivatives such as resistant malto-dextrins (soluble fibre with less viscosity) and glucose. However, lower level of starch content in SM hinders their use in Industrial level production of starch and allied products including ethanol. Suitable SM germplasm with desirable carbohydrate characteristics need to be identified for this purpose.

Roller drying, spray drying and vacuum shelf drying

All these are contemporary technologies being applied to SM for the development of healthy, wellness, weaning and supplementary foods; milk-based beverages can be also prepared.

Another area that can be explored is the application of nanotechnology to prepare millet flours with characteristic physical and rheological properties for diverse food applications with enhanced bioavailability of nutrients and phytochemicals. Milk and millet-malt-based fermented probiotics indicate a possible use for millets in synbiotic foods (Chaudhary and Sreeja, 2020).

The glycaemic properties of SM preparations range from low to high (Shobana et al., 2013); however, the information on varietal variations is still lacking. Identifying suitable germplasm for high amylose and soluble dietary fibre contents and complex-rigid protein and starch architecture may help in identifying low glycaemic index (GI) millet products.

Identifying suitable germplasm for specific foods, developing value chains for consistent supply of such varieties to processing industries and creating public awareness about the health benefits of SM need to be better addressed.

Conclusion

SM are important alternative cereal grains to help mitigate malnutrition, of both the under and over kind. Millet varieties with higher protein, dietary fibre and micronutrient content, and better bioavailability of nutrients would be suitable for preparing holistic SM-based products, while high-amylose varieties could be

316 Vasudevan Sudha et al.

used for preparing foods with lower glycaemic properties. Germplasm characterization is essential for an effective selection of varieties to unleash the benefits contained in SM, which are otherwise still largely underutilized.

References

Bergamini, N., Padulosi, S., Ravi, S.B. and Yenagi, N. (2013) Minor millets in India: A neglected crop goes main stream. In Fanzo, J., Hunter, D., Borelli, T. and Mattei, F. (eds.) *Diversifying food and diets: Using agricultural biodiversity to improve nutrition and health* (pp. 313–325) Routledge, London.

Chaudhary, J.K. and Sreeja, V. (2020) Effect of incorporation of Finger millet (*Eleusine coracana*) on the antimicrobial, ACE inhibitory, antioxidant and antidiabetic potential of a milk-millet composite probiotic fermented product. *Indian Journal of Dairy Science*, 73(3), pp. 222–230.

De Benoist, B., Andersson, M., Egli, I., Takkouche, B. and Allen, H. (2004) *Iodine status worldwide. WHO global database on iodine deficiency.* Geneva: World Health Organization.

Delost-Lewis, K., Lorenz, K. and Tribelhorn, R. (1992) Puffing quality of experimental varieties of proso millets (*Panicum miliaceum*). *Cereal Chemistry*, 69(4), pp. 359–365.

Devi, P.B., Vijayabharathi, R., Sathyabama, S., Malleshi, N.G. and Priyadarisini, V.B. (2014) Health benefits of finger millet (*Eleusine coracana* L.) polyphenols and dietary fiber: A review. *Journal of Food Science and Technology*, 51(6), pp. 1021–1040.

Desai, A.D., Kulkarni, S.S., Sahoo, A.K., Ranveer, R.C. and Dandge, P.B. (2010) Effect of supplementation of malted ragi flour on the nutritional and sensorial quality characteristics of cake. *Advance Journal of Food Science and Technology*, 2(1), pp. 67–71.

FAO, IFAD, UNICEF, WFP and WHO (2020) The state of food security and nutrition in the world 2020. Transforming food systems for affordable healthy diets. Rome, FAO. DOI: 10.4060/ca9692en

Jaybhaye, R.V., Pardeshi, I.L., Vengaiah, P.C. and Srivastav, P.P. (2014) Processing and technology for millet based food products: a review. *Journal of Ready to Eat Food*, 1(2), pp. 32–48.

Kaur, A., Kumar, K. and Dhaliwal, H.S. (2020) Physicochemical characterization and utilization of finger millet (*Eleusine coracana* L.) cultivars for the preparation of biscuits. *Journal of Food Processing and Preservation*, 44(9). DOI: 10.1111/jfpp.14672

Kulkarni, D.B., Sakhale, B.K. and Giri, N.A. (2018) A potential review on millet grain processing. *International Journal of Nutrition Science*, 3(1), pp. 1–8.

Malleshi, N.G. (1989) Processing of small millets for food and industrial uses. In A. Seetharam, K. W. Riley, and G. Harinarayana (eds.) *Small millets in global agriculture*, pp. 325–339. New Delhi: Oxford and IBH Publishing Company.

Malleshi, N.G. (2007) Nutritional and technological features of ragi (finger millet) and processing for value addition. In *Food uses of small millets and avenues for further processing and value addition* (pp. 9–19). Bangalore: Project Coordination Cell, All India Coordinated Small Millets Improvement Project, ICAR, UAS, GKVK, India.

Malleshi, N.G. and Desikachar, H.S.R. (1979) Malting quality of new varieties of ragi (*Eleusine coracana*). *Journal of Food Science and Technology, India*, 16(4), pp. 149–150.

Ministry of Agriculture and Farmers Welfare, Department of Agriculture, Cooperation and Farmers Welfare, Govt. of India (2018). Gazette notification on millets dated 10th April 2018 (F.No.4-4/2017-NFSM (E), The Gazette of India.

Mirza, N., Sharma, N., Srivastava, S. and Kumar, A. (2015) Variation in popping quality related to physical, biochemical and nutritional properties of finger millet genotypes. *Proceedings of the National Academy of Sciences, India Section B: Biological Sciences*, 85(2), pp. 507–515.

Pasha, K.V., Ratnavathi, C.V., Ajani, J., Raju, D., Manoj Kumar, S. and Beedu, S.R. (2018) Proximate, mineral composition and antioxidant activity of traditional small millets cultivated and consumed in Rayalaseema region of south India. *Journal of the Science of Food and Agriculture*, 98(2), pp. 652–660.

Saha, D., Gowda, M.C., Arya, L., Verma, M. and Bansal, K.C. (2016) Genetic and genomic resources of small millets. *Critical Reviews in Plant Sciences*, 35(1), pp. 56–79.

Saleh, A.S., Zhang, Q., Chen, J. and Shen, Q. (2013) Millet grains: nutritional quality, processing, and potential health benefits. *Comprehensive Reviews in Food Science and Food Safety*, 12(3), pp. 281–295.

Shivran, A.C. (2016) Biofortification for nutrient-rich millets. In: U. Singh, C. S. Praharaj, S. S. Singh, and N. P. Singh (eds.) *Biofortification of food crops* (pp. 409–420). Springer: New Delhi.

Shobana, S., Krishnaswamy, K., Sudha, V., Malleshi, N.G., Anjana, R.M., Palaniappan, L. and Mohan, V. (2013) Finger millet (Ragi, *Eleusine coracana* L.): A review of its nutritional properties, processing, and plausible health benefits. *Advances in Food and Nutrition Research*, 69, pp. 1–39.

Shobana, S. and Malleshi, N.G. (2007) Preparation and functional properties of decorticated finger millet (*Eleusine coracana*). *Journal of Food Engineering*, 79(2), pp. 529–538.

Shobana, S., Selvi, R.P., Kavitha, V., Gayathri, N., Sudha, V., Anjana, R.M. and Mohan, V. (2018) Development and evaluation of nutritional, sensory and glycemic properties of finger millet (*Eleusine coracana* L.) based food products. *Asia Pacific Journal of Clinical Nutrition*, 27(1):84–91.

Upadhyaya, H.D., Ramesh, S., Sharma, S., Singh, S.K., Varshney, S.K. and Sarma, N.D.R.K., Ravishankar, C.R., Narasimhudu, Y., Reddy, V.G., Sahrawat, K.L., Dhanalakshmi, T.N., Mgonja, M.A., Parzies, H.K., Gowda, C.L.L. and Sube Singh (2011a) Genetic diversity for grain nutrients contents in a core collection of finger millet (*Eleusine coracana* (L.) Gaertn.) germplasm. *Field Crops Research*, 121, pp. 42–52.

Upadhyaya, H.D., Ravishankar, C.R., Narasimhudu, Y., Sarma, N.D.R.K., Singh, S.K., Varshney, S.K., Reddy, V.G., Singh, S., Parzies, H.K., Dwivedi, S.L. and Nadaf, H.L. (2011b) Identification of trait-specific germplasm and developing a mini core collection for efficient use of foxtail millet genetic resources in crop improvement. *Field Crops Research*, 124(3), pp. 459–467.

Ushakumari, S.R., Latha, S. and Malleshi, N.G. (2004) The functional properties of popped, flaked, extruded and roller-dried foxtail millet (*Setaria italica*). *International Journal of Food Science &Technology*, 39(9), pp. 907–915.

Vetriventhan, M. and Upadhyaya, H.D. (2018) Diversity and trait-specific sources for productivity and nutritional traits in the global proso millet (*Panicum miliaceum* L.) germplasm collection. *The Crop Journal*, 6(5), pp. 451–463.

Vetriventhan, M. and Upadhyaya, H.D. (2019) Variability for productivity and nutritional traits in germplasm of kodo millet, an underutilized nutrient-rich climate smart crop. *Crop Science*, 59(3), pp. 1095–1106.

Vetriventhan, M., Azevedo, V.C., Upadhyaya, H.D., Nirmalakumari, A., Kane-Potaka, J., Anitha, S., Ceasar, S.A., Muthamilarasan, M., Bhat, B.V., Hariprasanna, K. and

Bellundagi, A. (2020) Genetic and genomic resources, and breeding for accelerating improvement of small millets: Current status and future interventions. *The Nucleus*, 63, pp. 1–23.

Vetriventhan, M., Upadhyaya, H.D., Azevedo, V.C.R., Allan, V., Anitha, S. (2021) Variability and trait-specific accessions for grain yield and nutritional traits in germplasm of little millet (*Panicum sumatrense* Roth. Ex. Roem. & Schult.). *Crop Science*, 2021, pp. 1–22. https://doi.org/10.1002/csc2.20527

27

MILLETS IN FARMING SYSTEMS IN SUPPORT OF NUTRITION AND SOCIAL SAFETY NET PROGRAMMES

Priya Rampal, Aliza Pradhan, Akshaya Kumar Panda, Sathanandham Raju and R.V. Bhavani

Introduction

Agriculture and allied activities are a major source of livelihood for almost 60% of the Indian population. Nearly 85% are small and marginal farmers with less than five acres of land (GoI, 2011). Within this scenario, leveraging agriculture for nutrition is a strategic pathway to impact that should be decisively promoted. There are six main pathways linking agriculture to nutrition. Three are related to the nutritional impacts of farm production, farm incomes, and food prices. The other three are related to agricultural linkages with gender, i.e., how agriculture can influence women's empowerment, childcare and feeding, and women's and children's nutritional outcomes (Kadiyala et al., 2014). In spite of a major share of population dependent on agriculture, and agriculture's potential to improve nutritional outcomes, the nutritional indicators of India's population tell a different tale. In 2015–2016, 38.4% of India's children below the age of five were stunted, 35.7% were underweight, and more than 50% of women and children were anaemic (IIPS and ICF, 2017).

Millets promoted as a part of leveraging agriculture for nutrition through nutrition-sensitive agriculture policies and programmes have the potential to improve nutrition outcomes. Farming System for Nutrition (FSN) is an example of nutrition-sensitive agriculture policy, defined by Prof. M. S. Swaminathan as "the introduction of location-specific agricultural remedies for nutritional maladies by mainstreaming nutritional criteria in the selection of farming system components involving crops, animals and wherever feasible fish" (Nagarajan et al., 2014). Nutrition awareness is an integral component of the FSN approach. The M. S. Swaminathan Research Foundation (MSSRF) undertook a study between 2013 and 2018 to demonstrate the feasibility of an FSN approach to addressing undernutrition under the research programme "Leveraging Agriculture for

DOI: 10.4324/9781003044802-31

Nutrition in South Asia". The study was undertaken in a core set of seven villages (658 households, 2,845 people) in the Koraput district, Odisha, and five villages (556 households, 2,254 people) in the Wardha district, Maharashtra, both areas of millet cultivation and consumption (Bhaskar et al., 2017).

The promotion of nutrient-dense millets (sorghum in Wardha and finger millet in Koraput) represented a key crop intervention under the approach. The National Food Security Act (NFSA), enacted in 2013, provides for the inclusion of millets in government food distribution programmes like the public distribution system (PDS) and supplementary nutrition programme (SNP) under the Integrated Child Development Services (ICDS). Karnataka was the first state in the country to introduce the procurement of millets for supply under the PDS.[1] This chapter examines the acceptability and viability of millets as a component of the FSN approach and in the PDS.

Millets under FSN

Finger millet and sorghum are hardy, nutrient-dense, largely rainfed crops that have been a part of the traditional diet of Indigenous communities in many parts of the world. In India, different millet species are cultivated in different regions of the country. Overall, their cultivation and consumption has been on steady decline (Raju et al., 2018).

Finger millet (*Eleusine coracana*) is among the major crops cultivated in the Koraput district, Odisha. It is grown in the kharif season (June–September), on marginal lands in the upland and hilly regions, with few external inputs, either as a pure crop or under mixed-cropping systems.[2] It is sold directly as a grain or is brewed into local beer for sale in local markets. However, the area under finger millet cultivation in the district declined by 55%, from 144,480 ha in 1980 to 65,160 ha in 2013, leading to reduced consumption (GoO, 2015). Due to traditional cultivation practices, the grain yield is as low as 4 q/ha under the broadcasting method; even with traditional transplanting methods, the yield is only 9 q/ha (Adhikari, 2014). The primary reasons for low production are poor crop management (use of low-quality seeds, broadcasting method of sowing leading to low plant population, lack of nutrient and/or weed management practices, etc.) and the replacement of existing millet fields with commercial plantations of eucalyptus, leading to reduced consumption of this food by local communities (Pradhan et al., 2017). One possible way to improve the food and nutrition security of the population is to improve the yield of this native crop that is highly adaptive to the local climate, has a high nutrient value, and can efficiently withstand biotic and/or abiotic stresses. Therefore, under the FSN study, one intervention was to focus on increasing the productivity and profitability of finger millet, alongside nutrition awareness initiatives (Pradhan et al., 2019).

The traditional practice of cultivation included broadcasting local landraces of finger millet at a high seed rate (25 kg/ha), carried out by farmers as a safeguard measure against poor seed quality and uncertain soil moisture; this practice results in a very dense crop, causing crowded plant populations and high competition for water and the scarce nutrients present in the soil. Inadequate fertilization coupled with poor crop management practices results in low yields. Seeds saved from these fields also turn out to be of inferior quality, with considerable physical mixing, which further adversely affects the yield potential. The FSN study included on-field demonstrations (OFD) in the fields of seven farmers over 0.52 ha in kharif (2015–2016), in order to select the most suitable variety and appropriate agronomic practices for higher productivity. GPU-67, a high-yielding finger millet variety selected under a participatory varietal selection programme conducted earlier, was taken along with farmers' varieties for comparison (Mishra et al., 2014).

GPU-67 is a semi-dwarf (non-lodging) medium-duration variety (114–118 days) with desirable traits like good panicle shape, no grain shattering, and a yield potential of 30–35 q/ha. Improved agronomic practices adopted in the intervention included: (a) nursery raising with seed rate of 5 kg/ha^{-1}; (b) line transplanting of three to five week old seedlings; (c) population density at 20 cm × 10 cm; (d) application of recommended doses of fertilizer in the order of 40:20:20 kg nitrogen (N), phosphorous (P), and potassium (K) per ha; and (e) timely weeding and need-based plant protection measures. Each farmer's field in the OFD was considered as a replicate, and the allotted area under each farmer was split into four treatments for yield comparison. The details of the treatments with farmers' variety and improved variety are provided in Table 27.1.

GPU-67 under improved agronomic practices (T1) produced the highest grain yield of 2,067 kg/ha, 31% higher than that with farmers' variety and with

TABLE 27.1 Comparison of yield-contributing parameters, grain yield, and economics of production among treatments (2015–2016)

Treatments	No. of productive fingers per plant	Finger length (cm)	Grain yield (kg/ha)	Total cost of cultivation (Rs./ha)	Gross return (Rs.)	Net return (Rs.)
T1: GPU-67 + recommended agronomic practices	4.4	7.4	2,067*	20,800	41,340	20,540
T2: Farmers' variety + recommended agronomic practices	4.4	6.6	1,832	20,670	36,640	15,970
T3: GPU-67 + traditional farming practice	4.3	6.6	1,740	17,980	34,800	16,820
T4 (control): Farmers' variety + traditional farming practice	4.0	6.5	1,579	17,850	31,580	13,730

*$P < 0.05$; market price of finger millet @ Rs. 20.00 per kg

traditional practices, providing a larger quantity of nutrient-rich food to farmers' households. It was, thus, chosen for promotion and upscaling. The farmers' adaptability to the recommended practice was so good that by kharif 2017–2018, it was being cultivated by 167 farmers across 23.5 ha in the seven core study villages, and by 87 farmers in 11 ha in 18 other neighbouring villages. For sustainability, triple-layered bags were distributed among the farmers for the safe storage of seeds. A village-level seed bank was also established for timely availability of quality seeds and to help farmers in distress. Small-scale, village-level millet processing mills were installed to encourage farmers to process and consume their produce. Furthermore, several nutrition awareness initiatives were conducted about the nutrient content of the crop, its consumption benefits, and recipe demonstration, in addition to exposure visits to finger millet fields and trainings on improved agronomic practices. The end-line food consumption survey revealed an increase of 13% in the average intake of finger millet over the baseline (70 g/person/day). In addition, the number of households consuming finger millet daily increased from 172 during the baseline to 187, with 68% sourcing it from their own production versus 47% during the baseline.

Akin to finger millet, the area under sorghum has been also falling in Wardha (Rukmani and Manjula, 2009). A similar approach was followed to promote the cultivation of that crop and its consumption. The experience in the five core study villages was analogous to that of finger millet in Koraput, with 26% of households cultivating and consuming sorghum at end-line as compared to 6% during baseline; there was also an increase of 64% in average sorghum intake over the baseline (150 g/person/day).

Millets in social safety net programmes

Given that millets are naturally nutrient-dense cereals, making them available through the PDS can help address the problem of micronutrient deficiency among poorer households (Raju et al., 2018). After the enactment of the NFSA in 2013, the Karnataka government began to procure millets for distribution through the PDS – finger millet in south Karnataka and sorghum in north Karnataka as traditionally the cultivation and consumption of these two millets happens in these respective geographical areas. The scheme, titled "Anna Bhagya Yojana," had the double objectives of buying millets from farmers, which supported income generation for rural farm households, and allowing households with PDS cards to gain access to nutritious food grains at low prices (KAPRICOM, 2014).

As part of the study by MSSRF, primary surveys were carried out in four districts of the state. Districts with the highest numbers of procurement and distribution of finger millet through the PDS (Mandya and Tumkur in south Karnataka) and of sorghum (Gadag and Dharwad in north Karnataka) were selected for the primary survey. Two hundred farmers (50 in each district) were interviewed to collect information on issues related to production, pricing, and procurement. A consumer survey was conducted across 50 rural and 50 urban households in each district, after ensuring that they were either BPL (below poverty line) or Antyodaya Anna Yojana cardholders and eligible to receive grains under the PDS (Rajshekar and Raju, 2017).

The state government had offered attractive procurement prices, including a bonus over and above the Minimum Support Price (MSP) announced by the Central Government (Rs 2,250 per quintal for finger millet and Rs 2,300 per quintal for sorghum), to encourage farmers to grow millets and ensure sufficient quantities for procurement. Interviews with farmers showed that the MSP did cover the actual cost of production (A1) and imputed cost of family labour, making it attractive for them to cultivate millets. However, the time taken for payment after the government's procurement ranged from about three to four weeks, as against payment received in one to two days in the case of open market sales in all the surveyed districts. This time-lag offset the benefit of the higher price offered by the government. If announced on time and sustained, and if the procurement window is increased, this price should encourage farmers to switch to millets from cotton and maize.

On the consumption side, 87% of households in Karnataka reported consuming finger millet or sorghum. For 48% of the population of rural and urban Karnataka, millets accounted for 20%–40% of total cereal consumption. The average per capita monthly consumption of finger millet by rural households fell from 1.8 kg in 2004–2005 to 1.2 kg in 2011–2012, and from 1 kg to 0.8 kg for urban households. Similarly, the per capita monthly consumption of sorghum for rural households declined from 2.3 kg in 2004–2005 to 1.4 kg in 2011–2012, and from 1.2 kg to 0.7 kg for urban households in the same period. Wheat and finger millet or sorghum were consumed in equal quantities in urban areas, while in rural areas, millet consumption was higher than that of wheat. The majority of the respondents, when asked about desired changes in the PDS, did not want an increase in the quantity of millets supplied if it came at the cost of a lower quantity of rice. The main reasons given were that the quantity of rice supplied was already insufficient and that farmers could grow finger millet and sorghum if required, but could not grow rice as easily. Reasons given for the strong preference for rice were its taste, ease of preparation, and popularity with children. Finger millet and sorghum were preferred for their nutritive value and by those engaged in physical labour. Consumer preferences for different kinds of millet in different parts of the state call for a decentralized procurement and distribution mechanism. Consumer preferences and cultural factors need to be taken into account, and awareness needs to be created regarding the benefits of consuming millets (Raju et al., 2018).

Discussion and recommendations

The Green Revolution dominated by rice and wheat was backed by public policy and price support. Similar support is required for millets. The NFSA (2013) is a major step in this direction. The FSN study by MSSRF showed the potential for

324 Priya Rampal et al.

increasing the consumption of millets by increasing their production and processing, and promoting awareness on their nutritive value. The study on introduction of millets in PDS highlighted the need for streamlining the government procurement and payment mechanisms and for decentralized processing. Millet farmers in a block or district could be linked with the PDS and institutional feeding programmes, providing them an incentive to cultivate and supply their crops for a ready market. Across India, in many states, different millets are grown in small pockets of lands. Such pockets can be identified and mapped for local procurement in the distribution system to improve food and nutrition security, while also increasing the procurement window and reducing the time between procurement and payment.

Millets are generally regarded as women's crops and do not receive the attention given to rice and wheat. This has to change. Agriculture department officials and extension staff need to be sensitized on how to guide farmers to follow improved cultivation practices and use improved varieties of seeds. Millets should be promoted as climate-resilient crops and more R&D should be done by agricultural universities and research institutes to produce high-yielding varieties that are profitable for farmers. A few initiatives are underway and need to become more widespread. The government of Odisha launched the Odisha Millet Mission in 2017, in partnership with NGOs and research institutes, to increase production and consumption of millets.[3] Decentralized cluster-level millet processing units managed by self-help groups or farmers' groups in villages can reduce the drudgery in millet processing faced by women and also boost consumption by producer households.

Awareness-raising has to happen at two levels: first of all, at the level of production and production techniques (e.g., line sowing, timely weeding, and proper application of fertilizers); secondly, nutrition awareness under the Prime Minister's Overarching Scheme for Holistic Nutrition, also called the POS-HAN Abhiyan[4] of the National Nutrition Mission should actively highlight nutritive value of millets and promote their consumption. Nutrition awareness workshops should also be conducted at the local level by the ICDS centre workers and ASHA (Accredited Social Health Activists) in villages at regular intervals, highlighting the nutritive value of millets and providing recipe demonstrations. They can be trained and incentivized to conduct such workshops at least once a month, which would increase awareness about the nutritive value of various food groups and the combinations of food to be eaten for better nutrient absorption. Sharing recipes can also be promoted along with the organization of cooking competitions; millets can be cooked in ways that would be more attractive to children (such as snacks) and this can raise their interest in this food instead of viewing it as something they are forced to eat and don't enjoy.

The Covid-19 pandemic has turned the spotlight onto the already vulnerable and is expected to further worsen their nutrition status. The closure of schools and ICDS centres has, for instance, put a halt to the noon-meal served to children attending them. At this time, the government issued directives for distribution of Take Home Rations (THR) in lieu of the meal (Bhavani and Rampal, 2020). Millets procured can be a nutritive and integral component of this THR. In fact, millets can also be used for free distribution along with other grains and pulses in food distribution programmes during the pandemic and also during other disasters in future. This would provide income to the farmers while also providing additional nutrition to vulnerable populations.

Notes

1 For more insights on the introduction of millets in the PDS of Karnataka, the reader is directed to Rajshekar and Raju, 2017.
2 In India, a crop sown in early summer such as rice, maize is to be harvested in autumn or at the beginning of winter.
3 http://www.milletsodisha.com/ accessed in June 2020
4 http://poshanabhiyaan.gov.in/ accessed in June 2020

References

Adhikari, P. (2014) 'Pragati, Koraput experiences in system of ragi intensification'. http://sri.cals.cornell.edu/countries/india/orissa/InOdisha_Pragati_SCI%20_Ragi14.pdf,

Bhaskar, A.V.V., Nithya, D.J., Raju, S. and Bhavani, R.V. (2017) 'Establishing integrated agriculture-nutrition programmes to diversify household food and diets in rural India', *Food Security*, vol. 9, no. 5, pp981–999.

Bhavani R.V. and Rampal, P. (2020) 'The lockdown and its aftermath'. https://www.orfonline.org/expert-speak/the-lockdown-and-its-aftermath-67550/

Government of India -GoI (2011) 'Census of India 2011, Registrar General and Census Commissioner of India, Ministry of Home Affairs', New Delhi.

Government of Odisha - GoO (2015) 'Odisha agriculture statistics (2013–2014). Directorate of Agriculture and Food Production', Odisha.

International Institute for Population Sciences (IIPS) and ICF (2017) 'National Family Health Survey (NFHS-4), 2015–16: India.' Mumbai. http://rchiips.org/nfhs/NFHS-4Reports/India.pdf

Kadiyala S., Harris J., Headey D., Yosef S. and Gillespie S. (2014) 'Agriculture and nutrition in India: mapping evidence to pathways', *Annals of the New York Academy of Sciences* vol. 1331, no. 1, pp43–56.

Karnataka Agricultural Price Commission (KAPRICOM) (2014) http://kapricom.org/crop_production_statistics.html

Mishra, C.S., Taraputia, T. and Suchen, B. (2014) 'Policy advocacy for climate smart agriculture in Millets: An initiative for ensuring food security in tribal communities of Koraput tract, Odisha, India', *Global Advanced Research Journal of Agricultural Science*, vol. 3, no. 7, pp179–185.

Nagarajan S., Bhavani R.V. and Swaminathan, M.S. (2014) 'Operationalizing the concept of farming system for nutrition through the promotion of nutrition-sensitive agriculture', *Current Science*, vol. 107, no. 6, pp959–964.

Pradhan A., Panda A.K. and Bhaskar, A.V.V. (2017) 'Crop based demonstrations and trials under farming system for nutrition study in Koraput (2013–16)—a report'. Research Report MSSRF/RR/17/.

Pradhan, A., Panda, A.K. and Bhavani, R.V. (2019) 'Finger millet in tribal farming systems contributes to increased availability of nutritious food at household level: Insights from India', *Agriculture Research*, vol. 8, pp540–547.

Rajshekar, S.C. and Raju, S. (2017) 'Introduction of millets in PDS lessons from Karnataka'. Research Report MSSRF/RR/17/41.

Raju S., Rampal, P., Bhavani, R.V. and Rajshekar, S.C. (2018) 'Introduction of millets in PDS lessons from Karnataka', *Review of Agrarian Studies*, vol. 8, no. 2, pp120–136.

Rukmani, R. and Manjula, M. (2009) 'Designing rural technology delivery systems for mitigating agricultural distress: A study of Wardha district'. Research Report MSSRF/RR/10/25.

28

THE SMART FOOD APPROACH

The importance of the triple bottom line and diversifying staples

Joanna Kane-Potaka, Nigel Poole, Agathe Diama, Parkavi Kumar, Seetha Anitha and Oseyemi Akinbamijo

Food system solutions incorporating a Smart Food Triple Bottom Line approach

'Food security' was the key focus in developing countries while mass starvation was a real threat. Alleviating hunger was a driving force for the Green Revolution (Behera, 2017). Awareness of hidden hunger then surfaced, and *'nutrition security'* was added to the rhetoric. More recently, the UN and other organizations have underlined the imperative for *'sustainable diets'*, defined as "diets with low environmental impacts which contribute to food and nutrition security", and the urgency to set targets to strive towards this.

The next critical step is to cater to all these needs and go one step further with solutions that are not only good for you and the planet, but also for the farmer. The Smart Food Triple Bottom Line advocates for solutions that approach all these three areas in unison. This is recommended as a framework for food system solutions. It will also help break down both discipline and sector silos.

Applying these solutions with crops that are 'smart foods', that is, foods that are inherently good for you, the planet and farmer, will strengthen our ability to achieve the 'Smart Food Triple Bottom Line'. Many NUS may fit the criteria of being a smart food. They may be good for the farmer and environment because they bring diversity to the farm, are more suitable crops for varying agro-ecologies, are crops that need fewer inputs and are resilient to the vagaries of climate change. However, without well-developed value chains that are sensitive to consumer awareness and demand, it is challenging to make them financially viable for the farmer.

It is paramount that having less-developed value chains do not become the excuse for continuing to support the same few major crops. It is regularly expressed that we need to transform the food system. A purposeful and consistent strategy

DOI: 10.4324/9781003044802-32

328 Joanna Kane-Potaka et al.

for the said transformation becomes imperative and the 'Business as usual' will not achieve this. Changing where we invest resources and supporting policies are needed, and strengthening value chains of smart foods so as to mainstream them is an opportunity for us to contribute to many of the UN's Sustainable Development Goals (SDGs) in unison.

A project in Kenya applied the 'Smart Food Triple Bottom Line' approach with NUS, tackling diversity in diets, diversity on farms and diversity in incomes, with the aim of crops being commercially viable as well as being consumed by the local community to improve diet diversity and nutrition. Six smart foods including millets, sorghum and legumes were selected and focused on. Families of over 60,000 children below the age of five were reached through volunteer Smart Food Ambassadors, who spread nutrition messages and conducted fun activities like cook-offs. The integration of education, health, nutrition and a fun approach in conveying the same message imparted strong knowledge of millets, sorghum and legumes.

> In just one year, the behavior patterns of the women and children changed significantly towards adopting a more micronutrient-rich diet, indicated by an increase of 15% in dietary diversity score for women and of almost 80% in the children's dietary diversity score. Similarly, consumers showed a considerable change in buying patterns. Rich in iron and fiber, both cowpea and pearl millet sales at the farm level more than doubled. Production also increased for all the smart food crops except finger millet. Consumption of four of the smart food crops increased. Households became more commercially oriented and sales of four of the crops increased.
>
> *(ICRISAT, 2018)*

Diversifying staples with smart foods for big impacts

To complement this approach of all food solutions having a Smart Food Triple Bottom Line, there is also a specific objective under the Smart Food initiative to diversify staples. Big impacts can be achieved by focusing on diversifying staples, given that across Africa and Asia staples can typically constitute as much as 70% of what is on the plate, and are often refined and low-nutrient carbohydrates, with approximately 60% of calories in developing countries coming from cereals – a number that can even be more than 80% in the poorest countries (Awika, 2011; Anitha et al., 2019a).

The diversification of staples with foods that fit the smart food criteria of being good for you, good for the planet and good for the farmer will require dissolving the boundaries of the '*Food System Divide*', where the largest investments have for decades gone into the Big 3 staples – rice, wheat and maize – including government support, private industry investment, R&D, product development and even development aid.

The smart food approach **329**

NUS can regain their popularity and enter the mainstream through concerted multi-pronged efforts across the whole value chain. Lessons can also be learnt from the successes of the Big 3, but approaches must be applied in an appropriate, sustainable and healthy way.

Some steps key to diversifying staples being pursued as part of the Smart Food initiative are:

1 *Dedicated effort on just a couple of smart foods*: Breaking the food system divide will take a focused approach and significant investments to develop value chains. Hence, an approach focused on just a couple of foods at a time is required. This complements initiatives that work broadly on popularizing NUS to bring diversity to farms and diets, and also builds niche markets that can be the springboard for larger markets in the future.

2 *Selecting millets and sorghum first*: Millets and sorghum were selected as the first foods to focus on and mainstream as they fit the profile of a smart food. Moreover, they were already the staples across many countries in Africa and Asia, with different millets originating from many countries and continents and growing from the Sahel to the Himalayas. They also fit into many global health food trends – being a super food, an ancient grain, gluten free with a low glycemic index (GI), high in fiber, good for managing weight, and good with strong health management.

   In particular, millets and sorghum are highly nutritious and fulfil some of the biggest health needs. For example, a few millets are very high in iron and zinc, which are among the top three micronutrient deficiencies globally. Taking bioavailability into account, the right varieties can provide as much iron as white or red meat. Finger millet has three times the amount of calcium found in milk. Most millets have a low GI, which is extremely beneficial within the context of community/public health due to the growing incidence of non-communicable metabolic disorders like diabetes; they are also a good alternative to other food sources high in complex carbohydrates like white refined rice (Anitha et al., 2021). They also have high fiber, reasonable levels of protein and, when combined with legumes, create a complete diet of protein with good levels of all the essential amino acids (Longvah et al., 2017; Anitha et al., 2019b).

   From a sustainable resource management point of view, millets and sorghum have a low carbon footprint. They are typically grown and thrive with minimal inputs like pesticides and fertilizers. They tolerate high temperatures and survive with very little water. They are often the last crop standing in times of drought, are climate smart and are a good risk-management strategy for farmers. They have multiple uses, from food, feed and fodder, to brewing and biofuels (Tonapi et al., 2015; Davis et al., 2019).

3 *Create global commodities*: While the goal is to contribute to the SDGs and especially help poor and malnourished communities across Africa and Asia, in order to mainstream smart foods as staples, they need to be widely

adopted commodities globally. Focusing on portraying millets and sorghum as staples is also a key part of the plan for them to be affordable.

4 *The consumer comes first:* Most efforts and investments to date in millets and sorghum have been at the farm-production end. There is an urgent need to drive demand, by investing at the consumer end, changing perceptions, building awareness and creating a 'buzz' and desire around these smart foods. This is being achieved by working with food processors, governments and other key influencers.

Some key approaches that the Smart Food initiative has used to drive consumer demand include:

- **Working with the hidden middle**: Social entrepreneurs who genuinely want to change the food system for the better, and micro, small and medium enterprises (MSMEs) who are typically the pioneers in creating new consumer preferences, struggle as much as farmers do. Until MSMEs are equally supported, smart foods and NUS won't be available, affordable and, hence, accessible and demanded by consumers. They are often called 'the hidden middle' and need to be recognized as change-makers rather than only as operators in the value chain. Policy support is required to create a better enabling environment for MSMEs to thrive. The Smart Food initiative has launched a 'millet finder' that maps products around the globe to bring attention to the wide availability and silent revolution of millet and sorghum products being made available by MSMEs (smartfood.org, a).
- **Make it delicious, convenient and easy – the image and the reality:** For smart foods to be popularized, they have to be sought by the consumer. Although different foods are consumed for different purposes, to be popular in the mainstream and to reach the largest number of people, in general, food needs to be tasty, convenient and easy to make.
- **Promotion through chefs:** The catering sector is a conduit to taking new foods to the consumer as well as the way to change the food's image. The Smart Food initiative has engaged ambassador chefs, organized cooking master classes in West Africa, including with the President's chefs and a Smart Food Culinary Challenge for student chefs pan-India, and in Tanzania, chefs were introduced to street venders, who were trained on using millet and sorghum flour (smartfood.org, b).
- **Ambassadors and champions:** Influencers are important when perceptions and behavior need to be changed or significant awareness needs to be built. The Smart Food initiative has engaged VIPs to achieve this (smartfood.org, c), e.g., the First Lady of Niger, Dr. Malika Issoufou, became a Smart Ambassador, leading the way to a greater mobilization and commitment by the government for the cause of smart foods. She initiated an international millet festival (FESTIMIL), which captured a lot of attention among consumers, value-chain actors, farmers, processors and small and medium enterprises (SMEs) and served as a platform for a science and policy dialog on better developing value chains. This led

The smart food approach **331**

to Senegal announcing its interest in following suit to create an annual millet festival.

- **Media and social media outreach:** Outreach has been key in building awareness and reaching wider audiences (Diama et al., 2020). One example is the smart food reality show on Kenyan national television that reached 800,000 viewers (Vital, 2018).
- **International platforms**: Influencing researchers, governments, donors and industry are important and can be achieved through high-level panel discussions and international symposia (Diama et al., 2020).

5 **Scientific backing on nutritional benefits:** As far less R&D has been invested in NUS compared to the major staples, the field requires additional investment. The Smart Food initiative is currently collating and analyzing all existing nutrition studies on millets and sorghum, and is identifying research gaps. Some nutrition and consumer acceptance studies undertaken by the Smart Food initiative include:

- **India school feeding study:** A millet-based meal introduced for three months with 1,500 adolescent children had significantly higher nutritional levels compared to the control group of iron fortified rice-based meals, see Figure 28.1, and led to:
  − growth in terms of BMI and height, 50% more in the intervention group relative to the control group; and − high acceptance scoring ≥ 4.5 out of 5 for taste (Anitha et al., 2019a).
  Key lessons learnt on how to introduce millets to maximize nutritional benefits and acceptance, along with policy recommendations were identified and are shown in Figure 28.2.
- **Tanzania school feeding study:** Over 2,800 students in four boarding schools were introduced to finger millet and pigeon pea in their menu cycle in a participatory approach, taking into account cultural sensitivities. Fifteen months later, the schools were revisited and surveys identified that:
- 80% of the students changed their negative perception of finger millet;
- >95% of the students wanted to eat the finger millet dishes at school (Wangari et al., 2020).
- **Myanmar malnutrition and acceptance study:** This had a positive impact on the extent of wasting and underweight children between 2 and 14 months of age. Also sensory evaluations showed an average score of four out of five for all recipes and products (Anitha et al., 2019c).

Guiding the development of smart foods to keep or maximize their nutritional benefits is critical; this includes:

- **Popularize whole grain**: As most small millets have to be de-hulled, there is a risk they will also be polished to make them whiter and quicker to cook. It is important that consumers are exposed to the unrefined taste and

332 Joanna Kane-Potaka et al.

Laboratory tested nutrition composition of final meals: The typical school meal of fortified rice and sambar compared to millet based meals.
(Varieties used: Pearl millet *Dhanshakti*; Little millet *Phule Ekadashi*; and Finger millet used a range)

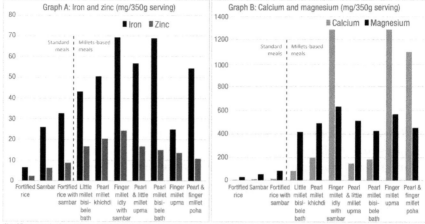

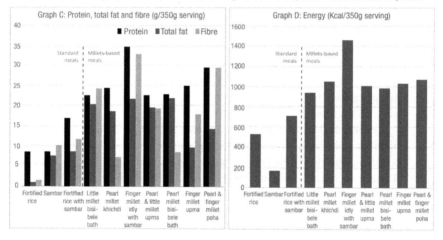

FIGURE 28.1 Nutritional value of fortified rice-based vs. millet-based meals. (ICRISAT, 2019).

convenient products made with the whole grain. Building awareness about the nutritional value of whole grain is essential.

- **Not ultra-processed or excessive added ingredients:** Efforts to diversify and popularize orphan crops will be to no avail if they are then over-processed and lose their nutritive value or if unhealthy ingredients like sugar, salt, saturated and trans fats and artificial additives are incorporated in high levels.
- **Selection of biofortified varieties:** Nutrition levels vary significantly by variety of the millets, so value chains from seed to consumer need to be

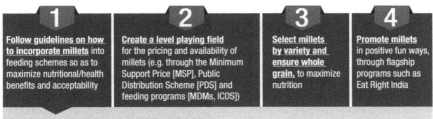

FIGURE 28.2 Policy recommendations for how to introduce millets into school meals.

developed and branded to recognize biofortified varieties in order to maximize the nutrition levels.

Conclusion

The Smart Food vision is a world where food is healthy, environmentally sustainable and contributes positively to the welfare of those who produce it, especially smallholder farmers. Studies have shown the positive nutritional benefits of millets and sorghum and high consumer acceptance for them. With 2023 declared by the UN as the International Year of Millets, this will be the turning point for millets to be globally recognized and popularized. Asia and Africa need value chains developed to be able to leverage the impending millet revolution. This can be the opportunity for millets and sorghum to return to their status as staples across many countries and be globally recognized, heralding their reach as a major staple, showing the potential for smart foods and NUS to gain popularity and acceptance and move into the mainstream.

References

Anitha, S., Govindaraj, M. and Kane-Potaka, J. (2019c) 'Balanced amino acid and higher micronutrients in millets complements legumes for improved human dietary nutrition', *Cereal Chemistry*, vol 97, pp74–84.

Anitha, S., Htut, T. T., Tsusaka, T. W., Jalagam, A. and Kane-Potaka, J. (2019b) 'Potential for smart food products in rural Myanmar: Use of millets and pigeonpea to fill the nutrition gap', *Journal of the Science of Food and Agriculture*, pp1–7. DOI: 10.1002/jsfa.10067

Anitha, S., Kane-Potaka, J., Tsusaka, T. W., Tripathi, D., Upadhyay, S., Kavishwar, A., Jalagam, A., Sharma, N. and Nedumaran, S. (2019a) 'Acceptance and impact of millet based mid-day meal on nutritional status of adolescent school going children in India', *Nutrients*, vol 11, no 2077, pp1–16

Anitha, S., Kane-Potaka, J., Tsusaka, T., Botha, R., Rajendran, A., Givens, I., Parasannanavar, D.J., Subramaniam K., Kanaka, P., Mani, V., Bhandari, R. (2021) 'A systematic review and meta-analysis on the potential of millets and sorghum for managing and preventing diabetes mellitus', *Frontiers in Nutrition*. UIN number: reviewregistry1094. www.researchregistry.com

Awika, J.M. (2011) 'Major cereal grains production and use around the world', in Awika, et al. (eds) *Advances in Cereal Science: Implications to Food Processing and Health Promotion*. American Chemical Society, Washington, DC, pp1–13.

Behera, M. J. (2017) 'Assessment of the state of millets farming in India', *MOJ Ecology & Environmental Science*, vol 2, no 1, pp16-20.

Davis, K.F., Chhatre, A., Rao, N.D., Singh, D., Ghosh-Jerath, S., Mridul, A., Poblete-Cazenave, M., Pradhan, N. DeFries. R. (2019) 'Assessing the sustainability of post-Green Revolution cereals in India', *Proceedings of the National academy of Sciences of the United States of America*, vol 116, no 50, pp25034–25041.

Diama, A., Anitha, S., Kane-Potaka, J., Htut, T., Jalagam, A., Kumar, P., Worou, O.N. and Tabo, R. (2020) 'How the Smart Food concept can lead to transformation of food systems and combat malnutrition – different approaches globally and a case study from Myanmar with lessons learnt for creating behaviour change in diets', in H.K. Biesalski (eds) *Hidden Hunger and the Transformation of Food Systems. How to Combat the Double Burden of Malnutrition? World Review of Nutrition and Dietetics*. Karger, Basel, vol 121, pp149–158. DOI: 10.1159/000507494

ICRISAT (2018) Stories of Impact from the drought tolerant crops value chain. Project Report ICRISAT. http://oar.icrisat.org/11205/1/Stories%20of%20Impact_%20 high%20res_7_12_2018.pdf

ICRISAT (2019) How to include millets in menus to maximize both nutrition and likeability. Smart Food brief 2. https://www.smartfood.org/wp-content/uploads/2019/12/How-to-include-millets-in-menus-final.pdf

Longvah, T., Ananthan, R., Bhaskarachary, K. and Venkaiah, K. (2017) 'Indian Food Composition Table', National Institute of Nutrition, Indian Council of Medical Research, India, pp1–578.

Smartfood.org (a) 'Millet Finder', www.smartfood.org/smart-food-products/

Smartfood.org (b) 'Activities', www.smartfood.org

Smartfood.org (c) 'Ambassadors', www.smartfood.org/smart-food-ambassadors/

Tonapi, V.A., Mahala, R.S., Elangovan, M., Yogeswara Rao, Y. and Yadav, O.P. (2015) 'Public-private partnership with special reference to seed industry', in Tonapi, V., Dayakar Rao, B., Patil, J.V. (eds) Millets promotion for food, feed, fodder nutritional and environmental security, *Proceedings of Global Consultation on Millets Promotion for Health & Nutritional Security, Society for Millets Research, ICAR-Indian Institute of Millets Research, India*, pp55–64.

Vital, M. (2018) 'Reality TV for a cause – Brining back Smart Food', Food Tank, May 2018. Available at: https://foodtank.com/news/2018/05/reality-tv-bringing-back-smart-food-icrisat/

Wangari, C., Mwema, C., Siambi, M., Silim, S., Ubwe, R., Malesi, K., Anitha, S. and Kane-Potaka, J. (2020) 'Changing perception through participatory approach by involving adolescent school children in evaluating Smart Food dishes in school feeding program – Real time experience from Central Tanzania', *Ecology of Food and Nutrition*, vol 59, no 5, pp472–485.

PART V

Stakeholders and global champions

29

VOICES FROM THE COMMUNITIES

The custodians of neglected and underutilized species

Aiti Devi, Bibiana Ranee, Biswa Sankar Das, Girigan Gopi, Indra Bai, Kamla Devi, Loichan Sukia, Loknath Naure, Maganbhai Ahir, Manikandan, Maruthan Ganeshan, Maruthi, Meera Bai, Melari Shisha Nongrum, Nanchiyamma, Phool Bai, Prashant Kumar Parida, Rakesh Kumar, Ramesh Makavana, Rami, Ramkali Bai, Ridian Syiem, Sangeeta Devi, Shalini Devi, Sharad Mishra, Sunadei Pitia, Sunamani Muduli, Tukuna Burudi, Usha Devi

Introduction

Farming communities have played a crucial role in the conservation and management of agrobiodiversity, including the rich diversity of nutritive crops that their ecosystem supports. Communities and farmers are the key stakeholders in the conservation of NUS, which plays a crucial role in their livelihoods and wellbeing. This chapter records the perspectives of custodian farmers, and the community-based and non-governmental organizations that support them, across the Indian landscape. Frequently identified bottlenecks in the cultivation and consumption of neglected and underutilized species (NUS) include lack of awareness, inadequate skills, drudgery (including limited access to available technology), inadequate support price mechanisms and procurement arrangements by the state, and limited marketing arrangements.

To address these issues, key interventions included working on all aspects of millet cultivation. The formation of core groups of millet farmers and self-help groups, and participatory assessments of traditional knowledge related to millets were important initial steps to ensure continuity. The technical aspect included the nutritive value assessment, seed collection from farming communities, quality seed production, distribution to farmers, revival of seed storage systems along with exchanges through community seed banks, demonstration of modified methods of cultivation including line sowing, intercropping to increase yield and additional income, and the provision of machinery to reduce drudgery in de-husking millets. The commercial aspect included training on value-added

DOI: 10.4324/9781003044802-34

products preparation, the creation of suitable rural infrastructure, facilitating enterprise units, and identifying and linking producers with potential markets.

Small millets: perspectives of custodian farmers of Koraput, Odisha

Smt. Sunadei Pitia and Shri Loichan Sukia of Machhara Village, Smt. Tukuna Burudi, Khiloput and Shri Sunamani Muduli of Janiguda Village of Koraput District

Facilitator: Prashant Kumar Parida, MSSRF

Millet has a long history of cultivation in the high lands of this region, so much so that Koraput is called the island of *mandia* (finger millet). Millets, especially *mandia*, are the staple food of the tribal people of Koraput. They begin their days, especially in the summer, with the consumption of *mandiapejo* (ragi gruel).

Views with regard to NUS?

Though millets are still part of subsistence agriculture in Koraput, the industrialization of the recent past has had a major impact on their cultivation. The area under millets has reduced drastically as farmers have shifted to commercial crops, with eucalyptus tree plantations and maize replacing large areas previously occupied under millets cultivation.

Major obstacles for NUS enhanced use

Most NUS crops like finger millet, little millet, foxtail millet, pearl millet, niger, horse gram, red gram, dolichos lablab, wild vegetables, etc. were mainly part of shifting cultivation practices (*Podu Chaso*). Now, the Forest Department or community forest groups like the Vana Surakshya Samiti (VSSP) prohibit such practices. Moreover, people are not so interested in cultivating NUS crops in their meagre land, replacing other crops that may fetch better returns. These NUS crops are mostly suited for cultivation on the uplands rather than other land categories. Other factors for the decline included the lack of marketing opportunities for NUS crops with remunerative prices. Since no processing facilities for NUS are available in rural areas, there are very low-level value-addition opportunities, leading to distress sales of surplus grains. Modernization has also influenced the present generation a lot, especially children who are not so interested in consuming millet in their daily diet, with the supply of rice in Public Distribution System (PDS) by the state government having changed the dietary preference of most tribal people.

Interventions to be pursued for their effective promotion:

- Institutions should create opportunities for processing at the Village/Gram Panchayat level, along with value addition, marketing, and remunerative prices, which will encourage farmers to cultivate more NUS crops.
- Researchers have to revisit the crop species available in the region and focus on seed purification and quality seed production of traditional varieties,

blending with improved varieties, and must make available quality seeds at reasonable rates at the right time.

- There is a need to shift from monocropping systems to traditional cropping systems such as mixed cropping, intercropping, multiple cropping, and crop rotation practices, with the introduction of modern technologies and sustainable agricultural practices and a focus on capacity building of farmers.

Custodian farmers' perspectives from Rayagada, Odisha

Mr. Loknath Naure, Kearandiguda, Bisamacataka, Rayagada, Odisha
Facilitator: Mr. Biswa Sankar Das, WASSAN.

Views with regard to NUS?

NUS such as millets, pulses, and oilseeds are very important. They are very healthy for humans and for soil. This also reduces pest incidence and provides different foods for consumption.

Major obstacles for NUS enhanced use

Previously, people used to grow different crops for meeting their food security needs. But after PDS has come in, people have slowly stopped growing different crops. In addition, the younger generations are more used to the taste of rice, and prefer not to consume millets and coarse cereals. Without the need for consumption, there will be reduction in production.

Interventions to be pursued for their effective promotion

The government should create awareness and build prestige around these crops. They should also be included in schools and hostels so the youth does not lose interest. Some incentives should be given to promote them.

Perspectives from custodian farmers of Madhya Pradesh

Ramkali Bai, Meera Bai, Magartagar Village, Dindori District
Phool Bai, Bhilai Village, Indra Bai, Dhiravan Village, Dindori District
Facilitator: Mr. Sharad Mishra, Action for social Advancement, Bhopal

Views with regard to NUS?

Kodo and kutki millets are well known in Madhya Pradesh and many farmers maintain their seeds. These crops are the lifeline of the tribal communities, and they are eager to learn of improved practices for their cultivation and use. The availability of improved varieties of millets, better linkages of farmers with markets, and improved processing units to reduce the drudgery of women are key interventions in our view.

Major obstacles for enhanced use

- Poor knowledge of new technology, and very weak links with concerned research and extension departments
- Lack of seeds
- Poor price for millets in local markets and unfair weighing of produce

Interventions to be pursued for their effective promotion

- Continuous supply of good quality seed production.
- Linking millet growers with markets through Farmer Producer Companies (FPC).
- Elimination of processing difficulties in de-stoning and de-husking operations.
- Introduction of minor millets in school meals and Anganwadi schemes to strengthen nutrition and raise demand.

Conserving Indigenous crops by custodian farmers in Uttarakhand

Kamla Devi, Paini Village, Usha Devi, Saldhar Village, Sangeeta Devi, Saloor Village, Shalini Devi, Poona Village, Aiti Devi, Mirag Village, Chamoli District
Facilitator: Dr Rakesh Kumar, HESCO, Dehradun

Views with regard to NUS?

NUS have always been very important sources of food in the Himalayas – in the Chamoli, Rudraprayag, and Tehri Garhwal regions, we use millets both for food and feed. We appreciated their resilience in the face of difficult soils and climates; millets are important for our culture and our nutrition security. Over the years, many NUS like Amaranths (chaulai), foxtail millet (kauni), finger millet (kauda), horse gram (gehath), rice bean (lobiya), and buckwheat (kuttu) are being cultivated less and less, and I believe they should be brought back. To do so, we need better varieties, better cultivation methods, and more favorable market conditions (like minimum support price).

Perspective of Anmol Desi Beej Sanvardhan Khedut Sangathan (seed saver farmer association)

Mr. Maganbhai Ahir, Ningal, Anjar Block, Kachchh, Gujarat.
Facilitator: Mr. Ramesh Makavana, Satvik

Views with regard to NUS?

Traditional varieties of NUS are well adapted to difficult conditions, performing better than major crops in resource-scare areas. However, the reduced availability of trusted traditional seeds has adversely affected their continued cultivation.

Farmers in the Kachchh District are conserving diverse traditional seeds, which they use to cope with climate variability. For example, if there is an early monsoon, they grow bajra (pearl millet), but if the rains are delayed by a month, they opt for sorghum. Similarly, they choose white sesame in the case of early rain, and prefer brown sesame for mid-late rains or brown type for late rain. This is to say that the more crops and varieties we keep in the communities, the more options we have to adapt to climate variability and satisfy other livelihood needs.

Major obstacles for NUS enhanced use

Today, only few farmers still maintain traditional seeds, with low genetic purity. This is because traditional crops have been widely replaced by modern ones, promoted aggressively by the government and private seed companies. Extension agents denigrate NUS as being inefficient crops and this is not helpful.

Interventions to be pursued for their effective promotion

- All NUS should be documented for their traits like drought tolerance or disease resistance, and their germplasm should be protected against bio-piracy.
- The government should change their attitude and promote NUS for their special features such as nutritional value, drought tolerance, etc.
- Incentives should be provided to farmers for the conservation and cultivation of NUS.

Perspectives of custodian farmers of Attappdi, Kerala on NUS

Mrs. Nanchiyamma & Manikandan, Kallakkara Village & Rami and Maruthi, Marrappalam Village, Sholayur Panchayath, Attappadi Block; Maruthan Ganeshan, Nakkupathi Village, Agali Panchayat
Facilitator: Mr. Girigan Gopi, MSSRF
Attappadi is valley in the Western Ghats inhabited mostly by tribal communities. People here appreciate traditional crops like NUS for its adaptation to harsh climates and difficult environments and for being an important source of food. The area under NUS has shown drastic reduction since the 1980s for various reasons, including the abandonment of agriculture by youth, changes in traditional diets, high competition from subsidized rice, laborious processing of NUS, and preference towards modern foods.

Major obstacles and suggested solutions

Millets and other NUS would be very helpful to fight the growing malnutrition recorded in Attappadi; however, farmers feel that it will be very difficult to bring NUS back into cultivation. We believe, however, that some interventions from the government can help towards that goal, e.g., the promotion of collective ownership of land, more credit opportunities to farmers, dissemination of

342 Aiti Devi et al.,

improved know-how, more irrigation, and protection from wild animals. As for younger farmers, we believe that if the cultivation of NUS were shown to be profitable, they would be happy to remain in their farms without migrating. In this regard, the creation of local markets for NUS will help realize fair prices for these products and improve their commercialization. Many varieties of NUS have been lost and we need support to better safeguard their seeds; we also need improved varieties, more irrigation for our fields, promotion of collective actions, improved processing (especially for millets), and application of the PDS also for these local crops.

Perspectives from custodian farmers of Meghalaya, India

Ms. Ridian Syiem, Khweng, Ms. Bibiana Ranee, Nongtraw
   Facilitator: Dr. Melari Shisha Nongrum, North East Slow Food & Agrobiodiversity Society (NESFAS), Meghalaya

Wild edible NUS

We belong to Indigenous communities that altogether form a numerically small minority of India's population. The state of Meghalaya is inhabited largely by the Khasi people (including *Khynriam, Bhoi, War, Pnar,* and *Lyngngam* sub-groups) and *Garo* Indigenous communities. Our forests are rich in biodiversity and agrobiodiversity, with many wild edibles (green leafy vegetables, fruits, tubers, shoots, nuts and seeds, and mushrooms) used in traditional food preparations. Wild edibles are important to us as Indigenous communities and they are an integral part of our ancient food culture. These edibles are safe foods as compared to most market foods, which contain a lot of toxins due to the heavy use of chemical fertilizers and pesticides. They can be found both in the wild and in home gardens, and are precious foods to supplement our diets, especially during lean seasons. Even during the lockdown period due to the COVID-19 pandemic in April 2020, many of these plants (wild and cultivated) helped households with a secure source of food.

Major obstacles for NUS enhanced use

Though wild edibles are integral to the traditional foods of the Indigenous People, we are seeing a change in their availability and consumption. Due to growing populations in the Villages, there is increased pressure on land and forest resources. There is exploitation of forests for various purposes, either for fuel wood, the heavy collection of non-timber forest products, wood collection for making charcoal, stone quarrying, mining of coal, or limestone mining, which destroys both land and forests. Also agricultural activities such as cash crop cultivation can be deleterious for wild edibles, due to massive land clearing associated with them. Because of these facts, we need now to walk increasingly longer distances to gather wild edibles, which is affecting regular access to and consumption of these plants. Also, these days, much of the youth and many young mothers regard wild edibles as a not-so-nutritious food. Major crops dominate markets and the difficulties in accessing these for wild edible sellers are also important limiting factors for their popularization.

Interventions to be pursued for their effective promotion

Despite these challenges, the translation of research evidence of the high nutritive values of wild edibles into local and simple languages by organizations like the NESFAS has really helped the youth and people at large regain confidence in wild edibles. More of such awareness-raising is required for school children and youth so that they understand the importance of wild edibles in enriching their diets.

Knowledge sharing among farmers is key to sustaining and continuing the existence of this knowledge. This sharing can be done across different platforms such as Agroecological Learning Circles, a platform of farmers that NESFAS has facilitated in many communities; the Indigenous Food Communities Alliance among the youth; and also in the school gardens where children can grow and learn about the wild edibles. From our experiences, this sharing of knowledge has led to sharing seeds. "If we lose our seeds, we have lost our food," said one custodian farmer. Preserving and promoting the cultivation of underutilized crops is very important. Thus, it is important to preserve traditional seeds as individual farmers or in community seed banks.

We can also allow wild edible plants to grow in kitchen gardens and cultivated fields. This will enhance access to these edibles as we frequent the fields regularly. This will enhance the access and consumption of wild edible greens.

30

VOICES FROM THE PRIVATE SECTOR ENGAGED IN THE USE ENHANCEMENT OF NUS

Ana Luiza Vergueiro, Daniel Kirori, Jacqueline Damon, Jose Alfredo Lopez, Leon Kenya, Mahesh Sharma, Ram Bahadur Rana, Margaret Komen, Meghana Narayan, Michael Ngugi, Rohan Karawita, Sergio Vergueiro, Serkan Eser, Shauravi Malik, Simon Nderitu, Sohini Dey, Sridhar Murthy Iriventi, Vikram Sankaranarayanan

Introduction

The participation of the private sector is widely considered critical in order to achieve a truly sustainable conservation and use of agrobiodiversity. The role of the private sector is particularly relevant for the use enhancement of neglected and underutilized species (NUS) (Padulosi et al., 2013, 2019) whose promotion requires knowledge of the proper functioning of value chains –from farm to fork – that is rarely available to communities, community-based organizations (CBOs) or even government organizations. Experiences and competences from the private sector are highly complementary to those of other stakeholders; thus, canvassing the cooperation among all actors should be central to all projects and programmes focusing on NUS. This is not a minor task though. Different are too often the sensitivities of private firms with regard to the methods, approaches and tools for the sustainable use of NUS, in terms of type of biodiversity to deploy in production systems (e.g., landraces vs. uniform, high-yielding varieties), participation of vulnerable groups, interest in safeguarding traditional knowledge, fair and transparent sharing of benefits among value-chain actors, capacity building of local actors, especially youth, women or Indigenous People, to just mention a few. The 11 contributions included in this chapter (from Guatemala, Brazil, India, Nepal, Sri Lanka and Kenya) are brief testimonials of private companies engaged at various levels in business activities dealing with NUS. The most common messages across all contributions include calls for the greater involvement of private companies in NUS projects, greater support from governments for upgrading the processing technologies of private firms engaged

DOI: 10.4324/9781003044802-35

in the promotion of NUS, greater support from governments in raising peoples' awareness of the many livelihood benefits associated with NUS, greater sharing of research findings with the private sector, capacity building where the private sector can be both a recipient and giver of knowledge, and enabling policies in support of entrepreneurship for the cultivation, processing and marketing NUS. General perspectives about NUS are provided along with some examples on specific issues related to the cultivation, processing and marketing of minor millets or other emblematic NUS.

National Food Promotion Board, Sri Lanka

Dr. Rohan Karawita, National Food Promotion Board,
   Ministry of Agriculture, Colombo, Sri Lanka

Company's views with regard to NUS

The main task of Sri Lanka's National Food Promotion Board (NFPB) is to promote local and indigenous foods (there are many NUS among these) as our citizens have become increasingly dependent upon poorly nutritious starchy foods and junk foods rich in fats, and have become used to consuming their meals in short eats. These food habits are not healthy as they cause the spread of non-communicable diseases among the population. As a way to counter these negative trends, the NFPB has been carrying out several programs and projects aimed at increasing peoples' awareness about Sri Lanka's valuable, rich biodiversity and promoting its greater use. Thanks to the Biodiversity for Food and Nutrition Project (BFN),[1] NFPB was able to strengthen these efforts and successfully develop three underutilized fruit drinks, sold in 80 ml ready-to-serve (RTS) bottles (namely 'Ceylon olive', 'Sour soup' and 'Ceylon mango'), all very natural and free from artificial additives and preservatives. These drinks are now being commercialized by the Peradeniya Agriculture Department's Food Research Unit in Gannoruwa.

Major obstacles in the use enhancement of NUS

Though NFPB has manufactured RTS drinks in the past, the demand for such products has always exceeded our capacities. Due to a number of factors – including the lack of novel technology, insufficient labor force, high unit costs and limited fruit pulp preservation capacity – we were, in fact, unable to manufacture RTS products regularly throughout the year. Not helpful either have been NFPB's efforts in disseminating relevant technologies to local communities in the hope that they would pick them up through cottage-based productions.

346 Ana Luiza Vergueiro et al.

Top 3 solutions for the effective promotion of NUS

What is much needed to broaden the use of NUS in Sri Lanka is the strengthening of their commercialization through the close involvement of the private sector. Although at NFPB we already sell RTS products made out of NUS through the company 'Thriposha Limited' and our two outlets in Narahenpita and Dehiwala, we do recognize the need to invest more in marketing efforts. To that regard, NFPB has now taken all the steps necessary to transfer RTS technology to a fruit-processing factory, the 'V and J Industries',[2] located in Pallekele (Kandy). All is ready for the new arrangements to take off and as soon as the COVID-19 pandemic eases up the pressure on our work, we are ready to start this new joint venture. The legal agreement between this company and NFPB will allow the mobilization of the company's know-how and physical capacities to support the wider promotion of NUS products across the country. This is expected to create greater demand of NUS raw products for farmers and, in so doing, boost local incomes, open up job opportunities and ultimately contribute to empowering smallholder farmers as per our vision.

Mace Foods Ltd., Kenya

Margaret Komen, MACE FOODS Ltd., Eldoret, Kenya

General considerations

The increasing awareness of the pro-health properties of non-nutrient, bio-active compounds found in fruits and vegetables has directed immense attention towards vegetables as vital components of our daily diets. For Kenya and sub-Saharan Africa in general, such an attention is very significant, as leafy vegetables have long been indispensable ingredients in the traditional sauces that accompany carbohydrate-rich staple foods. As a result, demand for traditional vegetables such a black nightshade, spider plant, common amaranth and cowpea leaves have been on the rise in local, urban and regional markets across Kenya. These vegetables have been typically growing in the wild, or in disturbed areas such as wheat farms, disused cow sheds and other farm areas.

In 2009, Mace Foods Ltd. (MFL) embarked on an ambitious project to produce, process and market five varieties of African leafy vegetables (ALV) for the East African diaspora and medium-income markets in urban areas. Mace Foods Ltd aimed at bridging the demand and supply gap between rainy and dry seasons as well as delivering high-quality, hygienic and nutritious food options to busy working-class women, who have little time to spare for selecting, plucking, cooking and serving these vegetables on a regular basis.

Since its inception, MFL has mobilized, recruited and trained many farmers in the production of ALV. So far the enterprise has piloted black nightshade (managu), common amaranth (dodo), spider plant (saga), cowpea leaves (kunde) and kales (sukuma wiki[3]).

Challenges in promoting NUS

Many are the obstacles that our company has been facing in promoting NUS. Following are some of the most relevant ones:

- Lack of access to high-quality seed varieties, leading to poor yields and poor quality of final products from farmers
- Producers and processors not able to tap the NUS market potential due to:
  - Unsustainable and unreliable production seasons, leading to inconsistent supply of the product to the market
  - Low quality products in the end-market due to poor harvests, post-harvest management, packaging and presentation of the final products
  - Low value-for-money for end-consumers due to poor product presentation, ease of use and convenience
  - Non-compliance with food-safety standards and regulations for ethnic food currently offered in the domestic market.

The solutions adopted

Our company aims at providing solutions for the above problems by:

- Providing a high-quality, hygienic, nutritious option on the table for the busy urban consumer
- Bridging the demand and supply gap between the rainy and dry seasons
- Lowering the cost of the end-product and increasing consumption of the ethnic foods in our retail chains
- Improving our business operations and the efficiency of supply chains, through continuous improvement systems, adherence to standard/certifications and other technologies relevant to our processes.

D.K. Engineering Ltd., Kenya

Daniel Kirori, D. K. Engineering Company Ltd., Food Processing Equipment Sales & Services, Nairobi, Kenya

Major obstacles in and recommendations for enhancing the use of NUS

- There has been a dramatic change in eating habits of the population, especially among younger ones who have abandoned traditional foods in favor of modern 'fast foods'. A better presentation of our traditional foods in the market is, thus, highly needed; this includes the better packaging of NUS products, to make them more appealing to consumers

348 Ana Luiza Vergueiro et al.

- More information be provided to consumers about the health benefits associated with the consumption of NUS. Both governmental and non-governmental organizations should provide platforms to educate the public on this matter, for example through media, road shows, trade fairs, schools and organized visits to research institutions
- Major institutions like ministries of agriculture, should lobby for policy bills in support of the wide dissemination of food processors like flour mills and snack processors that can help people make better use of local crops, to produce foods like 'ugali' from composite flour (i.e., maize plus millets flour, rather than only maize flour) or snack bars and breakfast cereals from NUS

One of the main technical challenge encountered in processing NUS had to do with the popping machine. Through numerous trials using modified parts of the machine, we were able, in the end, to address successfully such a problem and our company is now able to use the right pressure for the popping any type of cereals, including NUS.

Eurotropics S.A., Guatemala

Ing. Jose Alfredo Lopez L., EUROTROPIC, S.A., Guatemala Central America

Our views about NUS

Eurotropics S.A. is a Guatemala-based import–export company established in 2003, dealing with exporting food and ornamental plants. In 2011, we were contracted by the National Council of Science and Technology, the University of San Carlos, the University Del Valle de Guatemala and the University Mariano Galvez, to characterize endemic species of high nutritional and medicinal value. After five years of research, planting, germinating, cultivating, pruning, harvesting and analyzing roots, seeds, stems, leaves and flowers of numerous species, grown from 14 m a.s.l. to 3,240 m a.s.l. in the high mountains of Los Cuchutamanes, we discovered plant species with amazing properties for pregnant women, kids, young, adults and elderly people! We decided, thus, to rescue those amazing plants to help Guatemala strengthen food security and the health of its people. This country has a shameful 49% level of malnutrition among children under five years old! Almost half the population has low corporal weight, and low neuronal development. This is no longer acceptable and we wanted to help change that.

Major obstacles in the use enhancement of NUS

First of all there is a lack of education related to Mesoamerican plants that are much underutilized in our current food systems. About 25,000 years ago, these plants grew wild in this marvelous, biodiversity-rich region, and we have documents indicating that such plants were already being widely used by the Mayan civilization some 4,000 years ago. Though still used today by some Mayan tribes, they are rarely known by the young people. Very sadly, knowledge about these plants is fast disappearing among common citizens. Universities do not teach ancestral knowledge regarding these plants, and schools do not educate students on their importance. Governments and policy makers do not pay attention to them at all.

Top 3 solutions

The Ministries of Health, Education, Agriculture, Environment and Development are not spending enough efforts to leverage NUS knowledge to improve peoples' livelihood. There is a need to establish national programs to educate students about NUS, starting from kindergartens to the doctoral degree level. Chefs, cooks, farmers, industry managers and food manufacturers have to be all involved in such programs too; they should be mobilized to help create a *'Certificate of Origin of Mayan Superfood in Mesoamerica'* for all kinds of NUS foods, such as drinks like *atole*,[4] crackers, powders, cookies or tortillas, and people ought to learn about those amazing nutritional and medicinal properties that are helpful in strengthening our health. The importance of these traditional species is even more relevant today, as we are frantically struggling against the COVID-19 pandemic. It is realistic to say, in fact, that these neglected medicinal plants from the NUS basket may well harbor important biochemical compounds that could boost our immune system for better fighting this disease as well; hence, more research is needed on them! Unfortunately, humankind spends more efforts in building hospitals rather than preventing diseases through the education, promotion and cultivation regarding nutritious local plants as well as in processing them into high-quality products that are cheap and available to national, regional and international markets. I strongly believe that greater focus on NUS would reduce malnutrition, especially among pregnant women and children, and it would create new jobs, reduce youth migration and protect weak national currencies from depreciation. Guatemala is the eighth richest megadiverse center of biodiversity on Earth and such natural wealth is not sufficiently put to work! I hope my plea for a better future will be heard!

Slurrp Farm, India

Ms. Meghana Narayan, Shauravi Malik & Sohini Dey
Slurrp Farm, New Delhi

Our views about NUS

Food is a fundamental need in our lives. But food isn't merely a substance for staying alive, but rather a means to enhance the quality of life by assuring the required nourishment for our body. The lack of dietary diversity is well known to be a major cause for malnutrition, not just in India but across the globe. With the long-standing popularity of rice and wheat as the main foods in peoples' diets, experts are calling for urgent changes in our meals: more diverse crops are needed to fill the nutritional void left by the overconsumption of just few grains in our diet.

At Slurrp Farm we believe that millets are indeed among those underutilized species that need to be better promoted, and through our work, we hope to play a role in bringing them back to the plate. We believe that millets, with their inherent high nutritional benefits, can be part of the solution in tackling food and nutrition insecurity in India. Lately, we have seen a rapid rise in the use of oats and quinoa, a South American staple that has truly become a world commodity in just few years, and minor millets could well follow the same trend.

Millets are rich sources of fiber, vitamin B-complex and minerals, as well as polyphenols, lignans, phytosterols, phytoestrogens and phytocyanins, which act as antioxidants, detoxifying agents and immunity modulators. Additionally, millets are gluten-free grains, an advantage in the midst of increasing cases of celiac disease and gluten intolerances that restrict or entirely prohibit the consumption of wheat for lots of people. The nutritional benefits of various millets are elucidated in other chapters of this book and here we just want to stress that the value of these grains is based not only on their nutritional merits, but also on the important contribution that they can make to supporting sustainable agricultural practices and resilience in times of a continuously changing climate. Millets are highly resistant to adversities of cold, drought and salinity, and are, thus, highly suitable for cultivation in dry and arid lands (Padulosi et al., 2009).

In India, farmers are struggling with the effects of drought on their cultivations, and this is raising great concerns over the sustainable provisioning of the food necessary to feed 1.2 billion people. To that regard, the capacity of millets (including small millets) to thrive in harsh climatic conditions offers an immense potential to rescue agricultural communities ridden by drought and poverty. The revival of underutilized and neglected crops such as millets can exert significant influence on the cultural traditions and diversity and contribute to the self-identity, self-esteem and visibility of local communities.

Major obstacles for the use enhancement of NUS

Limited awareness and nascent consumers' demand are the major challenges. For customers used to decades of diets based only on rice and wheat, transitioning to millets will be a challenge. Low awareness accompanied by a lack of interest in taste can be a deterrent for consumers, who may be unable to understand the value of these nutritious food in their diet

Struggles in post-harvest processing

Post-cultivation drudgery remains one of the greatest challenges for millet farmers and a major obstacle in the popularization of these and other NUS. Apart from finger millet, pearl millet and sorghum, all small millets must be threshed and de-hulled before they are ready for consumption (more details on millet processing technologies are provided in Chapter 26 of this)

Lack of research and development

As emphasis has always been placed on improving a few grains for large-scale commercial use, the consumption of millets has declined steadily as a result, which has impacted negatively their production and potential for growth

Top 3 solutions for effective promotion

We live in times of ever-changing food trends and rapidly mingling culinary cultures. Globally, chefs are continuously pushing the bar for innovation, combining far-flung ingredients, reinventing traditional recipes and presenting food not simply as a means of subsistence but also as a multi-sensorial experience. This new trend is very helpful for the revitalization of NUS and the time has come for millets to be put back on the global food map. To that end, we believe the following are three most robust solutions currently needed:

A   Develop innovative millet products that are tasty, convenient and affordable
Using millets to develop products that are tasty, affordable and in a convenient easy-to-use format is most critical. Finger millet (ragi), which was among the lesser-known crops until a few years ago, is now making its way back into mainstream markets as an ingredient in a variety of products, from pre-packaged dosa and pancake mixes to chips. Similar product development will also increase the popularity of other small millets.
B   Launch a nationwide awareness campaign
Millets are often called forgotten foods because the communities that once consumed them regularly are now out of touch with their benefits and associated food cultures. A national advertising campaign similar to the 'Egg Campaign' or the 'Operation Flood Milk Campaign' are the need of the hour if we are to push the agenda for nutritious millets, good for our health and the planet. Bollywood stars like Alia Bhatt, Anushka Sharma and Virat Kohli have all shown ways in which they include millets in their diets and it would be great to also include top nutritionists, sports stars and movie celebrities in such campaigns too.
C   Making R&D work for millets
Millets have remained out of the R&D sphere for decades, with only a handful of institutes investing time and resources into studying and promoting them. Having turned to mainstream crops in recent years, many farmers are unaware of the many advantages of growing millets. Giving farmers access to information on best practices and facilitating small-scale testing can help them assess the conditions that are best suited for each millet species and variety. Leading research into the nutritional content of millets is imperative

352 Ana Luiza Vergueiro et al.

in order to better understand the numerous health benefits of these crops. Making scientific information accessible to millet-based product manufacturers, as well as consumers, can also have a positive impact. With a number of independent companies around the country working on millet-based products, instead of conventional food grains, active support from research organizations can aid product development and enhancement.

African Forest Ltd., Kenya

Simon Nderitu, Leon Kenya and Jacqueline Damon
African Forest, Nairobi, Kenya

About our work

African Forest is a holistic agro-forestry social enterprise established in 2006 and located on the edge of the Soysambu Conservancy in Kenya, a 22,000 ha ranch surrounding Lake Elmenteita, in the Rift Valley. African Forest has an indigenous organic plant nursery, seed center and trial plantations. We produce and promote, in conjunction with local communities, a wide range of non-timber forest products. These include medicinal plants, vegetables and other food crops, skincare products, bee products, fibers, oils, tannins and dyes. African Forest also offers services such as plantation establishment and management, reforestation, gardening and sustainable development planning and implementation, as well as assistance in raising 'green' finance.

Our 'Planet Positive Forestry' program involves planting a mixture of indigenous climax and pioneer tree/shrubs, to create a bio-diverse forest. These are intercropped with foods, fruits and herbs.

Promoting private-sector involvement in the development of value chains for NUS

The livelihoods of the world's population depend on food systems and food value-chains. There is a dire need to make our food systems more sustainable, especially in view of climate change, cultural changes and the many adversities affecting production systems. We strongly believe that food systems that are socially inclusive should be promoted, as opposed to the current ones that are controlled by corporations. Currently, it is estimated that 50% of the world's population cannot meet its food needs. Research also estimates that to meet the population's food requirements in 2050, measures need to be implemented to facilitate a 70% increase in current production. And these measures have to be inclusive of women, youth and Indigenous People.

Out of the 1.4 billion people in the world living in extreme poverty, 70% live in rural areas and are the most affected by drought and famine. Climate change has created unpredictable weather conditions, presenting challenges for Indigenous producers who depend on natural conditions for farming.

With the emphasis on Western lifestyles, critical Indigenous knowledge has been lost and is no longer being transmitted from one generation to the next. This is depriving our societies of knowledge on vital foods, while at the same time new lifestyles are compromising people's health.

We have, nevertheless, registered lately an increased awareness of the pro-health properties n indigenous vegetables, rich in micronutrients and bio-active compounds. Vegetables should be, therefore, core components of peoples' daily diets. Scientific findings are also confirming that local species and varieties are rich in compounds that have antioxidant properties, such as β-carotene, as well as in iron, calcium and zinc.

More demonstrations of the production, processing, cooking, marketing, distribution and consumption methods of NUS should be carried out in communities, in order to leverage Indigenous People's knowledge about these species and help promote their wider use. Critical publications of research findings on these species have been made, but these are not readily accessible to the Indigenous People. More should be done to disseminate relevant findings to strengthen farmers' knowledge of sustainable production and the use of these species.

Main obstacles in promoting NUS

The main obstacles for the enhanced use of indigenous vegetables includes access to production means, including quality seeds, land tenure, soil additives, water, etc. Indigenous farmers barely get access to knowledge on best practices. The perishability of many NUS vegetables and fruits, caused by their short shelf-life, is also a great challenge in their effective storage and marketing.

Some recommended solutions

- Train traditional farmer communities on the sustainable production and use of NUS to empower them. Training packages should cover inter alia small-scale, diversified farming and the use of environmentally sustainable technology
- Implement participatory guaranteed systems, giving greater role to women, youth and farmer groups, building on trust, social knowledge and networks
- Provide support and outreach programs including extension, helping farmers establish their own community structures. Interventions could start with one ward in a county (e.g., Mbaruk ward in Nakuru County, Kenya) and 1–2 villages for demo production sites (e.g., the Kasambara and Echaririe villages of the Mbaruk ward, Kenya)
- Provide farmers with quality seed for the focus NUS and demonstrate potential for including other species in value chains

- Support community-based seed banks and promote community seed bank registrations
- Build resource centers for knowledge sharing on research findings, product development, demo-site trials and seed sharing among Indigenous farmers
- Reduce perishability through drying, fermentation, canning in glass bottles and implementing temporary storage measures. Be a point of collection, washing and storage
- Conduct research on methods of mixing maize and wheat flours with flours from traditional grain crops
- Negotiate contracts and tenders to support the local production and consumption of NUS, including public procurement for schools and hospitals
- Carry out more donor-funded projects and promote more research on the nutrition and health benefits of NUS, as well as the dissemination of findings to raise awareness and build capacity
- Commit to partnerships through multi-stakeholder platforms and foster the joint implementation of activities through constant meetings and advocacy actions.

Anamol Biu Pvt. Ltd., Nepal

Mahesh Sharma and Ram Bahadur Rana, Chief Executive Officer, Anamol Biu Pvt. Ltd., Chitwan, Nepal

Our views about NUS

We believe traditional crops, especially nutritionally dense cereals and vegetables, are going to be one of the strategic sources of food for the ever-increasing number of urban consumers around the world. In Nepal, NUS are playing an instrumental role in meeting the food and nutrition security needs of households, particularly those residing in hilly and mountainous areas, where farming systems, to a large extent, rely on traditional crops and varieties. Needless to say, growing the use of NUS contributes to a healthier environment, as these plants require less chemicals and pesticides. Some NUS are also known to be climate resilient, meaning they perform reasonably well even in harsh growing conditions (they are cold and drought tolerant, need low external input environment, etc.), thus contributing to household food security in rainfed farming systems. Popularizing NUS among health-conscious urban consumers may spur demand for these traditional crops, which in turn would contribute to improving smallholder farmers' livelihood.

Our NUS work

Anamol Biu was established 10 years ago and, since its inception, has been working closely with rural farmers in the seed supply chain, where community seed banks (CSBs) have played pivotal role in preserving and promoting NUS crops. We source our vegetable and cereal seeds primarily from CSBs under a buy-back guarantee system. With technical support provided through LI-BIRD, Anamol Biu has diversified its seed portfolio with the inclusion of several NUS into its activities. Amaranth, coriander, faba bean, fenugreek, broad leaf mustard, local beans, cucumber and many other vegetables are included in our 'Vegetable Seed Composite Kit', a family-nutrition-targeted product, that has become highly popular among development efforts such as the SUAAHARA Project[5] and the Home Garden Project,[6] both implemented in Nepal. In recent years, this tool has become a standard approach for distributing seasonal vegetable seeds to farmers by development organizations and government agencies alike, thereby increasing the potential for mainstreaming NUS in seed supply systems. In order to capitalize on the increasing demand for selected NUS products (including minor millets, amaranth, mountain beans, sticky rice, etc.), Anamol Biu, with the support of several agencies, is investing in the establishment of a 'Packaging House', planned to come into operation in early 2021. The Packaging House is expected to aggregate, clean, grade and package NUS products in a particular brand before marketing them in domestic and international markets. We will be starting a line of composite vegetable packets comprising NUS vegetables for rooftop urban gardens in the near future, as we have received an increasing number of requests during agricultural fairs/*melas*.

Unfortunately, the mainstreaming of NUS varieties into formal seed systems, securing a niche for themselves in the current seed market is still rather low. Many factors are responsible for such slow progress, including unfavorable government policies and programs, limited research and development support, lack of awareness on the part of consumers and lack of best practices knowledge among the growers. It is, therefore, highly imperative that responsible actors increase their investments in the R&D of NUS, if we are to really help these crops emerge from their marginalization.

Although the promotion of NUS features in strategic policy papers of the Nepalese government, in reality this area has garnered the least priority in research investments, resulting in negligible progress towards the identification of promising species and their varietal enhancement. On the other hand, the private sector in Nepal is rather weak, in financial and technical terms, to be able to invest in R&D and is interested in acting only at the stage of popularization where heritage crops are concerned. As a matter of fact, a market system for NUS crops including access to quality seeds is yet to be developed in the country. So even if a NUS variety is registered, there remains a serious problem in the supply chain. When the market demands a significant amount of a particular NUS crop, the reality is that producers cannot actually be located to meet such a demand. This is the case for instance of the local amaranth variety 'Ramechhap Hariyo Latte', which was successfully registered at the national level by the Seed Quality Control Centre in 2018, upon which Anamol Biu was able to purchase 17 kg of the seeds from a farmers-managed CSB in Jugu, Dolakha district; the following year, however, we required at least 10 kg of that seed but the CSB in Dolakha was not able to provide such amount. We met the farmer representatives and the program staff from LI-BIRD, and learned about problems that the producers were experiencing, which related mostly to price and financial issues. Anamol Biu agreed to meet the farmers' price expectations, and LI-BIRD helped as well in multiplying the seeds in the Sindhupalchok and Lamjung districts.

The irony is that when producers have something to sell, they complain about market availability and accessibility issues, which is a reflection of dysfunctional market for NUS crops. The government is not very helpful because it has failed to come up with any longer-term monetary and non-monetary schemes to promote NUS.

Two factors hindering the promotion of NUS are the limited number of food recipes available for their popularization among consumers and the persistence of incorrect socio-cultural beliefs regarding NUS, often termed 'crops of the poor' or 'famine foods'.

As mentioned already, R&D throughout the value chain, from production to consumption, is one critical area policymakers should think seriously of investing in, so as to promote NUS effectively. This will pave the way towards solving production limitations and marketing challenges.

On a positive note, we should say that in Nepal, the government has made the registration of local crop varieties in the formal seed system a much easier task, something that is expected to benefit the popularization and commercialization of these neglected crops. But more essential interventions are still needed from the government, such as access to soft loans, support in machinery and tools for production and post-harvesting operations, and financing crop insurance to lure producers and make the cultivation of NUS crops more competitive and profitable for farmers. Furthermore, policymakers need also to help promote NUS through the national education system: NUS education for healthy diets and improved nutrition should be promoted in schools, college cafeterias and even office canteens.

Public–private partnerships (PPP) often prove to be a fruitful and sustainable venture for promoting the value chains of products such as NUS, where women and youth can be targeted at large through attractive schemes.

Lastly, we foresee a potential alternative of promoting NUS as organic crops, targeting upper economic classes and regions. Such an approach could prove very helpful: once the supply chain is firmly established it can then gradually cater to the needs also of middle- to low-income classes and thereby contribute to the wider popularization of these nutritious crops.

Simlaws, Kenya

Michael Ngugi
  Simlaw Seeds, Nairobi, Kenya

Our views about NUS

The importance of these neglected species of plants cannot be over-emphasized. Here in Kenya, NUS have been used as source of food for generations by many communities. Sadly, government research bodies have given them little attention insofar, focusing instead only on main food crops such as maize, wheat and beans.

Our NUS work

We at Simlaws do appreciate the importance of NUS in food security and, most importantly, in nutrition. These species can grow well with minimum use of inputs. They have stable yields in local environments where they are highly adapted and able to thrive where exotic crop species are challenged, such as in marginal lands where rainfall is scarce and irrigation is not affordable.

Most communities would normally go to pick NUS vegetables in the wild, because these species are not easily found under cultivation due to the lack of seed systems dedicated to them. Consequently, there has been erratic supply of these food crops in the market. This despite their inherent superiority in terms of resilience, disease and pest tolerance. Unfortunately, private seed companies have not been availing NUS seed in the formal market either.

For the successful promotion of NUS there should be a concerted effort in funding projects involving both national governments and the international community. This will ensure that these species are collected, safeguarded, characterized, selected, improved and massively produced for wider use. Training should be undertaken especially in the fields of breeding, conservation, seed production, agronomy, nutrition and value addition. The private sector should be also involved in training, so as to leverage its knowledge, which is helpful in promoting NUS at affordable prices more widely. Moreover, communities will need also to be educated through demonstrations on how to grow and use these nutritious crops more efficiently.

Bilgi Sistem Yönetimi – İthalat & İhracat, Turkey

Serkan Eser, Bilgi Sistem Yönetimi – İthalat & İhracat, Serik Antalya –Turkey

Our views about NUS

We are located in the Mediterranean region and we have good knowledge about the biodiversity there, including those species occurring in the wild that have not been properly researched yet. We support several research projects involving scientists, who are directing the attention to all the promising leafy greens NUS growing in this region.

Major obstacles for the use enhancement of NUS

NUS need to be better considered by both household users as well as people involved in the so-called 'horeca' sector (hotel/restaurant/café) who should devote continuous attention towards their use enhancement. Unfortunately, today this is not happening, and such a situation is limiting the wider use of NUS by chefs, which, in turn, is hindering their wider promotion. With regard to home users, these traditional crops need to be suitable for kitchen use and food cultures. The many good qualities of NUS should be highlighted and leveraged through novel, versatile and easy recipes in order to promote their popularization as home foods; to that end, more research on cooking aptitudes and methods is needed.

Top 3 solutions for an effective promotion

A   We need to do more for the promotion of NUS, especially among chefs
B   We believe that geographical indications would play a highly strategic role in the marketing of NUS
C   It would be extremely helpful to create an ad hoc brand, as has been done for other products. For instance, in the case of coffee: the majority of Turkish people refer to standard American coffee as 'Nescafe' so as to distinguish it from the classic Turkish coffee; so this commercial brand is applied widely to refer in fact to many types of coffee. When thinking about that, I realize that advertising is extremely powerful and could be, indeed, leveraged for the promotion of these nutritious but little-known foods.

GoBhaarati Agro Industries and Services Pvt. Ltd., India

Mr. Sridhar Murthy Iriventi, GoBhaarati Agro Industries and Services Pvt. Ltd., Hyderabad, Telangana

Our views about NUS

Consumers' food behavior is mostly cultural. Throughout human history we see cultural evolution has been a function of local contexts, including the policy framework implemented in that place. Like lifestyles, food styles have their own comfort zones ingrained into the unconscious pattern of choices of items on people's plates.

For the last ten years or so, our company, made up of first generation of entrepreneurs, has been at the forefront of food innovation and of introducing 10 varieties of millets, including sorghum, in three states of India. Over the years we have observed a significant rise in people's awareness with regard to NUS, especially millets, and this has been very helpful in promoting the consumption of these foods. We see NUS as the food of the future – good for the consumer, for the planet, for the farmer and for the economy too. Over the last decade, we have interacted with more than 50,000 customers, in hundreds of road shows, conferences, cooking competitions and other events, and the lessons we learned during these interactions can be summarized as follows:

- Consumers wouldn't mind using NUS in traditional recipes. For example, a traditional Indian recipe like *idli*, can well be made with little millet too, and this innovative food will be as palatable as that prepared with conventional ingredients like rice
- Entrepreneurs/private sector cannot carry out NUS promotion alone. Ideally, the private sector should focus on marketing NUS products while research institutions and governments should focus instead on studying the health or nutrition benefits associated with NUS; thus, the outcome of such work can inform the private sector for their campaigns. I think this would be an ideal partnership for the promotion of NUS
- NUS, especially minor millets, contain many important nutrients that can be strategic in achieving a balanced diet, but more needs to be done to highlight this fact in promotional campaigns
- During interactions with our customers, we register comments ranging from, "We don't know millets", to "We are Millet Ambassadors". While we are happy to inform customers on the value of millets and contributing, thus, to filling the gap in awareness, there are a number of obstacles in doing that, viz.:
  - Food is first and foremost a cultural experience that goes back to what we ate in our childhood when our mothers fed us with love. Most of the current adult generation has been fed by mothers who had access to basic food grains through the Public Distribution System (PDS), which focused (until very recently) only on rice and wheat.
  - The millet basket is made of many species and, owing to the poor knowledge that people have about these crops, it is normal to register some kind of hesitation in consumers when they need to make a decision on which type of millet to buy.
  - NUS products change continuously and do not last on the shelf for more than a year or so. This is not good for their popularization, as clients fall back into the 'comfort zone', i.e., purchasing the usual foods made out of wheat or rice.
- 2018 was declared by the Indian government as 'The Year of Millets'. This was a good move, but unfortunately the awareness campaigns on millets that followed did not do it justice in presenting in a convincing manner the many potentials of these crops. Much more and better needs to be done to promote NUS!
- Policies in favor of millets need to be better implemented to bring change! What we observe, unfortunately, is that most work for the promotion of NUS is led by scientists and bureaucrats who don't appreciate entrepreneurial efforts. For instance, our company has made a breakthrough in packaging,

which has increased the shelf-life of de-hulled millets by 300%! Yes, we have been appreciated for this outcome, but were not supported beyond that in scaling-up such novel processes.

A three-pronged approach for promoting NUS

A   Instead of encouraging Farmer Producer Organizations to do everything from farm to fork, governments should encourage instead centers of excellence to take care of specific activities along the value chain

B   Develop a 45-day off-season business education program for farmers on how NUS can improve their standard of living and the quality of their jobs. Such capacity building sessions can be steered by business professionals with inputs from both academicians and administrators. We believe that this business education can be very helpful in promoting the establishment of effective 'micro business models' at the farm-gate level

C   Reduce bureaucracy and let government and research institutions invest more in NUS promotion through well-crafted public awareness campaigns, while entrepreneurs focus more on marketing aspects. NUS products must be included in government canteens, PDS and other programs, with entrepreneurs participating through healthy competition and not bidding/auction models.

Agropecuária Aruanã Ltda., Brazil

Ana Luiza Vergueiro and Sergio Vergueiro
    ECONUT Comércio de Produtos Naturais Ltda., Brazil

The case of Brazil nut

Brazil has abundant natural resources and, among those, is the majestic Brazil nut tree (*Bertholletia excelsa* Bonpl.), notable for its size and its edible seeds (the Brazil nut), which are the greatest natural source of selenium known. Even though it is a traditional product in the Brazilian export line, the Brazil nut is one of the main NUS. The tree grows along the entire Amazon rainforest, but it is mainly found along the Brazilian share of the region. Considering the vast area covered by these long-living trees, there is no precise data regarding its production, and our hypothesis is that 90% of the natural nut production is not harvested, left abandoned on the forest floor.

Even though Brazil nut exploitation is a traditional activity in the country, the main obstacle to its improvement is organizing a value chain. The process starts with the harvesters that live in the forest, who should receive a better salary for their hard work. At the end of the chain are the consumers that demand – rightly so – that quality and safety requirements are followed during processing operations. To that end, there is a need for clearer quality and food-safety regulations for the processing of Brazil nuts, along with appropriate packaging of the final product for proper preservation during its shelf-life.

In light of the above, and in order to achieve the effective promotion of the Brazil nuts, we propose three main measures tackling environmental, socioeconomic and commercial viability as follows:

- Environmental viability: the development of species cultivation, recovering deforested areas in the Amazon rainforest and considering viable logistics to exportation control points
- Socioeconomic viability: support for harvests taking place in the forests, with minimum regional prices along with the establishment of classification, drying and packaging stations in the producing regions, in addition to tax exemption for the entire production chain
- Commercial viability: promotion and marketing of the product and its qualities (regarding production and nutrition) to national and international consumer regions, as well as the simplification of bureaucracy for exports.

SanLak Agro-Industries Pvt. Ltd., India

Vikram Sankaranarayanan, Director, SanLak Agro-Industries Pvt. Ltd. Coimbatore,

General considerations

Given the current debates surrounding food and nutrition security under the threat of climate change, stakeholders engaged in food value-chains, both public and private, are exploring new sources of food from resilient and nutritious crops, so as to complement what is already available in the market. Another push in that direction is also coming from the increasing interest of consumers, which is shifting from food-fortification approaches to a more effective deployment of biodiversity, in order to achieve a more sustainable diet. These trends are contributing to the growing interest in NUS.

From an entrepreneurial perspective, NUS, under the aforementioned context, offer myriad market opportunities to tap at a time when consumers are also keen to explore wellness and nutrition through alternative plant-based diets. Our company, through the guidance and mentorship of Ms. Santha Sheela Nair, began working on minor millets, crops that have been long forgotten due to overdominance of rice and wheat, promoted under the Green Revolution.

Following are some examples of our work on NUS, focusing on millets and amaranth, which are central to our public awareness campaigns and which are needed to inform consumers about the multiple nutrition and health benefits of these species, so much marginalized also for the pejorative labeling of them as 'low-status' foods. Fortunately, perceptions are changing gradually but steadily, with even the nomenclature being rightly changed from 'coarse grains' to

362 Ana Luiza Vergueiro et al.

'nutri-cereals'. Millet consumption is gradually gaining momentum in our food systems also in reaction to the rise of non-communicable diseases and the need to make healthier food choices for our own wellness.

Why promote minor millets?

Minor millets are a group of cereal crops that are climate change resilient and low resource-intensive. They require one-sixth of the amount of water necessary for rice (some 3,000–4,000 liters on average, depending on the rice variety). They are capable of growing well in degraded soils, under drought conditions and can withstand temperatures up to 48°C. They are gluten-free grains and offer a range of micronutrients naturally produced, vital in combatting the rise in non-communicable diseases as well as malnutrition, present in numerous areas of India, especially among women and children, and afflicting poor and vulnerable members of society.

There is certainly a great opportunity to be seized in millets, in terms of widening their consumption by all citizens. Such an outcome is, however, contingent on proper institutional intervention, needed to ensure that millet value-chains receive the same attention that has been directed for decades at other commodities.

Key obstacles and solutions

Numerous factors have diminished and constrained momentum in the growth of millets use, from farm to fork. These factors, grouped under institutional, market and consumer behavior, are summarized as follows:

A   Institutional factors

Although the Government of India, in its official ordinance in favor of millets (called 'nutricereals' therein) has requested the Indian states to include these crops in their respective PDS, such a request has not been implemented by most states. The failure to do so seems most likely to have been caused by the lack of proper coordinated efforts, and sadly, it did not help the fact that millets are being acknowledged as vital crops for our food and nutrition security. Furthermore, given that minor millets do not fall under the category of cash crops and are not given a 'minimum support price' (MSP), the lack of guaranteed pricing has discouraged farmers, traders and processors from entering this space. Double-digit variation in pricing has been observed between harvests, which has rendered minor millets difficult commodities to trade.

Some states such as Karnataka have taken the lead in value-chain development for minor millets and establishing farm-to-market business-to-business (B2B) and business-to-consumer (B2C) linkages. Under the previous Hon'ble Minister for Agriculture, Govt. of Karnataka, Mr. Krishna

Byre Gowda, minor millets gathered momentum with wide market reach and demand pull, catalyzing a subsequent product push from the farm-gate. Such strong political support is very much needed, as it strategically addressed the politics of food. In Tamil Nadu, the Ex-Vice Chairman of the State Planning Commission, Ms. Santha Sheela Nair, undertook similar initiatives, which resulted in a number of successful state schemes, such as the 'Mission on Sustainable Dryland Agriculture'. Indeed, a continuous thrust from the political establishment and administration is much needed in order to circumvent macro-barriers for the introduction of NUS.

B   Technological factors

In India, one of the main bottlenecks in millets value-chains is the lack of availability of evenly distributed processing infrastructures. Processing machinery developed so far for these crops has, in fact, not been successful at all, lacking proper food-grade standards and processing efficiency. In view of this prominent gap in technology, our company has invested a lot in R&D, and these efforts have led us to developing 'Borne SMART' (Small Millet Aleurone Retention Technology), which allow the effective processing of millet grains. This technology includes machinery for carrying out several processing steps at the same time: pre-cleaning, de-gluming (to remove the glume layeron barnyard, kodo and browntop millet [Brachiaria ramosa]) as well as de-hulling, fine separation and gravity separation. Borne SMART machines are available in different dimensions, with processing capacities of 100 kg/hr, 250 kg/hr, 500 kg/hr, 1 ton/hr or 2 ton/hr. Though these machine could help process many NUS, they are being requested by our clients especially for millets.

C   Market factors

Market-based factors are to be addressed from a value-chain perspective, in a holistic manner. Starting at the farm-gate, cluster formations of millet growers are imperative if we are to gain economies of scale. Such an approach was developed through the 'Raitha Siri' program in Karnataka, wherein dry-land clusters were selected, and a sum per acre was distributed to farmers within these clusters, with each cluster focusing on the cultivation of a specific variety of minor millet. It was a successful initiative at the farm-level but, unfortunately, it was not effective in ensuring that millets reached the market effectively, as the forward integration of traders, processors and companies were not developed. Understanding and proper coordination of value chains are vital in ensuring that these crops reach the targeted consumers effectively. Such a major bottleneck is prevalent in various parts of the country, and needs to be addressed on war footing.

A key link between producers and the market is processing and, although machinery is now available for all the end-to-end processing steps needed for minor millets, this technology is yet to reach stakeholders at a mass level. Previous initiatives undertaken between 2009 and 2012, whereby machinery was distributed, failed miserably due to the low quality of machinery processing output, lack of food-grade standards and breakdowns encountered. Although improvements in processing technology and protocols have been established, such as those of the Borne SMART line mentioned earlier, their uptake is slow, due to lack of investor confidence and previous bad experiences faced by clients.

D   Consumers' behavioral factors

Though the private sector has acknowledged the scope for growth in minor millet consumption, there is still limited awareness among consumers, regarding the nutritional virtues of the range of minor millets. Robust public awareness campaigns are much needed in order to catalyze attention and promote wider consumption of these NUS.

Notes

1 See Chapter 13 for more about this project
2 ISO certified company; for more see http://www.vandjindustries.com/
3 Var. acephala of cabbage (*Brassica oleracea*), characterized by the lack of the close-knit core of leaves (a "head") like other cabbages
4 Corn-based traditional hot drink
5 https://www.usaid.gov/nepal/fact-sheets/suaahara-project-good-nutrition
6 http://www.b4fn.org/case-studies/case-studies/home-gardens-in-nepal/#:~:text=-This%20project%20has%20worked%20to,increase%20income%20in%20poor%20households.

References

Padulosi S., Bhag Mal, Bala Ravi, S., Gowda, J., Gowda, K.T.K., Shanthakumar, G., Yenagi. N. and Dutta, M. (2009) Food Security and Climate Change: Role of Plant Genetic Resources of Minor Millets. *Indian J. Plant Genet. Resour.* 22(1): 1–16.

Padulosi S., Thompson J. and Rudebjer, P. (2013) *Fighting Poverty, Hunger and Malnutrition with Neglected and Underutilized Species: Needs, Challenges and the Way Forward: Neglected and Underutilized Species.* Bioversity International, Rome, Italy.

Padulosi S., Phrang Roy and Rosado-May, F.J. (2019) *Supporting Nutrition Sensitive Agriculture through Neglected and Underutilized Species - Operational Framework.* Bioversity International and IFAD, Rome, Italy. 39 pp. https://cgspace.cgiar.org/handle/10568/102462

31

VOICES FROM THE AGENCIES: RESEARCH FOR DEVELOPMENT (R4D) AGENCIES SUPPORTING EFFORTS TO CONSERVE AND PROMOTE THE SUSTAINABLE USE OF NUS

Claudio Bogliotti, Generosa Calabrese, Hamid El Bilali, Isabella Rae, Marco Platzer, Mario Marino, Mary Jane Ramos de la Cruz, Paul Wagsta, Pietro Pipi, Shantanu Mathur, Stineke Oenema, Tobias Kiene

Introduction

Research for Development (R4D) agencies have historically played an important role in promoting the conservation and sustainable use of neglected and underutilized species (NUS) around the world by providing investments and assistance to vulnerable communities that has helped build research partnerships and capacity, which has enabled them to benefit from sustainable livelihoods, improved diets and nutrition, food security and enhanced resilience and adaptation to pest and diseases, water scarcity and changing climate. The historical development of the NUS movement, the key agencies and actors involved as well as landmark events and publications are highlighted in Chapter 2 of this volume. Furthermore, many R4D agencies are profiled in Chapter 33, and it is largely through the efforts of these agencies that financial and other needed resources have been, and continue to be, directed towards fundamental research partnerships on NUS, though the gap in research budget allocated to the top few staple crops alone and the wider NUS portfolio remains huge. These agencies have also been important in the development of research frameworks, building capacity, promoting greater awareness and creating better enabling policy environments for NUS conservation and use. Certain R4D agencies such as International Fund for Agricultural Development (IFAD) have played a critical role in ensuring ongoing contributions and support for NUS, helping catalyse research collaborations and partnerships among organisations, communities and farmers interested in advancing the conservation and sustainable use of NUS. It is the purpose of this chapter to showcase the vision, perspectives and work of a select group of these R4D agencies.

DOI: 10.4324/9781003044802-36

366 Claudio Bogliotti et al.

Bringing *NUS* back into *Use*: why IFAD champions pro-poor crops of the future

Shantanu Mathur, Lead Adviser, Global and Multilateral Engagement, IFAD, Rome, Italy

Introduction

IFAD promotes a holistic and pluralistic approach to inclusive rural transformation and agricultural food systems as a part of its lending programme, which currently exceeds USD 1.5 billion, annually. NUS are viewed by IFAD as an intrinsic part of its arsenal of nature-positive, biodiversity-friendly approaches. They are increasingly promoted by IFAD for their promise in contributing to an economically remunerative and environmentally sustainable future – for rural communities and their habitats – depending on the socioeconomic, cultural, institutional and biophysical context. IFAD pursues people-centred approaches and, in the context of the NUS–sustainable livelihoods nexus, we believe that the theory of change is really about positioning rural communities (including Indigenous Peoples as the key custodians of NUS) to make the most out of the promising nature of rich, nutrition-sensitive markets and the values chains they are linked to.

The role of NUS in inclusive and sustainable rural transformation

IFAD investments in the NUS-based food systems of rural (Indigenous) communities (and their local knowledge embedded in those) are strongly associated with multiple positive relationships with incomes, nutrition and health of humans and their habitats. They drive financially attractive investment projects with outcomes that are poverty reducing, nutrient rich, diet-diverse, environmentally friendly and climate smart. It would be hard to find a better win–win strategy in any inclusive rural transformation agenda pursued by the international development community.

Over the past two decades, IFAD has developed experience in supporting NUS and has used key lessons learned to develop guidance on how, when, and why to incorporate considerations of NUS into nutrition-sensitive agricultural projects, with due attention paid to the inclusion of Indigenous Peoples and the enhancement of their livelihoods and wellbeing. The focus has been on a multitude of truly participatory rural development and transformation approaches incorporating the importance of NUS and their contribution to biodiversity conservation, improving nutrition (in the context of dietary diversity), quality traditional diets and Indigenous Peoples' food systems and their knowledge on these.

IFAD and the genesis of NUS

The advent of the very recognition of so-called NUS in formal sciences and by international agricultural research for development systems started with IFAD's pioneering grant support back in the late 1990s. The Fund's first grant was to establish a strategic, global, multi-stakeholder partnership led by the erstwhile International Plant Genetic Resources Institute (now the Alliance of Bioversity International and International Center for Tropical Agriculture (CIAT)).

This initiative benefitted immensely from the passionate, scientific leadership of a few individuals, and IFAD's investment scaled-up the so-called NUS agenda and placed it in the global research and development spotlight – pursuing the multiple objectives of "conservation-through-utilisation" and, later, a more composite goal of "sustainable utilisation".

With the strong conviction that NUS were neglected and underutilized for counter-intuitive reasons, a global constituency became established in the early part of this century. This community of practise views NUS as the "crops of and for the future" and as multi-dimensional "weapons of poverty alleviation" (ironically) in the hands of the world's most marginalised communities – who are the true custodians of NUS and most of the world's biodiversity.

In early IFAD-sponsored international conferences, we conceived a new, global multi-stakeholder initiative. For instance, in 1999, the IFAD-supported workshop organised in Chennai, India, by the CGIAR Plant Genetic Resources Policy Committee, focused specifically on NUS. This landmark meeting drew broad-ranging support from participating CGIAR centres and donors. It is interesting to note that this meeting represents the first time ever that the CGIAR discussed NUS in a formal way.

This milestone conference received the intellectual guidance of Professor MS Swaminathan, who cautioned us early partners and practitioners against the use of negative nomenclatures connoting the neglect and underutilisation of such a promising set of crops and commodities.

Indeed, on the other hand, these have been often regarded as strategic assets of rural poor communities (especially Indigenous communities) – resources that have helped these otherwise marginalised communities build resilience and cope with the challenges of food and nutrition insecurity, health calamities and pandemics, climate change and in their struggle against economic and social disempowerment.

Since the early 2000s, IFAD has actively helped build new alliances with global initiatives such as the Global Forum on Agricultural Research and Innovation's (GFAR) Collective Action on Forgotten Foods. Given the multi-stakeholder and interdisciplinary nature of GFAR and the power of the IFAD-investment portfolio in supporting community-based, participatory workshops aimed at enriching beneficiary knowledge on resilient cultivation practices, financing rural infrastructure and small and micro-enterprises for improving processing and consumption options has helped improve small-scale producer links to NUS-driven value chains and niche markets.

Incorporating NUS into food systems affords a more dynamic, *in situ* "people-centred" approach to addressing the biophysical challenges of modern agriculture. The above meetings provided a deeper understanding of the incentive structure of farming communities in managing risks and how strategic production choices

368 Claudio Bogliotti et al.

(e.g., of cultivating NUS) can strengthen resilience, especially of resource-poor farming systems.

Informed by these important scientific exchanges, IFAD-grants financed multi-locational research programmes in all the developing regions of the globe and have generated a series of evidence-based innovations, which have been incorporated into multi-million dollar loan project designs.

Not orphans at all, but Cinderella crops and hidden gems

The innovative deployment of nature-positive and environmentally sensitive components in agricultural development projects, is bringing to light the immense value of the orphan crops of the past – from relative obscurity into prominence. They exhibit attractive transformational characteristics and elite traits – a metaphorical reference to the tale of "Cinderella" (hence the label).

Neglected no more, the increasing production and trade of erstwhile NUS (quinoa, amaranth, finger millets and other [mislabelled] "coarse" grains like sorghum, maize and kiwi fruit to name a few) represent a vast spectrum of hidden gems.

They have demonstrated improved and multiple beneficial outcomes while simultaneously including robust local adaptability, better incomes, improved livelihoods and nutrition and ecosystem benefits – all of which are designed to improve the resilience of smallholders to biophysical and socioeconomic adversities and shocks. The range of the potential of NUS is so extensive that it is now becoming an increasingly inherent part of conventional food production systems – making them valuable in so many respects, well beyond the local context and the niche markets that they were associated with in the past – now, they are a part of the economic mainstream.

The multipurpose species of the future

NUS are increasingly better known as the promising multipurpose species for the future, resurrected from their general underutilisation and even disuse (as indigenous knowledge systems on their utility had eroded over the generations). A more development-outcome-led formal research system has brought to light cutting-edge knowledge on the multiple benefits and high value embedded in NUS. The agrobiodiversity associated with NUS "production" systems is also a direct contributor to the diverse range of micronutrients that NUS have made available to the diets of Indigenous and other rural communities. Such dietary diversity is known for addressing their micronutrient deficiency and under-nutrition, and also spontaneously targeting some of the root causes of stunting and wasting among children while improving the resilience of adults to health shocks.

The extensive use of NUS will, thus, ultimately contribute not just to SDG2, which is associated with zero hunger; reduction of all forms of malnutrition and sustainable agriculture, they will also help reach several targets under SDGs 7, 12, 13, 15 and 17, as well as relevant targets within the context of the CBD post-2020 global biodiversity framework. IFAD's Strategic Objective 1 (increase poor rural people's productive capacities), which focuses heavily on improving nutritional knowledge and behavioural change communication on healthy diets for all, has NUS elements. IFAD's Action Plan on Mainstreaming Nutrition-Sensitive Agriculture has a predominant NUS focus and the recently published Bioversity-IFAD operational framework in support of nutrition-sensitive agriculture through NUS will certainly help promote the adoption of NUS for sustainable food systems.

Yet, much more is needed to help countries mainstream NUS through participatory, pro-poor, gender-sensitive, multi-sector, multi-stakeholder and multi-disciplinary approaches. Further development and testing of NUS to promote replicability and transferability to other contexts; an advocacy role for policy change at the global level and the engagement of development practitioners (through participation in training and capacity building) are key. Innovative policy sensitisation is required to provide an enabling institutional environment so that the rural poor can achieve the twin objectives of income generation along with the positive nutrition potential currently locked inside their local NUS-based farming systems.

Going forward

We will continue to need new investments in innovations systems, including by bringing leading-edge science to build on local and Indigenous knowledge systems – for instance, developing appropriate food technologies to maximise the nutritional benefits from local crops. The international community needs to better appreciate the power of the crops of the future and unleash their potential in addressing the challenges of faltering food systems. We need to trigger a paradigm shift towards the adoption of more (bio-)diverse agricultural systems – and a departure from current food systems, which are precariously focused on an ever-shrinking portfolio of crops and commodities – ironically, where there are a large variety of robust but overlooked Cinderella species-in-waiting.

Self Help Africa – don't let an opportunity go to seed: commercialising neglected crops in practice

Paul Wagstaff, Senior Agriculture Advisor, Self Help Africa and Isabella Rae, Head of Policy, Research and Evaluation

What are your agency's views with regard to NUS?

NUS in Africa are a complex topic and one where limited generalisations can be made. Traditional fruit, vegetable and grain species have been neglected due to declines in demand, a decline in access to wild plant foods, low yields and poor adaptation to the livelihoods and workloads of farmers, particularly women. It is important to distinguish between NUS that are preferred "luxury" foods, and

370 Claudio Bogliotti et al.

those that are only eaten in desperation during times of famine. Though not traded internationally, NUS are uniquely adapted to their local environments and play a vital role in supporting diverse diets in sub-Saharan Africa.

Self Help Africa (SHA) focuses on improving family nutrition and household economy through diversifying incomes and linking farmers to markets, both of which have to take into account climatic factors and women's access to and control over resources. SHA, therefore, looks at the nutritional, crop diversity and market potential of NUS, including the possible impact on women's workloads related to the collection or cultivation of NUS.

One of the biggest drivers of the marginalisation of NUS has been the radical changes in diet, especially in East and Southern Africa, where "chicken and chips" have become the "aspirational" foods that have driven traditional foods out of local restaurants, with local foods being seen as a "poor person's food" (less so in West Africa and Ethiopia). In complete contrast, some neglected crops are enjoying a renaissance and are being considered "status foods" for the urban middle classes, and "comfort food" for the diaspora.

Several NUS with significant potential as resilient food crops are not more widely used due to the presence of toxic, anti-nutritive, excessively bitter or allergenic compounds, which must be taken into account when promoting NUS for home gardens. Some of the best-known examples include:

- Grass pea, a highly drought-tolerant legume that can cause lathyrism when eaten in significant qualities;
- Solanaceous crops that have not been exposed to generations of selection for low solanine levels can pose a health risk;
- Water lily roots, Nymphaea, an important famine food in the wetlands of South Sudan and Uganda, contain a toxic alkaloid, and young ferns, a delicacy in Sierra Leone, can be carcinogenic;
- Members of the Cruciferae can be goitrogenic in areas of low natural iodine while species high in tannins, oxalic acid and phytic acid can inhibit the uptake of iron, a disadvantage in countries with high levels of anaemia.

By definition, the term NUS implies little research interest. While this is true when compared to the level of research conducted on the main staple crops, a significant amount of research on NUS does take place but is often confined to MSc and PhD thesis and project reports, which receive little attention. The promotion of NUS can be politically sensitive, and NUS research and market development initiatives are sometimes treated with suspicion by some sectors of civil society.

A lack of investment and research into these crops has led to farmers planting them less and less, reduced access to high-quality seed and loss of traditional knowledge.

What do you see as the key opportunities provided by NUS?

We see a number of opportunities offered by NUS, in particular:

- The diversification of food production and of diets (diversified nutritional intake), adding new species to our diets that can result in better supply of particular nutrients, i.e., essential amino acids, fibre and proteins
- Economic and environmental benefits (domesticating wild crops, reducing pressure on the environment, disrupting the cycle of some pests and diseases by changing species in a crop rotation system, etc.)
- Cultural distinctiveness
- Urban middle class: as they get richer they tend to abandon "fast food" and return to local traditional foods

What do you see as major obstacles for their enhanced sustainable use?

Registration of varieties: A key obstacle relates to the difficulties in getting these varieties recognised and registered. Finding ways to simplify the registration process would be necessary. SHA has been working for many years with local seed producers, the Zambia Agriculture Research Institute and the Seed Control and Certification Institute (SCCI), to improve the availability of seeds for traditional bean varieties in Zambia. Though there is demand for bean seed getting landraces formally registered, based on the Distinct, Uniform, Stable (DUS) and value for cultivation and use (VCU) criteria, it is a long and expensive process, especially as the VCU of many landraces is their lack of uniformity. SHA worked with Crop Innovations at Bath University, UK, and SCCI to test the use of rapid genetic fingerprinting technology alongside traditional field trials as an alternative approach to characterising and registering landraces.

Promotion at the community level: Promoting certain varieties can be challenging at community level due to some of the difficulties mentioned above, including cultural biases and issues with toxicity.

Shortage of seeds: It is becoming increasingly difficult to obtain seeds for many traditional vegetables in East and Southern Africa, with only a few seed companies including NUS in their catalogues. Scaling up the production of NUS seeds will require a reliable supply of early generation seed (EGS), a major bottleneck even for mainstream commercial crops.

Trading to the EU: Any variety that hasn't been imported into the EU in the last 10 years now has to go through full toxicity tests, which would be prohibitively expensive and would require major subsidisation in order to be undertaken.

372 Claudio Bogliotti et al.

What are the key solutions you believe should be considered for better NUS promotion?

Creating demand: Researching niche markets and creating demand for NUS as "aspirational" foods, starting with the urban middle classes. Chia seeds are a well-known example, with affluent health-conscious consumers in Europe creating demand for chia seeds grown by farmers supported by SHA in Uganda.

Addressing NUS seed and germplasm shortages: This requires a vibrant, diverse seed sector, that supports farmer-to-farmer seed exchanges and SME- and farmer-owned seed enterprises as well as large-scale commercial seed production of a wide range of crops and varieties. Diversifying crops and varieties will require the adoption of newer techniques, such as "Crop Innovation's Genetic Finger Printing" for registering new varieties and recognising the intrinsic value of genetic diversity in landraces. Small- and medium-scale NUS seed production requires a reliable source of EGS. SHA's Edget Cooperative Union model in Ethiopia has proved that a union of seed producer cooperatives can produce EGS for its member co-ops, an approach also used by co-ops in the USA and Canada. Other EGS models that show promise include contracting out EGS production to specialist EGS companies (Brazil) and Agriculture Universities (USA and India), and setting up commercial spin-off EGS companies from National Agricultural Research Organizations (Nigeria, Uganda).

Plant breeding to improve the nutritional value and commercial viability of NUS: Using conventional plant breeding, plant breeders in India have produced varieties of grass pea with very low toxicity and there is considerable potential to use gene editing to disable genes for toxins and anti-nutritive compounds. The domestication of wild NUS is important to address the reduction in access to wild foods and can also reduce the time taken to reach maturity, especially for wild tree crops. Conversely, farmer-managed natural regeneration of degraded lands can increase the availability of wild fruits through the selective protection of wild fruit-trees with desirable traits.

The Italian Development Cooperation Agency (AICS): advancing the agenda of NUS

Pietro Pipi,
> *Head, Agriculture and Rural Development Department (AICS), Italy and*
> *Marco Platze, Senior Specialist, Agriculture and Rural Development Department,*
> *Italian Agency for Development Cooperation (AICS), Italy*

Agriculture, food security and nutrition are consolidated priority areas of intervention for Italian development cooperation, centred on family agriculture and small-scale farmers and contributing to the achievement of SDG2 and other related SDGs. In this context, the Italian Development Cooperation Agency

(AICS) approach entails: the promotion of sustainable agro-ecological practices that protect the environment and preserve biodiversity; strengthening networks that improve access to resources and local markets and increasing resilience to shocks and climate change by addressing the economic, social and environmental causes that underlie poverty, conflict and migration.

Moreover, AICS aims at strengthening and developing sustainable production chains, with particular emphasis on the enhancement of local/typical products with the objective to increase productivity and improve quality while also promoting good agricultural practices, aimed at the conservation of production areas. In this framework, AICS shares fully the growing interest around NUS and is highly aware of the strategic role that NUS can play in improving food and nutrition security, resilience to climate change, income generation and the empowerment of vulnerable people and communities.

In recent years, the yearly average disbursement of AICS for initiatives concerning agriculture and food security, rural development, including biodiversity conservation and protection of the environment, have amounted to around €150 million. AICS believes that in advancing the agenda of NUS at the global level, there are some aspects that should receive greater attention, with particular reference to the conservation of wild and cultivated diversity of such species, through *ex situ* and *in situ* methods, in order to allow countries to safeguard precious resource assets that are now under threat of being lost. Another relevant aspect is the enhancement of capacities of National Agricultural Research Systems in carrying out strategic research needed to tap effectively the potential of NUS for tackling the key challenges highlighted above. Lastly, the importance of strengthening appropriate information sharing tools to harness the wealth of knowledge currently scattered for the improved conservation and sustainable use of NUS needs to be understood.

The interventions promoted by the Italian Cooperation in this domain are aimed at enhancing the use of climate-resilient varieties, encouraging the production and consumption of high-quality and nutritious food and promoting healthy diets, identifying markets and raising awareness of communities. In this regard, AICS has supported several projects targeting biodiversity conservation and valorisation, both bilaterally and multilaterally through the Rome-based UN agencies (FAO, IFAD and WFP), Bioversity International, UNIDO and CIHEAM-Bari. The involvement of Italian research institutions and academia is also promoted.

Among the principal areas of intervention by AICS has been: the enhancement of Tunisian plant genetic resources (cereals, olives, fruit trees, food legumes) either through *in situ* or *ex situ* conservation actions, by strengthening the National Gene Bank of Tunisia through providing capacity building for Tunisian technical and institutional staff; value-chain development of local crops with high nutritional value in Ethiopia and the diversification of local communities diets in Oromia region; ongoing support to the FAO Mountain

Partnership to better utilise NUS and support to the production, productivity and value chains of moringa (M. *stenopetala* and M. *oleifera*) in Ethiopia. Finally, AICS recognises the pivotal role of women in NUS conservation and use through initiatives aimed at supporting women and youth to grow, process, market and cook healthy food.

International Centre for Advanced Mediterranean Agronomic Studies (CIHEAM): perspectives on NUS

Hamid El Bilali, Generosa Calabrese and Claudio Bogliotti, International Centre for Advanced Mediterranean Agronomic Studies (CIHEAM-Bari), Valenzano (Bari), Italy

Founded in 1962, International Centre for Advanced Mediterranean Agronomic Studies (CIHEAM) is an intergovernmental organisation focused on the sustainable development of agriculture, from food security to food quality and nutrition in rural and coastal areas, including the sustainability of local natural resources, agro ecosystems and fisheries in the Mediterranean. Its member countries are Albania, Algeria, Egypt, France, Greece, Italy, Lebanon, Malta, Morocco, Portugal, Spain, Tunisia and Turkey, and operations are carried out through its four institutes based in Bari (Italy, CIHEAM-Bari), Chania (Greece), Montpellier (France) and Zaragoza (Spain) and the headquarters in Paris. The *CIHEAM Strategic Agenda 2025* provides an innovative development framework that revolves around four pillars (viz. protecting the planet, food security and nutrition, inclusive development and crises and resilience) divided into 15 thematic priorities, the fifth one being *"enhancing agrobiodiversity conservation and agro-ecological practices"*.

The Mediterranean Basin is one of the world's 25 hotspots of global biodiversity, presenting a high level of diversity of different food species, forages, aromatic and medicinal species. CIHEAM-Bari provides education and knowledge building on agrobiodiversity conservation and agro-ecology in its M.Sc. course on Mediterranean Organic Agriculture. It also works on the conservation of Mediterranean biodiversity through projects on the protection of endangered species (e.g., MEDISLANDPLANT – ensuring the survival of endangered plants in the Mediterranean) and/or the conservation of local and ancient species (e.g., Crop Wild Relatives) and local crop varieties (e.g., Tunisian Phytogenetic Resources Better Conserved and Enhanced). Furthermore, it carries out institutional capacity-building and promotes networking among national plant conservation centres through programs such as GENMEDA (Network of Mediterranean Plant Conservation Centres). CIHEAM-Bari also carries out analyses of agriculture and natural systems sustainability, cropping systems management, soil quality and participatory approaches in the management of natural resources. Particular attention is also paid to biodiversity's role in safeguarding and promoting the Mediterranean diet.

CIHEAM-Bari believes that NUS can help in addressing several challenges such as climate change, food and nutrition insecurity, livelihoods vulnerability and poverty, biodiversity loss and ecosystem degradation. However, NUS development requires research and innovation investments. However, many constraints hinder the wider adoption of NUS, most of them relating to seed production and seed systems, agronomic practices, genetics and eco-physiology, use and markets. Furthermore, different factors hamper the promotion of NUS, among them: lack of sound data on their nutritional properties; low awareness by farmers, researchers and extension workers; poor economic competitiveness; inefficiencies in value chains (production, storage and processing) and limited availability of germplasm.

Research and innovation (R&I) are important for NUS to play an essential role in building a resilient and economically vibrant agricultural sector, able to sustain food and nutrition security needs and sustainable rural livelihoods under climate change. In this respect, cross-country collaboration is of paramount importance to set up an R&I strategy to bridge existing gaps and overcome the issues highlighted above. The strategy should address all relevant scientific disciplines, from natural and agronomical sciences (e.g., agronomic research on adaptation of NUS to changing climate, poor soils and dry environments) to social sciences and economics. Investigating the connection between NUS and food security, nutrition and livelihoods of rural communities could raise the awareness of all actors, both in the academic and policy arenas, on the benefits of mainstreaming NUS in farming systems and local diets. Attention should be also paid to adopting adequate policies and institutional/governance arrangements to enhance NUS in relation to agriculture, the environment, health and gender. In particular, concrete political support is needed to include NUS in national strategies for climate change adaptation and sustainable food systems, with the view of fighting environmental degradation and natural resource depletion that constrain the productivity of NUS in the context of agroecosystems, especially in remote rural communities.

CIHEAM-Bari pursues strengthening research on NUS within the Agricultural Knowledge and Innovation Systems by developing a long-term international agenda for research, innovation and development through concerted efforts involving relevant stakeholders from researchers to policy makers, farmers and consumers at different levels (from local to regional). An example of common engagement and shared efforts by different actors is given by the EU-funded project SUSTLIVES (Box 1). This project will play an important role in unlocking NUS potential in Burkina Faso, Niger and beyond (e.g., West Africa, Sahel), while also addressing the SDGs and the Paris Climate Agreement. In conclusion, it appears essential to build and support multi-actor partnerships that engage in raising joint funds for R&I on NUS, especially in the countries of the Global South.

376 Claudio Bogliotti et al.

Box 31.1 SUSTLIVES project on the promotion of NUS in Burkina Faso and Niger

The project SUSTLIVES (*SUSTaining and improving local crop patrimony in Burkina Faso and Niger for better LIVes and EcoSystems*) falls under the thematic priority "Vegetables, legumes, roots and tubers" of the DeSIRA initiative (Development Smart Innovation through Research in Agriculture). It is under the direct management of AICS and brings together partners including CIHEAM-Bari, Bioversity International, Italian National Research Council (CNR), University Roma Tre, the Natural Resources Institute Finland (LUKE), the University of Ouagadougou (Burkina Faso) and the University of Niamey (Niger). It has a budget of €6 million and a duration of four years. SUSTLIVES aims to improve food and nutrition security and livelihoods of rural communities in Burkina Faso and Niger through agrobiodiversity. Agrobiodiversity, and particularly the valorisation of NUS, constitutes the driving force behind our ability better meet nutritional needs of rural communities, to create income-generating opportunities for youth, farmers and women and to support national and local institutions to cope with climate change and its effects on agriculture. SUSTLIVES consists of three expected results: (i) identifying and assessing stress-tolerant crops, target areas, local stakeholders and their needs, and achieving a higher resilience of agroecosystems; (ii) raising awareness of communities and consumers about stress-tolerant NUS, and empowering local actors, especially youth and women, in inclusive NUS value-chains and (iii) ensuring policy-making and sustainable planning on agrobiodiversity, supported and in coordination with EU planning and priorities. The main direct beneficiaries are value-chain actors (farmers – especially smallholders – processors, etc.) and their organisations, women and young people, local administrations, research institutions and extension services and local seed producers and companies. Indirect beneficiaries include rural communities as well as consumers.

UN System Standing Committee on Nutrition (UNSCN): a nutrition-focused agency's perspective on NUS

Stineke Oenema, Coordinator, UNSCN, Rome, Italy

What are your agency's views with regard to NUS?

The UNSCN does not have a formal view regarding NUS. However, it does have a clear understanding that unhealthy diets these days are the major contributors to disease. Therefore, the promotion of healthy diets is an important part of the solution. A major characteristic of healthy diets is diversity. Diets are an outcome of food systems and the current dominant food systems do not deliver healthy diets. In addition, food systems are also major contributors to biodiversity loss (and to the pollution of water, soil and air, as well as to climate change). Our current food system utilises and promotes a very limited number of crops and species. During the Second International Conference on Nutrition (ICN2), in 2014, member states acknowledged the key role played by diversified and sustainable diets, including traditional ones, in reducing malnutrition. At the same time, the ICN2 acknowledged the importance of cultural identity in relation to diets. This resulted in formal commitments to, among other things, enhancing sustainable food systems and promoting safe and diversified healthy diets. It also recommended strengthening local food production and promoting diversification of crops, including underutilized traditional crops. Many studies have shown that NUS can play an important role in improving nutrition. Their nutritional value is often higher than corresponding exotic or imported foods. In addition, they are locally adapted and are often part of cultural heritages. In summary, they can play an important role as part of sustainable healthy diets, to confront the challenges described above.

What do you see as major obstacles for their enhanced sustainable use?

A major obstacle for the enhanced sustainable use of NUS are some of the very same positive characteristics listed above. They are so locally adapted and context specific that it is often difficult to bring them to the forefront in one solid overarching definition. One could say they are so contextualised that they are difficult to (sustainably) promote beyond their immediate context or locality. There is no common understanding of what NUS are – thus, the beauty of NUS is also its weakness. In addition to this, and possibly linked to it, is the lack of knowledge about the nutritional value of many species and varieties. The lack of a common understanding of what exactly NUS are has also contributed to their "neglect" or insufficient protection in (international) law. At the level of individual crops, additional issues can pop up, such as: consumers have lost the habit of preparing them or don't appreciate the taste, smell or colour (anymore), and producers don't see their potential (anymore) because they have become used to more dominant crops that have been promoted at liberty by companies, government services and policies. However, obstacles to the uptake of individual crops (as opposed to NUS in general) can be overcome more easily (relatively speaking) – see below, under Question 3. More worrisome are the obstacles that extend to the promotion of NUS overall. There is a lack of:

- recognition of NUS' potential roles;
- awareness on NUS potential among governments;
- adequate research (especially in the nutrition field);

378 Claudio Bogliotti et al.

- coherence and adequate protection in local and international legal frameworks;
- interest from the private sector to invest in NUS;
- funds and advocacy to local communities to implement all the above.

Solving these gaps requires political solutions as well as the collaboration of various sectors and stakeholders.

What are the key solutions you believe should be considered for better NUS promotion?

Solutions can be distinguished for individual crops and NUS in general. For individual crops, chefs or other champions (farmers, caterers, etc.) in various settings such as in school gardens (e.g., biodiversity project Brazil) or in the tourism sector (tourism in Uganda) can handle promotion. Whenever NUS are promoted in their endemic areas, it is important to take care not to over exploit a specific crop, for example, by establishing value chains abroad. The local population must continue to have access to it and benefit. To promote the uptake of more NUS in general, it is paramount to do more research and make an inventory of NUS and their nutrient content. NUS can also be promoted by advising people to include them in their diets. A very helpful tool for this is the national food-based dietary guidelines (FBDG), especially those that have sustainability criteria. FBDG take into account the actual consumption of the population, as well as cultural acceptability and preferences. FBDG also consider what is locally available and accessible. As such, FBDG can prove to be a very good tool to promote locally produced crops and species, including NUS!

The International Treaty on Plant Genetic Resources for Food and Agriculture (International Treaty): farmers saving and nurturing local crops for food security and rural economy

Mario Marino, Mary Jane Ramos de la Cruz and Tobias Kiene, ITPGRFA, Rome, Italy

The International Treaty on Plant Genetic Resources for Food and Agriculture (International Treaty)[1] was negotiated by FAO and adopted in 2001 to create a global system that provides farmers, plant breeders and scientists access to plant genetic materials. It aims at:

- protecting and promoting Farmers' Rights, recognising the enormous contribution of farmers to the diversity of crops that feed the world;
- promoting the exploration, conservation and sustainable use of plant genetic diversity in an integrated manner and
- establishing a global system to provide farmers, plant breeders and scientists with access to plant genetic materials, while ensuring that recipients also share benefits they derive from the use of these genetic materials.

What are your agency's views with regard to NUS?

NUS are local crop varieties — also known as landraces or farmers' varieties — providing basic daily food nutrition and livelihoods, but also socio-cultural and environmental values that are fundamental for sustainable agriculture and local economic development. Economically, they have enormous potential for fighting poverty, hunger and malnutrition. Around the world, smallholders and peasant farming communities, particularly those in remote rural and marginal areas, rely on a wide range of NUS for their food, nutritional and health security, and livelihoods. NUS not only provide diversified healthy food and nutrition, but most of them, with ancient origins, have been conserved and sustainably managed until today because of their unique social and cultural values, as well as environmental importance.

To date, the International Treaty is the only legally binding international agreement that recognises the enormous contributions that farmers from all regions of the world have made, and will continue to make, for the conservation and development of plant genetic resources as the basis of food and agricultural production. Article 9 of the International Treaty lists some measures that recognise and promote Farmers' Rights,[2] or the rights of farmers to crop genetic resources for food and agriculture. These measures are closely linked to one of the core objectives of the International Treaty — the conservation and sustainable use of plant genetic resources. Promoting the expanded use of local and locally adapted crops, varieties and underutilized species is specifically suggested by the International Treaty as a means of sustainably using plant genetic resources. The implementation can be realised through a list of measures at national levels that could be taken to protect, promote and realise these rights.[3] Farmers' Rights are critical to ensuring the conservation and sustainable use of plant genetic resources for food and agriculture (PGRFA) and, consequently, for food security — today and in the future.

Now, 15 years since the International Treaty came into operation, the need to realise Farmers' Rights is more relevant than ever. With an estimated 1.2 billion of the poorest people living in rural areas and depending largely on traditional agriculture and their own local crop varieties, promoting their rights to crop genetic diversity can be a valuable channel for the eradication of poverty, hunger and malnutrition. Many of these local crops are nutritionally dense, and, therefore, their erosion can have immediate consequences for the nutritional status and food security of the poor, while their sustainable use can bring about better nutrition and success in fighting hidden hunger. For most of these rural farmers, access to commercial varieties and the required production inputs, such as fertilisers and pesticides, are unaffordable. Thus, they depend on the diversity of their local traditional crops. Crop diversification can provide insurance against crop failure due to pests and diseases, or adverse climatic conditions, such as drought or floods, while the produce may be central to traditional local cuisine and specific dietary requirements. Furthermore, diverse crop genetic resources

380 Claudio Bogliotti et al.

are an important source of locally adapted genes for the improvement of other crops, to build climate-change-resilient agricultural production systems. Therefore, enabling farmers to maintain and develop this crop diversity, and recognising and rewarding them for their contribution to the global gene pool are basic prerequisites for the achievement of the Sustainable Development Goals (SDG1: no poverty; SDG2: zero hunger; SDG15: life on land).

What do you see as major obstacles for enhanced sustainable use of NUS?

To date, the enormous potential of NUS for food security and nutrition, hunger and local economic development have not been fully explored or utilised. This is despite increasing awareness and discussions at various fora highlighting the nutritional, economic, social, cultural and ecological values of NUS. The enabling environments, such as strategies for promotion and institutional mechanisms to support NUS, were always missing or were not mainstreamed in existing food and nutrition security policies, programmes and agendas. NUS are still not a priority for many governments, and investments in researching and improving the productivity, adaptability and utilisation are not yet in place. The lack of investments in local crops has meant that their potential benefits remain undervalued and untapped. If there are no enabling environments and supporting polices, and if there is no value and demand for NUS, farmers will ultimately give up the cultivation of these crops. In addition to this, the lack of institutional mechanisms to support NUS farmers, e.g., market incentives, marketing support for producers, access to financial and technical services, etc., is making it difficult for them to continue conserving and sustainably using NUS. The widespread adoption of high yielding uniform varieties has contributed to the diminishing exploitation of many NUS and other diverse local varieties, along with the knowledge systems associated with their cultivation and use. Therefore, appropriate infrastructures need to be in place to support NUS farmers to continue the cultivation of local traditional crop varieties – varieties that are genetically diverse due to repeated cycles of selection, seed-saving and re-planting, which has resulted in their adaptation to local environmental conditions. The promotion of local crop varieties, increasing knowledge and raising awareness are important elements in efforts to sustain their cultivation.

What are the key solutions you believe should be considered for better NUS promotion?

It all starts with the seed. Farming begins with seeds. Without seed, there can be no crops and no food production. Farmers and other crop maintainers depend on access to sufficient quantities of good quality seeds of their varieties of choice, and for these seeds to be available in time for planting when they need them. Seed systems – from production, through processing, storage and distribution – are, therefore, central in efforts to sustain local crop diversity. The top three solutions are:

1 Mainstream NUS into national and local policies and programmes to create an enabling environment and to support farmers by providing technical support and access to financial services
2 Create institutional mechanisms to support rural livelihood strategies and household income diversification, such as access to market and market facilities, develop market value-chains – markets for local products, linkage to tourism and other potential income-generation support for NUS farmers
3 Strengthen local seed systems and increase the number and quality of local seed banks to promote sharing and exchange of seeds among farmers.

These solutions might help address the challenges that confront smallholder farmers and Indigenous local communities in conserving and sustainably using local crops and NUS. No difference should be made in terms of "formal" and "informal" seed systems at the local level to facilitate access to plant genetic resources. A supportive legal and policy framework for the cultivation and marketing of local crop varieties is lacking in many countries. There are activities that can help enhance their unique value and support the creation of new *in situ* diversity. This may, in turn, serve to inform and influence the development of a more appropriate, supportive policy environment. This is particularly critical in the many developing countries where as much as 80–90 percent of seed requirements of smallholder farmers are sourced through local exchange networks, as well as from markets and household stocks.

A further tangible way to safeguard local crop varieties and to secure the seed supply for local communities are Community Seed Banks (CSBs). CSBs serve as repositories of local crop diversity that is often adapted to prevailing climate conditions, including biotic stresses, such as pests and other infestations, and facilitate farmers' access to seeds in time for planting. This is particularly necessary during periods of calamities, natural disasters or emergencies/crises. Many experiences, including through the Benefit Sharing Fund[4] of the International Treaty, operating in more than 60 countries around the world, highlights the importance of local seed banks for local food security and the empowerment of local communities, as well as for maintaining traditional knowledge and raising awareness of the value of local crop diversity. Therefore, supporting farmers and local communities in developing and maintaining these seed systems is important to sustaining diversity and ensuring local food and nutrition security.

Creating stable value chains for crop produce is one option that is widely considered to be central to efforts promoting the sustainable use of local crop varieties. Effective commercialisation and innovative marketing systems can promote NUS and maximise their economic value. This can involve a range of stakeholders, including farmers and farmers' cooperatives, local promotional associations and businesses, research and development organisations, food

382 Claudio Bogliotti et al.

processing companies and local government agencies. To enhance marketing options, value-adding measures should include the development of new products from raw sources, the use of high-quality processing methods and packaging and registration through schemes such as Geographic Indications (GI) and traditional knowledge. Products may be sold in local markets, grocery shops and supermarkets, and via internet-based outlets,[5] as well as to restaurants.

Lastly, knowledge management and information dissemination is crucial to the conservation of local crop diversity. The International Treaty promotes knowledge sharing and information dissemination through a Toolbox for Sustainable Use of PGRFA.[6] The toolbox is aimed at assisting countries in designing and implementing measures to promote sustainable use. It caters to those seeking information or guidance on policies, strategies and activities that can promote and enhance the sustainable use of PGRFA, particularly at the national and local levels. Intended users may come from a wide range of stakeholder groups, including those working in or associated with public research institutions and gene banks; government agencies; farmers' associations; agro-NGOs; local and Indigenous community enterprises; seed networks; educational establishments; international bodies; networks and services; private plant-breeding companies and the commercial seed and plant production industries, as well as independent plant breeders, farmers and seed producers.

To demonstrate the top three solutions to safeguard local crop diversity, below are some of the living examples extracted from the national measures and practices on the implementation of Farmers' Rights under the International Treaty.[7] These examples support the conservation of local crop diversity while enhancing food and nutrition and the livelihoods of farmers and rural communities.

Law for Agrobiodiversity, Seeds and Promotion of Sustainable Agriculture (2017)

In 2017, the Ecuadorian government adopted a Law for Agrobiodiversity, Seeds and Promotion of Sustainable Agriculture. One of the important components of the law is the conservation of native seeds. The main objectives of the law are to protect, revitalise and promote the dynamic conservation of agricultural biodiversity; ensuring production and free and permanent access to quality seeds, including by strengthening scientific research and promoting sustainable agricultural production models. It also respects the diverse identities, knowledge and traditions that guarantee the availability of healthy, diverse, nutritious and culturally appropriate foods to achieve food sovereignty and contribute to "Good Living" ("Sumak Kawsay"). It guarantees the free use and exchange of peasant seeds, and establishes rules for the production, certification and commercialisation of certified seeds. The law is a normative instrument that would allow the conservation of diverse agricultural biodiversity species of nutritional and economic importance, as well as with industrial potentials.

The Heirloom Rice Project

The Heirloom Rice Project, started in 2014, is supported by the Department of Agriculture of the Philippines and the International Rice Research Institute. It aims to enhance the productivity and enrich the legacy of heirloom and traditional rice varieties by empowering communities in rice-based ecosystems in the Philippines. Heirloom rice varieties, grown and handed down for generations by small landholders, have exceptional cooking quality, flavour, aroma, texture, colour and nutritional value. There is high demand for these varieties, and they command higher prices in both domestic and international markets. However, there are also challenges hindering farmers from seizing these opportunities, and some varieties are at risk of extinction. The Heirloom Rice Project takes a market and product development approach, from characterising the existing heirloom or traditional varieties alongside modern climate-resilient varieties to capacity development and enterprise building in farming communities, and identifying opportunities for value addition and market linkages. The project resulted not only in an almost 80% increase in the production of heirloom rice varieties in six years but also, more importantly, in providing opportunities for farmers to increase their income-generating capacity.

Creation of a micro value-chain for a local variety of rye, the "Lermana"

During the 2013–2014 period, *Agenzia Lucana di Sviluppo ed Innovazione in Agricoltura* (ALSIA), an agency of the regional government of the Basilicata region in Italy supporting agricultural development and innovation, carried out a survey among farmers of the region to identify rye growers. Four farmers cultivating a total area of two hectares with an old variety called "*Lermana*" or "*Germana*" were identified. In 2018, ALSIA succeeded in registering the local "*Lermana*" rye variety in the Italian National Seed Catalogue in the "Conservation Varieties" section. This was followed by activities for value-chain development, which increased the area cultivated with "*Lermana*" rye from 2 hectares to 15 hectares, and also increased the number of farmers cultivating the variety from 4 to 30. This success demonstrates the importance of the "conservation through use" approach for the conservation of local PGRFA, and also the role that local authorities can play in supporting local development through on-farm conservation of traditional varieties.

Ejere Farmer Crop Conservation Association and CSBs

CSBs have been established in many countries to safeguard local crop varieties and to secure seed supply for local communities. CSBs are commonly established and managed by farming communities but may also involve collaboration with NGOs or research institutes. The CSB in Ejere, Ethiopia, attracts many visitors each year, from Ethiopia as well as abroad, who wish to learn about

384 Claudio Bogliotti et al.

their achievements and success. Through conservation and the participatory improvement of local crop diversity and related activities, the CSB has significantly improved seed and food security, nutrition and livelihoods in the whole area. Initiated in 1990 by USC Canada[8] in collaboration with the erstwhile Plant Genetic Resources Centre, Ethiopia, the work was later taken over by the Ethio-Organic Seed Action, with support from the Development Fund of Norway. The initiative promotes sustainable climate-change-adaptation practices among farmer communities through enhanced capacity to sustainably manage, develop and utilise local agrobiodiversity as an adaptive mechanism. To bolster local crop diversity conservation, it supports implementation of practical actions on the ground, such as a reintroduction of traditional crops; participatory varietal selection to adapt promising crops to changing environmental conditions and improve desired properties; quality seed production and distribution; seed fairs and training in advanced organic production methods and income-generating activities.

Notes

1 The International Treaty on Plant Genetic Resources for Food and Agriculture was adopted by the Thirty-First Session of the Conference of the Food and Agriculture Organization of the United Nations on 3 November 2001. More information is available at: http://www.fao.org/plant-treaty/en/
2 Article 9.1: Recognition of the enormous contribution that local and Indigenous communities and farmers of all regions of the world have made and will continue to make for the conservation and development of plant genetic resources; Article 9.2(a): The protection of traditional knowledge relevant to plant genetic resources for food and agriculture; Article 9.2(b): The right to equitably participate in sharing benefits arising from the utilisation of plant genetic resources for food and agriculture; Article 9.2(c): The right to participate in making decisions, at the national level, on matters related to the conservation and sustainable use of plant genetic resources for food and agriculture; and Article 9.3: The right that farmers have to save, use, exchange and sell farm-saved seed/propagating material, subject to national law and as appropriate.
3 Inventory of national measures, best practices and lessons learned from the realisation of Farmers' Rights, as set out in Article 9 of the International Treaty http://www.fao.org/3/na906en/na906en.pdf
4 Projects under the Benefit-Sharing Fund are showing positive results, demonstrating that the greater the diversification of crops, the more food secure a community can become and the more resilient they find themselves in the face of current threats like climate change, pests and diseases.
5 Online marketing and selling of agricultural produce.
6 For more information on the Toolbox for Sustainable Use of PGRFA, refer to http://www.fao.org/plant-treaty/tools/toolbox-for-sustainable-use/overview/en/
7 National measures, best practices and lessons learned on the realisation of Farmers' Rights are compiled in a dedicated section of the International Treaty website: http://www.fao.org/plant-treaty/areas-of-work/farmers-rights/farmers-rights-submissions/en/
8 It is now known as Seed Change; for more information, refer to https://weseedchange.org/

PART VI
Building an enabling environment

32
EQUITY, GENDER AND MILLETS IN INDIA

Implications for policy

Nitya Rao, Amit Mitra and Raj Rengalakshmi

Introduction

The global resurgence of interest in neglected and underutilized species (NUS) like millets, including viewing them as a panacea for fighting food and nutrition insecurity during the COVID-19 pandemic, is welcome. These crops are being considered as alternatives to rice and wheat, especially for the poor in the peripheral areas of the Global South (Muthamilarasan and Prasad, 2020). However, the focus on millets will be counter-productive if social equity, including gender concerns, are not taken into account in their promotion, as happened in the case of major staples, especially during the Green Revolution.

Debates on male versus female farming systems, especially in the African continent (Boserup, 1970), have highlighted the social and technical reasons as to why particular crops are considered as men's or women's. Instead of sophisticated machinery, women often used hoes to cultivate; due to lower access to resources and assets they used fewer inputs than men; and, importantly, given their role in ensuring household food security, their cropping patterns often prioritized food and subsistence crops over cash crops. Therefore, highlighted the importance of recognizing women's roles and contributions to farming in national accounts and supporting them through better access to and provision of resources, inputs, services, and markets.

Women are often portrayed as risk-averse subsistence farmers, perhaps due to the choices they make or the constraints they confront. Policies, hence, ignore their crops (Devkota et al., 2014), as does research (Donatti et al., 2019). Despite their low social standing, these crops are crucial to household food and nutrition security, and the optimum and sustainable use of the ecosystem including soil management. Using counter-examples from Zambia and Zimbabwe, Jackson (2007) demonstrates how risk behaviours of women in farming are strongly related to the nature of marriage,

DOI: 10.4324/9781003044802-38

conjugality, and gendered notions of insurance and dependence. Further, most crop production systems involve both men's and women's labour, making it important to understand relational dimensions, including gendered contributions, in order to ensure gender equality and household food and livelihood security, alongside sustainable resource use (Rao, 2017; Doss et al., 2018).

Small millets and other NUS, like roots and tubers or wild fruits have been integral to farming systems, especially in the hills and mountains or their adjoining forests in rural India and elsewhere across the world. These traditionally included multiple crops, meant for both consumption and sale. Crop rotation, mixed/inter-cropping including of non-cultivated edible greens or fodder for livestock, and practices such as bund crops to reduce soil erosion on slopes, growing pulses, or mulching crop residues helped maintain soil fertility. Overall biomass productivity was enhanced by effectively sharing resources (water, sunlight, nutrients), ensuring regulation of pests and attacks from wild animals, and collective action to maintain a uniform sowing time, sharing seed resources, and maintaining informal seed networks and crop watching at harvest time. All these actions, directed at maintaining the ecosystem, also strengthened resilience to climate variability (cf. Voelcker, 1893). This strategy diversified sources of income and food and contributed to ensuring diverse diets and nutrition security. Additionally, the system helped generate and stabilize employment by extending the cropping season throughout the year.

Importantly, such diversified farming systems were premised on principles of social equity, including gender equity, involving elements of mutuality, albeit asymmetric, in production, processing, and consumption (Mitra and Rao, 2019). With changes in land tenures and agricultural modernization as well as the emphasis on agricultural production as a source of income, there has been a shift to mono-cropping of high-value crops. Single varieties are subject to higher risks from price fluctuations and climate change. They conform to more individualized forms of production, with little attention given to the collective management and maintenance of the larger ecosystem. Using the example of millets, we draw insights the broader dimensions of inequalities and the mechanisms of their production and reproduction over time. We focus particularly on the evolution of three dimensions of inequalities: access and control over resources including knowledge and skills; labour relations; and decision-making processes in the production, processing, sale, and consumption of millets. These dimensions impact gender and the wider relations between people, collective action, and the natural environment. The reflections in this paper come from the long-term engagement of the authors with communities in the Kolli Hills in Tamil Nadu and Koraput in Odisha. Tribals, historically seen as 'marginalized' and 'backward' communities in the Indian context, predominate in both sites.

Understanding social equity and gender relations

What do we mean by social equity? There are multiple levels of inequity prevalent in any society, based on many social and economic markers including class,

Equity, gender and millets in India **389**

landownership, caste, ethnicity, age, educational status, and gender. While our emphasis here is on gender, we understand gender identity as embedded in other social relations of caste, class and ethnicity. We adopt an intersectional perspective on gender, exploring the experiences of women – not as a category, but, rather, seeing women and men, as heterogenous groups, and how their experiences are shaped by their particular social position (Rao, 2017).

Gender analysis unpacks the power relations involved in resource access, especially land, a critical resource for agriculture, but equally reflected in both labour contributions and control over income and decision-making processes. We find variations in women's labour and time burdens depending on the quantity, quality, and the location of land in the ecosystem, and crop choices. Most tribal communities in the study areas manage cultivation with household labour. Hired labour is uncommon, though harvesting, threshing, and cleaning draw, at times, on reciprocal community labour (Mitra and Rao, 2019).

Women's knowledge of cultivation, especially in seed management and forming seed exchange networks, was valued and recognized in scenarios where millets were cultivated regularly. Additionally, women were free to choose the crop mixes in particular types of lands, especially the uplands. Plantation crops, like eucalyptus, in which agreements are mostly made by men and lumpsum payments received in cash, are now replacing millets in the uplands (Mitra and Rao, 2019). With market-driven cropping systems, including of millets, gaining ground, women have lost their say in crop choices, but equally in the kind of crops to be sold as well as the quantities. Their knowledge, needs, and preferences, including for domestic consumption and assuring basic nutrition, are no longer given due consideration.

The final issue relates to decision-making control or the exercise of agency. Priorities clearly are negotiated within the household (Sen, 1990). Changes in the types of land and the types of cultivation practiced, upland versus lowland, are accompanied by shifts in decision-making control. This is because uplands are often associated with women and their crops, used for subsistence, while lowlands are seen as better-quality lands, for paddy cultivation, which lies in men's domain (Rao, 2008). This normative distinction is linked to gender roles, with men considered 'providers', responsible for the staple crop, and women 'home-makers', responsible for the food accompaniments rich in micronutrients (Carney and Watts, 1990). With the growing need for cash incomes to meet needs beyond food, which is considered to lie in the public (male) domain, there has been a shift towards greater male control decision-making over land use and household expenditures. In the process, women's central role as farmers is forgotten.

Land management

In India, social position matters in shaping both access to agricultural land, and the nature of cultivation practiced. In Koraput, the experiences of the landed and slightly better-off Bhumiya tribe differ from the near-landless Parojas or the landless Scheduled Caste Doms. While for the Bhumiyas, their crop choices are increasingly based on the exchange value of particular species in the markets, the driving factor for the Parojas is still the use-value of different varieties for a range of purposes like food, liquor, fuel, and social rituals. Among the small millets, finger millets constitute the bulk of the production, as they can be fermented easily, and the straw used for fodder and housing material. Little millets are still offered to the gods. The Doms, dependent entirely on wage labour, are not necessarily aware of the intricate drivers of crop choices. While both the Bhumiyas and Parojas cultivate millets, the former are shifting to mono-cropping finger millet (ragi) in their lowlands due to growing market demand. The Parojas, mostly upland cultivators, continue to grow multiple millet varieties mixed with pulses and vegetables (Mitra and Rao, 2019). This suggests a growing gap between those with access to resources, particularly lowlands, and those without. Yet this gap in landownership does not necessarily translate into dietary diversity in household food consumption. In fact, higher incomes through sales of surplus production are invested in expenditure on education, capital assets, and at times health, rather than better food.

A second, related trend in land management is a shift from diversified cropping systems suited to particular land types towards cash crops, be it a shift from millets to tapioca/cassava in the midlands in the Kolli Hills, or to eucalyptus plantations in the uplands of Koraput, both crops that are contracted to industry. In the Kolli Hills, the local Malayali people regularly practiced mixed cropping, crop rotation, and inter-cropping (Vedavalli and Rengalakshmi, 2019). The *poramboke* lands (uncultivated commons), often interspersed with the rocky steep slopes (*kolla kadu*), provided fuel, fodder, uncultivated foods, and leaf mulch for millet cultivation in the uplands or paddy in the wetlands. Diverse farming systems also contribute to soil health, maintaining water resources and managing ecosystem services. While the new cash crops do bring in assured cash to the individual household, this is at the cost of environmental health and regeneration. Here too, the negative effects on household nutrition are visible, as is the shift in decision-making and control from women to men.

In terms of formal land tenures, the notion of private property was introduced by issuing *pattas* (recorded land titles) in men's names in the first half of the twentieth century in Koraput, a princely state until 1947. In the Kolli Hills, annexed by the British in 1792, the *ryotwari* system of revenue collection (where the state collected revenue directly from the cultivators) was introduced in 1797–1798 (Saravanan, 2010). While in Koraput many non-tribal outsiders, clerks, and moneylenders occupied tribal lands (Behuria, 1966) – similar to experiences in the tribal parts of colonial British India (Rao, 2008) – in the Kolli Hills, major land-use changes began with the introduction of commercial tapioca in the early 1970s (Vedavalli and Rengalakshmi, 2019).

Despite the changes in the local economy due to the land settlements and commercial crops, the *de-facto* situation did not change much for a long time on the ground as women continued to cultivate upland plots. While individual ownership was recognized over individual plots, the ownership and management of the ecosystem remained a collective responsibility. A farmer could not make changes or cultivate crops that would adversely impact plots downstream (Mitra and Rao, 2019). In both sites, contrary to the images of life being on the margins of subsistence, there was food self-sufficiency, considerable trade in forest products, and various arts and handicrafts. Surpluses might not have been large, but there was a balance and reciprocity with the natural environment (Mitra and Rao, 2019).

Divisions of labour

Each crop has a different set of timings for different activities from planting to weeding, irrigating, and harvesting. These are gendered, with men normally performing land clearing and ploughing activities, while women are involved in sowing and weeding. In Koraput, where households now cultivate both lowland and upland crops – the former usually for sale in the market and controlled by men – women's labour is diverted to this crop, with little time available for weeding or attending to their traditional varieties, often millets and pulses, grown on the uplands. This results in a decline in the production and consumption of NUS (Rao and Raju, 2019), with negative effects also on soil productivity.

An analysis of gendered time use across seasons in Koraput highlighted that it was major crops like paddy that were most time intensive for women, both during planting and harvesting. In fact, during these seasons, women had much less time available for domestic work, or household care, including self-care. Rest and leisure are not just critical for ensuring women's own health, but also the squeezing of time for domestic work directly impacts time spent on cooking and feeding, with potentially negative outcomes for children and, indeed, the entire family. Interestingly, women from the Gadaba community, who continued to practice a more diversified cropping and livelihood system, did not confront the severe time shortfalls or seasonal bodyweight losses during the peak cultivation seasons as did the Bhumiya women (Rao and Raju, 2019).

There is renewed interest in diversified farming systems on grounds of food and nutrition concerns. It needs to be highlighted that these systems also valued women's time and labour, alongside ensuring they had control over the crop and its use. In India, more research that accounts for women's labour in the calculation of costs and prices of different agricultural commodities is needed. As women are counted only as 'unpaid household helpers', their labour tends to remain undervalued, with negative effects on their decision-making control, nutrition, and indeed health.

Women's agency and decision-making

A key decision area relates to the choice of crops across different types of land – lowland, midlands, and uplands. Each of these land types and crop choices have

392 Nitya Rao et al.

different contours, water resources, and labour needs. The available market opportunities also shape household decision-making processes. This is evident in the case of Kadaguda (name changed), a village in Koraput district. Women from Paroja households used to cultivate a variety of crops, including small millets, niger, and pulses on the uplands, like all other villages in the area. Almost every household had one or two vegetable plots owned and managed by the women. Changes began about 15 years ago with the introduction of eucalyptus plantations on the uplands, and sweet potatoes on their vegetable plots. The cash income, however, officially accrues to the men, though it involves the labour of both men and women. It is not spent on the purchase of food, so dietary diversity has declined. The villagers now rely on cheap cereals from the Public Distribution System. Kadaguda was famous in the area for the quality of its fermented millet beer. This has now been replaced by cheap country liquor and Indian-made foreign liquor. Women used to drink the nutritious millet beer but now are discouraged from consuming alcohol. Their labour is appropriated, but they receive neither cash nor nutrition in return. Further, not only is crop diversity lost, but so too is women's knowledge in adapting to the microenvironment nullified.

This is not to say that diversified farming systems in these communities were gender equitable. Men have always held leadership positions in the community and the household, with women excluded from participation in village or community-level decisions (Rao, 2008). Yet, similar to the case of the New Guinea Highlands, despite their exclusion from the ritual, political sphere, women's participation in production, and the social acknowledgement of that, ensured that they had a voice both in the household and in the community (Strathern, 1988). Their voice, though negotiated, could not be ignored. If this happened, women found everyday forms of protest, including through withdrawing their labour from the men's crops (Carney and Watts, 1990; Rao, 2008).

The situation began to change with the introduction of the notion of private property and the issuing of pattas (Land Title) in men's names, seen as the 'heads of households'. As social relations, especially vis-à-vis the state, became more male-centric, this tendency moved down to the community level too. Bride price was gradually replaced by dowry, as women were seen as a burden, to be looked after, rather than as productive members of the household. Women's rights in terms of the choice of marriage partner, the right to divorce and remarry, and control over their bodies more generally started being restricted.

While women's freedoms were curtailed, their work burdens were not. They continue to hold the primary responsibility for land management, production, and processing, yet their ability to make decisions on how to do so is constrained. While some women do resist their marginalization, new forces of development, including education, organized religion, and a cash economy based on enhancing the productivity of single crops, have limited their opportunities for resistance. As many women noted across sites, they now have fewer children and aspire to educate their children. This means that they lack household labour to manage

Equity, gender and millets in India **393**

a diversified farming system, ultimately giving in to men's pleas to shift to cash crops or contract farming (Rao and Raju, 2019). Despite a weakening in their material basis in this process, they nevertheless hope that their hard work for the household will ensure a degree of jointness in both production and consumption.

Implications for policy

The above analysis has several implications for gender-sensitive policy. While clearly recognizing women as farmers is key, and ensuring their basic entitlements to assured credit, insurance, remunerative prices, value-addition technologies, and so on, this is not sufficient. Specific constraints confronting the cultivation of NUS and maintaining diversified farming systems need to be acknowledged and addressed.

An important emergent need in rural communities is that of enhancing cash earnings. This has led to the devaluation of NUS, largely seen as subsistence crops. To enhance the status and value of NUS, urgent research is needed to improve the productivity of local, resilient varieties, but equally to value addition of the entire plant biomass, including straw and husk and not just the grain. This has been attempted in case of paddy, and some plantation crops such as coconut and banana, but continues to remain neglected for millets. A second strategy to ensure minimum incomes from the cultivation of NUS is through public procurement at assured prices, provision of insurance, including at the warehouse level, and credit provision to women farmers. While minimum support prices are announced, very little of the produce is actually procured at these prices or, indeed, made available for mass consumption. In fact, due to the nutritional quality of millets, the elite, rather than the poor, are now consuming them.

From a social equity and gender perspective, it is important to recognize women's contributions, both in terms of their knowledge of seeds, soils, and water management, and their time and labour spent on the cultivation and processing of these crops. Women in the Kolli Hills told the first author that they did not like to cook millets at home, not just because their children did not have a taste for it, but also because it required hand pounding and was too labour and time intensive to prepare. Research and policy need to focus on the development of labour-saving technologies that value and support women's labour in relation to NUS.

Finally, promoting the cultivation of millets by adopting a cluster approach would help to enhance ecosystem services along with the collective management of pests that damage the crops. This is in the interest of all farmers, and will ultimately benefit not just women, but the entire community.

References

Behuria, N.C. (1966) *Final Report on the Major Settlement Operations in Koraput District, 1938–1964*, Cuttack: Government of Orissa.

Boserup, E. (1970) *Women's Role in Economic Development*, New York: St Martin's Press.

Carney, J., and Watts, M (1990) Manufacturing dissent: Work, gender and the politics of meaning in a peasant society. *Africa*, Vol 60, No 2, pp 207–241.

Devkota, R., M. Karthikeyan, H. Samaratunga, H. Gartaula, K. Khadka, S. Kiran, B.K. Nayak, and Nadhiya, M. (2014) *Increasing Gender Equality among Small Millet Farmers in South Asia, Stories of Change*, Ottawa: IDRC.

Donatti, C.I., Harvey, C.A., Rodriguez, M.R.M., Vignola, R., and Rodriguez, C.M. (2019) Vulnerability of smallholder farmers to climate change in Central America and Mexico: Current knowledge and research gaps, *Climate and Development*, Vol 11, No 3, pp 264–286.

Doss, C., Meinzen-Dick, R., Quisumbing, A. and Theis, S. (2018) Women in agriculture: Four myths. *Global Food Security*, Vol 16, pp 69–74.

Jackson, C. (2007). Resolving risk? marriage and creative conjugality. *Development and Change*, Vol 38, No 1, pp 107–129.

Mitra, A. and Rao N. (2019) Contract farming, ecological change and the transformations of reciprocal gendered social relations in Eastern India, *The Journal of Peasant Studies*. https:/doi.org/10.1080/03066150.2019.1683000

Muthamilarasan, M. and Prasad, M. (2020) Small Millets for Enduring Food Security Amidst Pandemics, *Trends in Plant Science*. https://www.sciencedirect.com/science/article/pii/S1360138520302557

Rao, N. (2008) *"Good women do not inherit Land": Politics of Land and Gender in India*, New Delhi: Social Science Press and Orient Blackswan.

Rao, N. (2017) Assets, agency and legitimacy: Towards a relational understanding of gender equality policy and practice. *World Development*. Vol 95, pp 43–54.

Rao, N. and Raju, S. (2019) Gendered time, seasonality and nutrition: Insights from two Indian districts. *Feminist Economics*. Vol 26, No 2, pp 95–125.

Saravanan, V. (2010) Agrarian policies in the tribal areas of Madras presidency during the pre-survey and settlement period, 1792–1872, *Indian Journal of Agricultural Economics*, Vol 65, No 2: April–June.

Sen, A. (1990) Gender and cooperative conflicts, in I. Tinker (ed.) *Persistent Inequalities: Women and World Development*. New York: Oxford University Press, pp 123–149.

Strathern, M. (1988) *The Gender of the Gift: Problems with Women and Problems with society in Melanesia*. Berkeley: University of California Press.

Vedavalli, L. and R. Rengalakshmi. (2019) *Crop Diversity in Peril: A Case of Kolli Hills in India*, New Delhi: Academic Foundation.

Voelcker, J.A. (1893) *Report on the Improvement of Indian Agriculture*, London: Eyre and Spottiswoode.

33

WHAT IS GOING ON AROUND THE WORLD

Major NUS actors and ongoing efforts

Stefano Padulosi, Gennifer Meldrum,
E. D. Israel Oliver King and Danny Hunter

Introduction

Through this updated list of key players and initiatives focusing on neglected and underutilized species (NUS) around the world, we hope to facilitate possible synergies, collaborations and partnerships among organizations interested in advancing the conservation and sustainable use of these resources. Far from aiming to be exhaustive, our analysis results from consulting various sources of information obtained from published literature, unpublished reports, the internet and LinkedIn profiles. The list includes research institutions (national and international), NGOs, universities, networks, consortia, major NUS conferences, private sector companies and donors who have been supporting/are supporting NUS activities. This chapter also complements Chapter 2, which highlights some of the historical landmark events related to NUS.

International players

CGIAR Centres

A number of centres of the Consultative Group on International Agricultural Research (CGIAR) are engaged in research and development of a crops, landraces, and other plant and tree species that can be considered NUS. Here we only highlight some of the key centres.

The Alliance of Bioversity International and CIAT

The recently launched Alliance is bringing together expertise of two centres, Bioversity International (formerly IBPGR, IPGRI) and the International Center

DOI: 10.4324/9781003044802-39

for Tropical Agriculture (CIAT), both with an active engagement in the promotion of NUS. With regard to Bioversity, its work on NUS dates back to its very establishment, at a time when the organization developed regional conservation strategies and priorities that included NUS (IBPGR, 1981). Currently Bioversity has just ended the fifth phase of an IFAD-EU supported programme on NUS, which was implemented from 2015 to 2020 in Guatemala, Mali and India (Padulosi et al., 2019), and a GEF-UNEP Project ended in 2019 and implemented in Brazil, Kenya, Sri Lanka and Turkey (Hunter et al., 2019) specifically addressing the nutritional value of a large portfolio of NUS. A three-year research project funded by GIZ on NUS has been also recently launched in Ethiopia and Kenya to help communities develop their own action plans to improve nutrition and diet diversity, harnessing local agrobiodiversity (Nowicki, 2019). CIAT maintains in its germplasm collection numerous species of NUS legumes and forages of tropical origin and involved in their use enhancement.

The World Agroforestry Centre, International Centre for Research in Agroforestry (ICRAF)

This CGIAR centre is active in promoting NUS, mostly tree species, such as moringa (*Moringa oleifera*), the safou plum (*Dacryodes edulis*) and the medicinal African plum (*Prunus africana*). The centre is a major partner in the African Orphan Crops Consortium (AOCC – see later) Website: http://worldagroforestry.org/

International Crops Research Institute for the Semi-Arid Tropics (ICRISAT)

ICRISAT as a centre actively involved in the promotion of millets, for which it holds a large *ex situ* germplasm collection. Currently engaged in a global initiative – the "Smart Food Initiative", described in Chapter 28 – which covers several NUS cereals and pulses (Smart Food Executive Council, 2019). Website: www.icrisat.org

Non-CGIAR international organizations

Crops for the Future (CFF)

Crops for the Future was established in 2009 out of the merger of the Global Facilitation Unit for Underutilized Species (GFU) and the International Centre for Underutilized Crops. In 2011, it launched the CFF Research Centre in Malaysia, which comprises laboratories, offices and a field research centre (Gregory et al., 2019), which is currently being relocated to another country. CFF and CFF-RC became one entity in 2014. Website: http://www.cropsforthefuture.org/

Royal Botanic Gardens (RBG), Kew, UK

In 2007, Royal Botanic Gardens-Kew launched Project MGU known as "The Useful Plants Project", which aims at enhancing the *ex situ* conservation of native

Major NUS actors and ongoing efforts **397**

useful plants for human wellbeing across Africa and Mexico by building the capacity of local communities to successfully conserve and use these species sustainably. Since its establishment, the project has been working with partners in Botswana, Kenya, Mali, Mexico and South Africa to conserve and sustainably use indigenous plants that are important to local communities (Ulian et al., 2016). More information on the project is provided in Chapter 9. Website: https://www.kew.org/science/our-science/projects/project-mgu-useful-plants-project. Worth mentioning here is also RBG Kew's latest publication by Ulian et al. (2020), released in conjunction with the RBG's State of the World's Plants and Fungi 2020 report (Antonelli et al., 2020), in which authors stress the urgency for the better promotion and recognition of NUS' role in improving the quality, resilience and self-sufficiency of food production.

The Food and Agriculture Organization (FAO) of the United Nations

The FAO has been actively involved in the promotion of NUS for some time. As a matter of fact, IPBGR (the first predecessor to Bioversity International) was established in 1974 within the FAO and from there it operated until its separation from the UN system in 1993. Through the Commission on Genetic Resources for Food and Agriculture, the FAO is active in promoting, inter alia, the genetic resources of underused species. The FAO's second Global Plan of Action for the Conservation and Sustainable Utilization of PGRFA is another important not legally binding multilateral framework that promotes NUS as well through its activities. The International Treaty on Plant Genetic Resources for Food and Agriculture (International Treaty) considers NUS one of its priorities (Article 6e) and efforts are deployed to promote their sustainable use through its Access and Benefit-Sharing Funding Mechanism (see Chapter 31). The FAO is also active in the use-enhancement of NUS through its efforts to safeguard Indigenous Peoples' food systems.[1] Within the FAO, the Global Forum for Agricultural Research Key Focus Area 1 contains a specific collective action (1C) dedicated to "Rediscovering Forgotten Foods and Ensuring Benefits to Smallholder Farmers Foods". Worth mentioning is the initiative implemented by the FAO's Regional Office for Asia and the Pacific on "Future Smart Foods", defined as NUS, carried out in the context of its efforts to achieve zero hunger. Its programme on non-wood forest products also covers many NUS in forest areas. With regard to nutrition, it is important to mention the work of its INFOODS programme, which includes information on the nutritional profile of several NUS crops and species. Website: https://www.fao.org

Global Crop Diversity Trust (GCDT)

The GCDT launched, in 2017, the Food Forever Initiative, a global partnership to raise awareness of the importance and urgency of conserving and using agricultural biodiversity to achieve SDG Target 2.5. This initiative is very active

398 Stefano Padulosi et al.

in promoting the wider use of crop diversity for improved nutrition, including NUS. Website: https://www.food4ever.org

The Lexicon, USA

This media company ("The Lexicon for Sustainability") was founded in 2009 to accelerate the adoption of practices that build more resilient food systems and help combat climate change; it is currently implementing – with the support of several organizations (including Bioversity) – the "Rediscovered Food Initiative", which focuses on 25 NUS from around the world. Website: https://www.thelexicon.org/rediscovered/

Centro Agronomico Tropical de Investigacion y Ensenanza, Costa Rica

This research centre has many activities dealing with NUS, including the maintenance of a germplasm collection of fruits, vegetables and palms (Ebert et al., 2007). Website: https://www.catie.ac.cr

The World Vegetable Centre (AVRDC)

This international nonprofit research and development institute, is committed to alleviating poverty and malnutrition in the developing world through the increased production and consumption of nutritious and health-promoting vegetables, many of which fall into the NUS category, such as traditional African leafy vegetables (see Chapter 17). In support of such a mission it maintains the world's largest public vegetable germplasm collection with more than 61,000 accessions from 155 countries, including about 12,000 accessions of indigenous vegetables, many of which are NUS. Website: https://avrdc.org/our-work/managing-germplasm/

Centre de coopération internationale en recherche agronomique pour le développement (CIRAD)

This French agricultural research and international cooperation organization works for the sustainable development of tropical and Mediterranean regions. In its key area dedicated to biodiversity, it works with partners to study the conditions in which conserving, restoring, mobilizing and exploiting biodiversity, including NUS, could help alleviate poverty and boost food security and safety. Website: https://www.cirad.fr/en

Deutsche Gesellschaft für Internationale Zusammenarbeit (GIZ)

With 50 years of service at the global level, the German Agency for International Cooperation works in the field of sustainable development and international education; it engages in a variety of areas, including the sustainable use of agro-biodiversity in which it funds numerous projects, including the recently launched "Improving Dietary Quality and Livelihoods using Farm and Wild Biodiversity through an Integrated Community-Based Approach in Ethiopia and Kenya", which will target several highly nutritious indigenous NUS (see also Chapter 6). Website: https://www.giz.de/en/html/index.html

OXFAM

This international NGO is very active in the promotion of NUS to tackle food insecurity, poverty and the marginalization of vulnerable peoples. Particularly relevant is its "Sowing Diversity=Harvesting Security" programme, which aims to improve access and use of crop diversity (including NUS) to change the current unsustainable and unequal food production systems. Website: https://www.sdhsprogram.org/

United Nations Educational, Scientific and Cultural Organization (UNESCO)

Among its efforts, UNESCO has numerous initiatives focusing on the safeguarding of wild and cultivated genetic food species, including NUS. The work of UNESCO for the documentation and valorization of traditional food cultures entrenched with NUS is particularly relevant; see, for example, some of the activities carried out in the context of "The Man and the Biosphere Programme". Website: https://en.unesco.org/

Food Tank

This USA-based advocacy NGO works to build a global community for safe, healthy, nourished eaters. It spotlights and supports environmentally, socially and economically sustainable ways of alleviating hunger, obesity and poverty and creates networks of people, organizations and content to push for food system change. Many of its initiatives focus on NUS and their sustainable promotion in today's food systems. Website: https://foodtank.com/

Slow Food International

Slow Food is a global grassroots organization founded in 1989 to prevent the disappearance of local food cultures and traditions, counteract the rise of fast food and combat people's dwindling interest in the food they eat, where it comes from and how our food choices affect the world around us. Through its Slow Food Foundation for Biodiversity, it works to defend local food traditions, protect food communities, preserve food biodiversity and promote quality artisanal products, with an increasing focus on the global south. Many of its projects focus on

400 Stefano Padulosi et al.

NUS (see also Chapter 18). Website: https://www.slowfood.com/what-we-do/preserve-biodiversity/

Pacific Agricultural Plant Genetic Resources Network (PAPGREN)

A network established in 2001 to promote the conservation and use of the genetic resources of crops of local importance in order to ensure long-term conservation and access to these genetic resources by Pacific Island populations, which in turn will contribute to sustainable development, food security and income generation. Some of the NUS that have been the focus of PAPGREN and CePaCT (below) are highlighted in Chapter 12. Website: https://lrd.spc.int/the-pacific-plant-genetic-resources-network-papgren

The Centre for Pacific Crops and Trees (CePaCT)

An internationally recognized gene bank established in Fiji to assist Pacific Island countries and territories in conserving the region's unique genetic resources and providing access to the diversity they need. It is a propagation material vault operated by the Pacific Community's (SPC) Land Resources Division. It conserves the region's major crops with over 2,000 accessions, including the world's largest taro collection (over 1,100 accessions). Website: https://www.spc.int/about-us

For further information on other organizations working directly or indirectly on NUS such as the World Wide Fund for Nature (WWF), the World Conservation Union (IUCN), Botanic Gardens Conservation International, United Nations Environment Programme (UNEP) and other crop-specific collaborations and thematic networks, including relevant civil society and Indigenous Peoples' networks, the reader is directed to Chapter 2 of Maxted et al. (2020).

Knowledge-sharing websites

NUS Community platform

An online platform for sharing research results, development news and policy advice regarding the use and conservation of NUS. It aims to support research and promote the use of NUS to strengthen food security; build more resilient, climate-smart agriculture and empower people through income generation and revitalized local food culture. It also provides access to information and lessons from projects that Bioversity International implements jointly with research and development partners around the globe. In addition, it serves as an information hub in support of human capacity development and synergy building among practitioners engaged in this field. Website: http://www.nuscommunity.org

Biodiversity for Food and Nutrition platform

Funded by the Global Environment Facility (GEF), the Biodiversity for Food and Nutrition (BFN) initiative (formally known as Mainstreaming Biodiversity Conservation and Sustainable Use for Improved Nutrition and Well- Being) commenced in 2012 and was led by Brazil, Kenya, Sri Lanka and Turkey, coordinated and executed by Bioversity International with implementation support from the UNEP and the FAO of the United Nations. The project prioritized nutrient-rich NUS in participating countries, demonstrating and providing evidence for their nutritional value by supporting significant research on food composition. Using this expanded knowledge-base, the project undertook the challenge of strengthening the enabling environment to better promote and mainstream NUS for improved diets and nutrition, including through the development of policy incentives, markets, public food procurement, school feeding programmes and food-based dietary guidelines. The initiative also undertook considerable awareness raising, working closely with chefs and gastronomy movements, school garden networks and consumer groups. Website: http://www.b4fn.org/countries/

Crops for the Future

Its website provides useful online resources on NUS, including relevant publications, portals, online libraries, weblinks and publications. More resources can also be found at the website (still accessible) of the CFF predecessor, GFU, at http://www.underutilized-species.org/MasksSearch/SearchInstitutionDetail_id_55.html

Website: http://www.cropsforthefuture.org/FutureCrop-@-LandingArticle.aspx#Other_Online_Resources

New Agriculturalist *(issues on underutilized crops)*

A journal that is sensitive to NUS and has been covering them in various articles. It has also published a number of special issues on various NUS. Website: http://www.new-ag.info/en/focus/on.php?a=423

Eat the Weeds

A useful site developed and maintained by Green Deane on wild greens and their foraging. It contains an online database with more than 1,000 wild edibles. Website: http://www.eattheweeds.com/

Plant Resources of Tropical Africa (PROTA)

An international foundation that networks with a large number of R&D organizations, within and outside tropical Africa to facilitate access to the existing wealth of information about useful plant species. Website: https://www.prota.org/

Plant Resources of South East Asia (PROSEA)

Similar to PROTA, this international cooperative programme has the main goal of documenting information on plant resources in South East Asia, and making it widely available for use by the education, extension, research and industry sectors, as well as for the end-users, for the betterment of the livelihoods of regional communities. Website: http://proseanet.org/prosea/

Introduction to useful plants (Pl@ntuse)

A collaborative space for the exchange of information on useful plants (including NUS) and their uses. It is not intended to duplicate existing encyclopedias (such as Wikipedia), but to offer additional features such as online resources, thematic literature, species lists, etc. Website: https://uses.plantnet-project.org/en/Introduction_to_useful_plants

Food Plant Solutions, Australia

A project of the Rotary Action Group designed to address malnutrition through the use of readily available and local food sources. It creates educational publications that help people understand the connection between plant selection and nutrition, and empowers them to grow a range of highly nutritious plants with differing seasonal requirements and maturities. It focuses on NUS, defined as plants that are growing in and adapted to their environment, which are high in the most beneficial nutrients. Website: https://foodplantsolutions.org/about-us/

New CROP, USA

This website, online since 1995, is a project of Purdue University's (USA) Center for New Crops and Plant Products and is associated with the New Crop Diversification project and the Jefferson Institute. It offers the following databases – (1) Help: Strategies for using New CROP; (2) CropINDEX: A list of scientific and common names of crops for information access; (3) CropSEARCH: for facilitate crop search; (4) CropMAP: A nationwide (US) location-specific crop information system; (5) Current: Projects, classes, presentations, websites; (6) CropREFERENCE: Books and manuals on crops; (7) CropEXPERT: Directory of new crop resource personnel; (8) IMPORT–EXPORT: Plant quarantine information and phytosanitation permits for all countries; (9) Famine Foods: A list of unconventional food sources and (10) NewCrop LINKS: Connections to related websites, external databases and libraries. Website: https://www.hort.purdue.edu/newcrop/

Moringa News, France

This is a free information-exchange platform devoted to the promotion of moringa and other NUS with a strong potential to improve the quality of life in developing countries. It offers consultancy services to help create and implement research, development and communication projects for NGOs and businesses. Moringa News offers through its website its knowledge, know-how and experience, as well as its still-evolving professional network of international partners. Moringa News implements research, development and communication projects (publications and seminars) with or without the partnership of other associations, businesses, farmer groups, research centres and universities. Website: http://www.moringanews.org/gb/

The Mediterranean Germplasm Database, Italy

A rich database used as reference for the germplasm collection of food crops of the Institute of Biosciences and Bioresources of Research Council (CNR), Bari, Italy. The collection contains some 56,000 accessions belonging to more than 100 genera and over 700 species and it devotes particular attention to landraces of major crops and NUS, including plants potentially useful for the extraction of bioactive or technological compounds. Website: https://ibbr.cnr.it/mgd/

National governmental research institutions

Kenya: Kenya Agricultural and Livestock Research Organization (KALRO)

KALRO focuses on a broad range of crops, including many considered to be NUS and supports research to generate and promote crop knowledge, information and technologies that respond to farmer needs and opportunities. KALRO also hosts the Genetic Resources Research Centre, which contains a number of NUS accessions. Most recently, through the earlier mentioned BFN project, KALRO has been active in the promotion and mainstreaming of African leafy vegetables (ALVs) in school feeding programmes and in organizing and linking local growers of ALVs to school markets.

India: National Bureau of Plant Genetic Resources (NBPGR)

One of the largest *ex situ* genebanks, it maintains genetic resources of many wild and cultivated NUS and it hosts a NUS-dedicated programme (All-India Coordinated Research Network on Potential Crops) focusing on the use-enhancement of 17 of them (see later on in this chapter). Website: http://www.nbpgr.ernet.in/

Mali: Institut d'Economie Rurale (IER)

IER is involved in collecting, conserving and characterizing NUS, as well as in carrying out research on agroecological adaptation, propagation, food

404 Stefano Padulosi et al.

technology, marketing and promotion. Currently, it is implementing a project on NUS in partnership with Bioversity focusing on fonio and Bambara groundnut (Mbosso et al., 2020). Website: http://www.ier.gouv.ml/

Pakistan: Pakistan Agricultural Research Council (PARC)

PARC holds large germplasm collections of wild and cultivated food crops, including NUS. Its plant genetic resources (PGR) Program implements a project entitled, "Conservation and sustainable utilization of agrobiodiversity of underutilized crops".

Website: http://www.parc.gov.pk/index.php/en/component/content/article/101-narc/pgri/818-pgrp-research-projects

Papua New Guinea: National Agricultural Research Institute

Works on a wide variety of crops, cropping systems and post-harvest technology, including NUS such as taro and traditional vegetable crops like bele (*Abelmoschus manihot*). website: https://www.nari.org.pg/

Sri Lanka: Plant Genetic Resources Centre (PGRC)

The PGRC of the country's Department of Agriculture, has the national responsibility for the conservation of all of Sri Lanka's crop varieties and their wild relatives. It actively promotes several NUS, including through germplasm collection, production development and processing. Website: https://www.doa.gov.lk/SCPPC/index.php/en/institute/35-pgrc-2

Sri Lanka: Industrial Technology Institute (ITI)

It carries out research to develop and promote the food industry also focusing on several NUS such as such as jackfruit, breadfruit, rambutan and durian. Website: http://iti.lk/en/

Vietnam: Fruit and Vegetable Research Institute (FAVRI)

It has a breeding programme for underused Vietnamese and exotic fruit and vegetable species. http://favri.org.vn/index.php/en/

Nepal: Nepal Agricultural Research Council (NARC)

It includes in its mandate work on a number of NUS such as buckwheat, minor millets and amaranth. Website:https://www.slideshare.net/apaari/country-status-reports-on-underutilized-crops-by-baidya-nath-mahto-nepal

Universities

James Cook University, Australia

The Agroforestry and Novel Crops Unit of the School of Tropical Biology covers a number of NUS studies related to genetic diversity, and the domestication and commercialization of indigenous fruits and nut trees in Papua New Guinea, the Solomon Islands and Australia. Website: https://www.jcu.edu.au/

Ghent University, Belgium

The Department of Plants and Crops and Department of Food Safety and Food Quality, Faculty of Bioscience Engineering, has been carrying out numerous research investigations on NUS, (e.g., cherimoya, baobab, tamarind). One of their recent publications is on dietary species richness as a measure of food bio-diversity and nutritional quality of diets (Lachat et al., 2018). Website: https://www.ugent.be/en

University of Abomey-Calavi, Benin

Its Faculty of Sciences and Technology is active in the survey, study and promotion of NUS (Dansi et al., 2012). Website: http://www.uac.bj/

Wageningen University & Research, The Netherlands

Currently leading an international project supported by The Bill & Melinda Gates Foundation and the UK Government's Department for International Development for studying fruit and vegetable consumption (including NUS) in Nigeria and Vietnam, in order to gather high-quality evidence linking agriculture, nutrition and health through systems-level approaches. This project is part of the A4NH Research Programme of the CGIAR. Website: https://a4nh.cgiar.org/

University of Nottingham, United Kingdom

The School of Biosciences has been working actively in Africa on several projects targeting Bambara groundnut (flagship project BamBREED). The school is closely associated with the CFF-RC. Website: https://www.nottingham.ac.uk/

University of Southampton, United Kingdom

The university's Centre for Underutilized Crops prioritizes high-quality inter-disciplinary research into underutilized crops contributing to poverty alleviation in an environmentally sustainable way. Some of its current and previous targets include: (1) sustainable agriculture (the cross-over between crop improvement

406 Stefano Padulosi et al.

and socioeconomics); (2) ecosystem services (the economic value of indigenous species and habitats); (3) climate change: crop acclimation and adaptation; (4) climate change: carbon sequestration and biofuels; (5) health and social development and (6) infrastructure and economic development. Website: https://www.southampton.ac.uk/cuc/about/index.page

Non-governmental organizations (NGOs)

Action for Social Advancement (ASA), India

One of the leading NGOs in the sector of farm-based livelihoods for the poor and for natural resource management. It is engaged in the field implementation of NRM projects with active community participation, supporting communities in financial inclusion, sustainable agriculture, establishing farmer producer companies, capacity building and institutional development. ASA is involved in the cultivation, promotion and marketing of minor millets, partnering also with Bioversity in the implementation of the IFAD-EU Global Programme on NUS. Website: https://asaindia.org

Archeologia Arborea, Italy

An Italian non-profit foundation, located in the Umbria region of Italy and focusing on the conservation, study and promotion of local fruit-tree diversity (some 150 ecotypes rescued so far, belonging to both popular species and NUS, like quince and sorb tree). Germplasm is maintained in a field collection near Perugia. Website: http://www.archeologiaarborea.org

Centre for Agriculture and Bioscience International (CABI)

CABI's mission is to improve people's lives worldwide by providing information and applying expertise to solving problems in agriculture and the environment. Several of its knowledge management resources are relevant for NUS, including books, e-books and multimedia tools, the Global Open Data for Agriculture and Nutrition Action and the mobile advisory services such as m-Kisan and m-Nutrition. Website: https://www.cabi.org/

Crops of the Future Collaborative (COTF), USA

Crops of the Future is a public–private collaborative established by the Foundation for Food and Agriculture Research to enhance US and global agriculture by developing the crops needed to feed a growing population. COTF aims to expand knowledge of the genes and traits that give rise to the characteristics crops need to adapt to a changing future. Website: https://foundationfar.org/cotf-crops-of-the-future-collaborative/

ECHO Community, USA

An international NGO engaged in promoting sustainable agricultural practices and providing assistance to local communities through training activities and other interventions. Involved in the promotion of NUS and advocating their greater use for improving the resilience, nutrition and incomes of local communities. Website: https://www.echocommunity.org

Fundación Promocion e Investigacion de Productos Andinos (PROINPA), Bolivia

The PROINPA Foundation promotes technical innovation to improve the competitiveness of Andean crops (including NUS), food security and the conservation and sustainable use of genetic resources, for the benefit of farmers and Bolivian society as a whole. Website: http://www.proinpa.org/web/

Local Initiatives for Biodiversity, Research and Development (LI-BIRD)

A Nepal-based organization that carries out numerous projects focusing on NUS, committed to capitalizing on local resources, innovations, and institutions for sustainable management of natural resources for improving livelihoods of smallholder farmers. Long-standing partner of Bioversity and other international R4D organizations. Website: http://www.libird.org

M.S. Swaminathan Research Foundation (MSSRF), India

The MSSRF was established in 1988 as a not-for-profit trust. MSSRF was envisioned and founded by Professor M.S. Swaminathan with proceeds from the First World Food Prize that he received in 1987. The Foundation aims to accelerate use of modern science and technology for agricultural and rural development to improve the lives and livelihoods of communities. MSSRF follows a pro-poor, pro-women and pro-nature approach and applies appropriate science and technology options to addressing practical problems faced by rural populations in agriculture, food and nutrition. It is involved in numerous projects focusing on NUS (see also Chapter 21). Website: https://www.mssrf.org/

NIRMAN, India

This Indian, Odisha-based NGO has worked with 20,000 tribal and farmer households on sustainable, biodiversity-based farming, community resilience and climate justice, land and forest rights, natural resource conservation and management and nutrition for the last two decades. It gives special emphasis to community-led processes for the conservation of agrobiodiversity and has extensively campaigned for mainstreaming millets since 2011. Website: https://www.nirmanodisha.org/

408 Stefano Padulosi et al.

Rete Semi Rurali, Italy

The Italian Farmers' Seeds Network was established in 2007 and, as of 2014, consisted of more than 30 associations conserving and promoting local agrobiodiversity, including NUS. RSR supports farmers, politically and scientifically, in the creation and dissemination of self- and truly sustainable organic farming processes. Its objective is to rebuild sustainable agricultural systems starting from the re-localization of agriculture, territorial re-contextualization of agricultural research and the technical, political and cultural centrality of farmers and rural areas. Actively engaged in the PGR advocacy policy arena. Website: https://www.semirurali.it/

Native Seeds/SEARCH, USA

A seed conservation organization based in Tucson, Arizona, whose mission is to safeguard and promote the arid-adapted crop diversity of the southwest USA in support of sustainable farming and food security. In its gene bank, it maintains about 2,000 varieties of crops adapted to arid landscapes, including many NUS and their wild relatives. The collection represents the cultural heritage and farming knowledge of over 50 Indigenous communities and recent immigrants. Website: https://www.nativeseeds.org

SAVE Foundation, Switzerland

It is involved in the conservation of plant varieties and breeds. SAVE is also engaged in the documentation and promotion of the genetic diversity of NUS, including fruits and berries in Europe. Website: http://www.save-foundation.net/en/

Society for Research and Initiatives for Sustainable Technologies and Institutions (SRISTI), India

SRISTI is a developmental organization aiming to strengthen the creativity of grassroots communities, including individual innovators. It supports eco-friendly solutions to local problems. It also nurtures ecopreneurs engaged in conserving biodiversity, including NUS, common property resources, cultural diversity and educational innovators. Website: http://www.sristi.org/

Private companies

Anamolbiu, Nepal

A company based in Bharatpur, Chitwan, its work is driven by a "triple bottom line" approach of pursuing economic, social and environmental objectives.

It provides good quality seeds to farmers, for both popular varieties as well NUS, such as broadleaf mustard and amaranth. Website: https://www.anamolbiu.com/

Oroverde, Switzerland

It operates, with selected tropical fruits (many of which are NUS), a trading business from the farmer to the industrial partner, encourages production models of ecological cultures, initiates the bio-certification of fruit culture regions and processing plants in Amazonia as well as of products in the consumer countries, implements quality promotion and quality assurance, develops new innovative products with and for customers and is characterized by its proximity to the producers, co-operatives and communities. Website: http://www.oroverde-fruits.com/EN/fruits.html

Naturally Zimbabwean, Zimbabwe

This is an online magazine for Zimbabweans and everyone else with a passion for our natural foods, natural health and our natural environment. It showcases news, stories, opinions, recipes, interviews, trends and profiles of companies to entertain and educate readers on local foods, many of which are NUS. Website: https://naturallyzimbabwean.com/2015/07/01/welcome-to-the-1st-edition/

Tulimara Specialty Foods, Zimbabwe

This company focuses on processing and commercializing NUS such as indigenous beans, wild fruits and herbal tea. Website: https://www.organic-bio.com/en/company/5173-TULIMARA-SPECIALITY-FOODS

UNILEVER

Its brand Knorr has teamed up with WWF-UK, plus leading scientists, nutritionists and agricultural experts, to compile the Future 50 Foods report, aiming at promoting the wider consumption of highly nutritious NUS from around the world. Website: http://www.knorr.com/uk/future50report

Donors

Australian Centre for International Agricultural Research (ACIAR)

ACIAR is Australia's specialist international agricultural R4D agency, whose purpose is to broker and fund research partnerships between Australian scientists and their counterparts in developing countries. It supports numerous efforts to research and promote NUS in the Pacific, Asia and Africa, including the promotion of traditional vegetables in Papua New Guinea, sustainable taro production

410 Stefano Padulosi et al.

in Samoa and other Pacific countries and the promotion and consumption of ALVs in schools and communities in Kenya. Website: https://www.aciar.gov.au/

International Development Research Centre

The International Development Research Centre (IDRC) funds research in developing countries to promote growth, reduce poverty and drive large-scale positive change. The agency works with numerous development partners and brings innovations to people around the world. Many of its supported projects focus on agrobiodiversity and NUS in particular, and aim at enhancing their sustainable conservation and use. They are based in Canada with five regional offices spread across South America, sub-Saharan Africa and Asia. Website: https://www.idrc.ca/en

International Fund for Agricultural Development (IFAD)

IFAD is an international financial institution and food and agriculture hub of the United Nations. It has been championing financial support for the use-enhancement of NUS since 2001, funding numerous projects (the most important being the Bioversity-led IFAD-NUS Programme). In 2019, it published jointly with Bioversity an operational framework for the promotion of NUS in its loans programme and through other R4D projects (Padulosi et al., 2019). Website: https://www.ifad.org/en/

European Union (EU), Belgium

The EU has supported, over decades, numerous projects focusing on NUS. Particularly worth mentioning is the DESIRA Programme (Development of Smart Innovation through Research in Agriculture) and the Framework Programme for Research and Innovation, Horizon 2020. Both cover NUS through numerous international efforts focusing on Europe and other regions of the world. Website: https://europa.eu/european-union/index_en

Global Environment Facility

The GEF is the financial mechanism of the Convention on Biological Diversity (CBD) and provides financial support to countries for the implementation of projects to help meet their commitments within the framework of the CBD. Mainstreaming biodiversity conservation and sustainable use into production landscapes is recognized as a key strategy to secure the objectives of the CBD, and as a major objective for projects supported by GEF. Over the last few decades, the GEF has provided much support for projects that actively promote the conservation and sustainable use of NUS (Hunter et al., 2019). For example, many of the projects undertaken by the UNEP as a GEF-implementing agency over the past 20 years, many with the execution support of Bioversity International, have provided a rich body of experience for ensuring the effective conservation and use of NUS. Fourteen projects in 36 countries have been implemented in diverse agricultural landscapes by a wide range of national and international partners, supported by civil society and in collaboration with local communities who continue to maintain and use globally important agricultural biodiversity (Mijatovic et al., 2018). These projects have made a substantial contribution to achieving GEF's mainstreaming strategy as well as contributing to the CBD's Aichi Biodiversity Target 13 (maintenance of genetic diversity of crops, animals and other socio-economically important species).

McKnight Foundation, USA

The goal of its Collaborative Crop Research Programme is to improve access to local, sustainable, nutritious food using collaborative research and knowledge-sharing with smallholder farmers, research institutions and development organizations, working to ensure a world where all have access to nutritious food that is sustainably produced by local people. Website http://www.ccrp.org/ Accessed 5 March 2020

The Deutsche Gesellschaft für Internationale Zusammenarbeit (GIZ), Germany

GIZ is Germany's development agency that provides services in the field of international development cooperation. It funds and/or supports several projects focusing on NUS to strengthen food and nutrition security, income generation and the empowerment of vulnerable groups. https://www.giz.de/en/html/about_giz.html

The Bill and Melinda Gates Foundation, USA

The Foundation supports projects dealing with agrobiodiversity, including NUS, and aims at the conservation of plant genetic resources, and the enhanced use of NUS for improved nutrition and income generation (see A4NH-led Project in Nigeria and Vietnam, the African Orphan Crops Consortium and the Crop Trust). Website: https://www.gatesfoundation.org/

USAID

Actively engaged in fighting hunger at the global level, through its programme "Feed the Future", it covers also the use-enhancement of NUS. One of its projects being implemented in Kenya (FOODSCAP) addressed the shortage of improved crop varieties and plant protection options for orphan crops or open-pollinated varieties. Website: https://www.usaid.gov/what-we-do/agriculture-and-food-security/increasing-food-security-through-feed-future

Winrock International, USA

Committed to empowering the disadvantaged, this organization works on providing solutions to some of the world's most complex social, agricultural and environmental challenges, partnering with communities to develop and implement strategies that boost food production and that are resilient in the face of a changing climate, including NUS as well in these interventions. Website: https://www.winrock.org/about/

Syngenta Foundation for Sustainable Agriculture (SFSA)

It provides financial and technical support to a number of projects focusing on NUS. Website: https://www.ipixel.ch/images/Syngenta_Review_2018.pdf

Networks and consortia

African Orphan Crops Consortium, Kenya

The consortium was established in 2011 to facilitate the genetic improvement of NUS, through the genomic characterization of 101 traditional African local food crops. Its core founding members are the World Agroforestry (ICRAF), Mars Inc., AUDA-NEPAD, the University of California – Davis (USA) and WWF. Website: http://africanorphancrops.org/

All-India Coordinated Research Network on Potential Crops (AICRN on PC), India

Operating under NBPGR (ICAR), New Delhi, this collaborative network has as its main objective the generation of improved technologies in selected crops of minor economic importance for food, fodder and industrial use. It coordinates and conducts research on 17 crops of food, fodder and industrial value through various centres across agro-climatic zones of India. Website: http://www.nbpgr.ernet.in/AICRN_on_PC.aspx

International Society for Horticultural Science (ISHS)

ISHS is actively engaged in promoting knowledge sharing on horticultural crops, including NUS, through conferences and symposia. It has a Work Group on Underutilized Plant Genetic Resources that meets usually during the Symposia on Underutilized Plant Species (see below). Website: https://www.ishs.org/

Millet Network of India (MINI), India

This was established in 2007 to promote community action for the revival of millet-based farming and food systems, placing control of food, seeds, markets and natural resources in the hands of the poor, especially women. Today, it is a large alliance of over 120 members representing over 50 farmer organizations, scientists, nutritionists, civil society groups, media persons and women. They represent over 15 rainfed states of India. Website: https://milletindia.org/

The Forgotten Foods Network, Malaysia

A global initiative launched by CFF that collects and shares information on foods, recipes and traditions in danger of being lost. The objective is to promote the rediscovery of foods that can transform the way people eat and that could better nourish future populations. Website: http://www.forgottenfoodsnetwork.org/

The Asia-Pacific Association of Agricultural Research Institutions (APAARI)

APAARI was established by the FAO in 1990, it is a membership-based, apolitical, multi-stakeholder and inter-governmental regional organization working to bridge gaps between national, regional and global stakeholders to bring about collective change in the agri-food systems of Asia-Pacific. They are involved in numerous initiatives for the promotion of NUS, including dedicated NUS conferences (see Tyagi et al., 2017). Website: https://www.apaari.org/

PELUM Association

This is a network of more than 250 members of civil society organizations/NGOs working with small-scale farmers in East, Central and Southern Africa engaged in promoting participatory ecological land use and management practices including promoting indigenous foods/NUS. Website: https://www.pelum.net/

Participatory Enhancement of Diversity of Genetic Resources in Asia (PEDIGREA)

This is a Southeast Asian initiative focusing on the conservation and use-enhancement of plant and animal genetic resources, including NUS (focus on local vegetables). Website: https://field-indonesia.or.id/en/participatory-enhancement-of-diversity-of-genetic-resources/

Conclusions

The many initiatives listed above are a genuine indication of the fact that concrete interest in NUS has become a reality at both the national and international levels. While this is an excellent development, more should be done to connect isolated efforts to avoid duplications, allow better use of limited funding and build a robust collaborative network at the regional and international levels to bring impact to scale. In particular, more resources should be directed in support of knowledge-sharing events (conferences, trainings) and mechanisms (e.g., one robust internet-based platform) to facilitate access to knowledge, especially from students, practitioners and personnel from GOs or NGOs from less-developed economies that too often are challenged in attending regional/international events focusing on NUS that are most promising for improving the livelihood of communities in their own countries.

Note

1 http://www.fao.org/indigenous-peoples/food-systems/en/

References

Antonelli, A., Fry, C., Smith, R.J., Simmonds, M.S.J. and Kersey, P.J. (2020) *State of the World's Plants and Fungi 2020*. Royal Botanic Gardens, Kew. DOI: 10.34885/172

Dansi, A., Vodouhè, R., Azokpota, P., et al. (2012) Diversity of the neglected and underutilized crop species of importance in Benin. *Scientific World Journal* 2012, 932947. DOI: 10.1100/2012/932947

Ebert, A., Astorga, A., Ebert, I., Mora, A. and Umaña, C. (2007) Securing our future: CATIE's germplasm collections – Asegurando nuestro futuro: Colecciones de germoplasma del CATIE

Gregory, P.J., Mayes, S., Hui, C.H., et al. (2019) Crops For the Future (CFF): An overview of research efforts in the adoption of underutilized species. *Planta* 250, 979–988. DOI: 10.1007/s00425-019-03179-2

Hunter, D., Borelli, T., Beltrame, D., Oliveira, C., Coradin, L., Wasike, V., Mwai, J., Manjella, A., Samarasinghe, G., Madhujith, T., Nadeeshani, H., Tan, A., Tuğrul Ay, S., Güzelsoy, N., Lauridsen, N., Gee, E. and Tartanac, F. (2019) The potential of neglected and underutilized species for improving diets and nutrition. *Planta* 250(3), 709–729. DOI: 10.1007/s00425-019-03169-4

Hunter, D., Borelli, T. and Gee, E. (2020) *Biodiversity, Food and Nutrition: A New Agenda for Sustainable Food Systems*. Issues in Agricultural Biodiversity, Earthscan/Routledge, London.

UKIBPGR (1981) *Revised Priorities among Crops and Regions*. International Board for Plant Genetic Resources, Rome.

Lachat, C., Raneri, J.E., Walker Smith, K., Kolsteren, P., Van Damme, P., Verzelen, K., Penafiel, D., Vanhove, W., Kennedy, G., Hunter, D., Odhiambo, F.O., Ntandou-Bouzitou, G., De Baets, B., Ratnasekera, D., The Ky, H., Remans, R. and Termote, C. (2018) Dietary species richness as a measure of food biodiversity and nutritional quality of diets. *Proceedings of the National Academy of Sciences* 115(1), 127–132.

Maxted, N., Hunter, D. and Ortiz, R.O. (2020) *Plant Genetic Conservation*. Cambridge University Press, Cambridge.

Mbosso, C., Boulay, B., Padulosi, S., Meldrum, G., Mohamadou, Y., Berthe Niang, A., Coulibaly, H., Koreissi, Y. and Sidibé, A. (2020) Fonio and bambara groundnut value chains in Mali: Issues, needs, and opportunities for their sustainable promotion. *Sustainability* 12, 4766.

Mijatovic, D., Sakalian, M. and Hodgkin, T. (2018) *Mainstreaming Biodiversity in Production Landscapes*. United Nations Environment Programme, Washington.

Major NUS actors and ongoing efforts **415**

Nowicki, M. (2019) Empowering communities to use agrobiodiversity for nutritious diets. https://www.bioversityinternational.org/news/detail/empowering-communities-to-use-agrobiodiversity-for-nutritious-diets/ Retrieved on 20 May 2020.

Padulosi, S., Phrang Roy and Rosado-May, F.J. (2019) *Supporting Nutrition Sensitive Agriculture through Neglected and Underutilized Species - Operational Framework*. Bioversity International and IFAD, Rome, Italy. 39 pp. https://cgspace.cgiar.org/handle/10568/102462

Smart Food Executive Council (2019) INSIGHTS. 21 January 2019 https://www.insightsonindia.com/2019/01/21/smart-food-executive-council/ Retrieved on 20 May 2020.

Tyagi, R., Pandey, A., Agrawal, A., Varaprasad, K., Paroda, R. and Khetarpal, R. (2017) *Regional Expert Consultation on Underutilized Crops for Food and Nutritional Security in Asia and the Pacific–Thematic, Strategic Papers and Country Status Reports*. Asia-Pacific Association for Agricultural Research Institutions (APAARI), Bangkok, Thailand.

Ulian, T., Diazgranados, M., Pironon, S., Padulosi, S., Liu, U., Davies, L., Howes, M.-J.R., Borrell, J., Ondo, I., Pérez-Escobar, O.A., Sharrock, S., Ryan, P., Hunter, D., Lee, M.A., Barstow, C., Łuczaj, Ł., Pieroni, A., Cámara-Leret, R., Noorani, A., Mba, C., Womdim, R.N., Muminjanov, H., Antonelli, A., Pritchard, H.W. and Mattana, E. (2020) Unlocking plant resources to support food security and promote sustainable agriculture. *People Plants Planet* 2(5), 421–445. ISSN: 2572–2611.

Ulian, T., Sacande, M., Hudson. A. and Mattana, E. (2016) Plant conservation for the benefit of local communities: The MGU - Useful Plants Project. In: *Botanists of the Twenty First Century: Roles, Challenges and Opportunities*. Based on the Proceedings of UNESCO International Conference, 22–25 September 2014, Paris, France. UNESCO, 2016. pp. 28–34, illus. ISBN 978-92-3-100120-8.

34

IN A WELL-NOURISHED WORLD, UNDERUTILIZED CROPS WILL BE ON THE TABLE

Marco Antonio Rondon and Renaud DePlaen

Ten thousand years of decline

In what could be considered a modern food paradox, current human diets have narrowed to a very small number of food items worldwide, even though globalization and people's mobility should increase the availability and integration of foods from almost anywhere into more diversified diets everywhere. Many factors have led us to this condition. Well before the onset of agriculture, the diet of early humans consisted of a wide variety of fruits, grains, leaves, seeds, nuts, tubers, mushrooms, lichens, flowers and even plant saps, collected over vast territories by communities of hunter gatherers. From the roughly 375,000 plant species known in the world today (Christenhusz and Byng, 2018), it is estimated that between 6,000 and 7,000 are edible for humans (FAO, 2019). Many of them may have been part of the diets of human population around the world. Sedentarization, made possible by the invention of agriculture around 10,000 years ago, brought more predictable and available food sources and accelerated human population growth. But it also had some unintended consequences. It triggered a gradual process of erosion of the number of plant species and food sources that provide the necessary nutrients for human survival. From the vast number of viable food sources, only around 200 plant species were subsequently cultivated, and the continued race for higher yields and selected plant characteristics further reduced that number to about 30 major crops that constitute today the large majority of our plant-based foods. Three of them – maize, rice and wheat – alone account for around 60% of our global energy intake (Seck et al., 2012).

Modern agriculture has undeniably improved food available but it has not yet succeeded in providing sufficient calories and adequate nutrients to everyone on the planet. Despite significant progress over recent decades, the number of undernourished people is growing again, rising from 784 million in 2015 to 820

DOI: 10.4324/9781003044802-40

A well-nourished world and NUS on the table **417**

million in 2018 (FAO et al., 2019). Insufficient food consumption is only part of the challenge; today, more than two billion people, many of them children, do not consume sufficient micronutrients. There is also a realization that the success in increasing agricultural productivity has come at a cost: the progressive reduction in nutrient content for most cultivated species, particularly the major staple foods (Willett et al., 2019). This constant impoverishment of food diversity is masked by the industrialization of food processing, which transforms a small number of plants and ingredients into a multitude of products and presentations, giving the false impression of a great abundance and diversity of food sources. Processed foods are typically high in energy and poor in nutrients. As our consumption of industrialized foods increases, the impact becomes apparent and difficult to ignore. More than two billion people in both developed and developing countries have become overweight or obese. This is now recognized as one of the major challenges to global health that societies have to address (WHO, 2018).

If we want to achieve the UN's Sustainable Development Goals by 2030, business as usual is not an option, and our food systems need a drastic and rapid transformation. Several changes are necessary, and part of the solution is at hand – bringing back more diversity to our diet and reintroducing or revalorizing some of the numerous crops that used be part of our ancestral diets (Mayes et al., 2012). We must use our knowledge on best agronomic practices to increase the production of these crops without losing their nutrient value, and must take advantage of existing and novel food technologies to add value and create new products based on those crop species. In the next sections, we would like to illustrate how this can be achieved.

There is a bright future for underutilized plants species

A growing body of scientific evidence (Ebert, 2014; Nyadanu and Lowor, 2015) shows that traditional but undervalued and underutilized food sources, frequently have a higher nutrient content and are particularly rich in minerals, vitamins and dietary fibre, which are the missing component in many "modern aliments". Recognizing the potential of traditional food, the Agriculture and Food Security Program[1] of the International Development Research Centre in Canada[2] launched, a decade ago, an ambitious research initiative to generate scientific evidence on the potential of underutilized crops for our food systems. The effort targeted communities that still consumed undervalued traditional foods and looked at ways to reintroduce or increase the production and consumption of plant species with high nutrient content from various regions of the world. Two large programs supported this work: the Canadian International Food Security Research Fund,[3] jointly funded by Global Affairs Canada, and the Cultivating Africa's Future program,[4] a joint initiative with the Australian Centre for International Agricultural Research. Results from these programs clearly showed the potential for underutilized crops and the willingness of local populations to produce and consume them, but also demonstrated that they could represent an attractive and viable business model for new agri-entrepreneurs, particularly for the youth and women. Scientific methods documented the potential of indigenous species to contribute to daily diets (as components of new foods or through new presentations of traditional foods), and as new sources of compounds for the pharmaceutical and nutraceutical industry. The following examples illustrate this potential.

Ancient crops, new terrains: In the narrow terraces of Nepal, the availability of arable land and crops species that can thrive on terraces present major constraints. The "Nepal Terrace Farming and Sustainable Agriculture Kits"[5] project tested an innovative public –private model to scale up low-cost and regionally relevant sustainable agriculture kits to address these limitations, reduce drudgery for farmers – the vast majority of whom are women – and increase farmers' income and food security. Implemented as a collaboration between LI-BIRD[6] in Nepal and the University of Guelph in Canada, researchers developed a set of options to make use of the terrace's vertical walls. The toolkit included species of local value such as climbing beans, and also introduced others such as chayote (*Sechium edule* – originally from Central America), and yams (*Dioscorea* spp. from Southeast Asia). Yams were planted in sacks at the base of the walls, which simplified the harvesting of the tubers and provided an important source of income for farmers, while chayote found a niche in local cuisines and as food for small livestock and poultry.

The toolkits included inputs and seeds for growing high-protein legumes and micronutrient-rich vegetables, as well as small technologies such as manual corn shellers and kneepads from which farmers could select based on their preferences, needs and opportunities. The project also took advantage of established local retailers to sell products that farmers identified as needed and for which they were willing to pay. Overall, the project reached more than 60,000 smallholder-farming households in nine districts in Central Nepal, impacting more than 173,000 people. Adoption of the toolkit earned families up to US$200 a season and reduced female drudgery (Chapagain and Raizada, 2017) while at the same time allowing households to significantly increase their dietary diversity. The project proved that when technologies are affordable enough and are properly explained using tools such as picture books (Raizada and Smith, 2019) designed for illiterate farmers, they can be quickly and widely adopted.

Greens to the plate: In Nigeria and Benin, the project "Microveg",[7] run by a consortium of universities from Benin, Nigeria and Canada, reintroduced and promoted several species of indigenous vegetables particularly rich in micronutrients, iron and zinc that were locally produced and consumed but not on a commercial scale. The project identified the best-yielding varieties with high micronutrient content, established seed multiplication systems by farmer cooperatives and trained farmers on the best agronomic practices. Four species of indigenous vegetables were particularly appreciated and were adopted by farmers in the two countries: fluted pumpkin (*Telfairia occidentalis*), African eggplant

(*Solanum macrocarpon*), amaranth *(Amaranthus cruentus)* and African basil (*Ocimum gratissimum*).

The project trained more than 250,000 farmers who increased, by ten-fold, the area cultivated with indigenous vegetables (from 8,090 to 82,000 ha in three years) (Kanyinsola et al., 2018). It increased the demand for indigenous vegetables, developed new products based on these crops such as syrups and bottled beverages, breads and cookies and expanded markets through promotional campaigns. Over 20,000 women were involved in the marketing of indigenous vegetable products. New business, mostly operated by young entrepreneurs from the community were initiated and have continued operating successfully. Novel research identified a variety of polyphenol compounds extracted from these indigenous vegetables, with potential applications in the nutraceutical industry (Moussa et al., 2019), and suggest that they can reduce blood pressure and control diabetes (Olarewaju et al., 2018). The potential of these products is immense and will hopefully spearhead research on other underutilized crops in Africa and beyond (Odunlade et al., 2019). The project organized the first International Conference on indigenous vegetables in Cotonou, Benin in 2017, which led to the publication of a special issue on African vegetables in the *Acta Horticulturae Journal* in 2019.

Pulses once again on the menu: Legume seeds (or pulses) are particularly rich in proteins, iron, zinc and micronutrients and have been an important part of traditional diets. They are not only good for human nutrition and health, but also play a role in improving soil quality, thanks to their symbiotic association with soil microorganisms (*Rhizobia*) that fix nitrogen from the atmosphere, making it available to plants. Legumes have dispersed from their sites of origin to reach almost every place where agriculture is practiced. Common beans from Mesoamerica are now cultivated globally, and lentils, faba beans, and chickpeas from the Middle East have become staple foods in South Asia and are widely consumed worldwide, particularly in developing countries. Despite this apparent success, the true potential of pulses for global food systems is far from being realized.

In Ethiopia, despite recent progress on food security, malnutrition remains a challenge. The levels of stunting ranks among the highest in the world and anaemia affects nearly 37% of children under the age of five. The project "Pulse Innovations: Food and Nutrition Security in Southern Ethiopia"[8] tested and adapted technologies to allow farmers to increase their incomes by planting pulses on land that was often left idle after the cereal harvest. The establishment of women's micro-franchises increased employment for women and their participation in production and marketing, improved household nutrition, and popularized pulse products. Consumers in 15 districts were introduced to ready-to-eat, pulse-rich products and more than 23,000 female-led farm households in 52 villages benefited from recipe demonstrations, complementary food preparation training, improved children feeding practices, and nutrition education (Kabata et al., 2017) and demonstrated that the dietary diversity of children and lactating mothers improved with pulse consumption. A unique partnership between male and female farmers, processors, consumers, universities, and the government transformed subsistence agriculture into a dynamic and market-oriented enterprise. There has been especial interest in expanding the use of the Hawassa Dume seed variety, which produced higher yields, is preferred by consumers, and has proven to be highly resilient to heavy rainfalls and flooding.

Another project on "Scaling Legume Technologies in Tanzania"[9] aimed to introduce soybeans and locally desirable common bean varieties in Tanzania where demand is growing and legumes are fetching good market value. Over a period of two years, an estimated 656,000 farming family members were provided consistent information on improved legume technologies through multimedia campaigns. Six radio series were created and broadcasted nationally, and interactive radio campaigns were launched to target large audiences. Community radio listening groups, with space for listeners to question and discuss information, proved particularly effective at reaching women and youth. A dedicated series of a comic book called *Shujaaz,* developed and distributed nationwide to reach young people in bean-farming families in target areas, reached more than 23% of Tanzanian youth. These multimedia campaigns, complemented by farmer-to-farmer exchanges and demonstration schools, resulted in an estimated 128,589 farming family members adopting at least one of the improved legume technology practices such as improved seeds, row spacing, fertilizing, weeding, and storage (Kansiime et al., 2018). The adoption of soya bean, introduced by the project, led to the rapid increase of production to feed small livestock and poultry and is being progressively integrated into local diets. The international market is booming, and the government is facilitating an increase in national production for export. Soybean is a good example of legumes that can and should play a role in local food systems. The project successfully influenced key policies in Tanzania to speed-up seed varieties registration, cut input costs, and expand community-based seed systems for the adoption of new varieties. The Agricultural Seed Agency (ASA) significantly altered its business model by stocking seeds for both soybean and common bean varieties, and private-sector seed producers are following suit. ASA developed a network that trained 75 agro-dealers on input business management.

The declaration of 2016 as the "international year of pulses" by the FAO brought to the forefront the potential of underutilized species, contributed to their dissemination, and opened new opportunities for farmers in many regions. The International Development Research Centre (IDRC) and its partners worked with a high-level group of public- and private-sector experts to develop an internationally coordinated pulse crop productivity and sustainability research strategy for the following ten years.[10]

Finger millets, from obscurity to supermarkets: Finger millet was domesticated more than 5,000 years ago in the highlands of Ethiopia. It spread within Africa and rapidly reached northern India, where it has been part of traditional dishes for centuries. The manual processing of small millets is tedious and labour intensive, which contributed to a rapid decline in their consumption. Lower yields compared to other cereals, weak supply chains, poor consumer awareness, inadequate or inefficient processing facilities, and policy neglect led to the quasi disappearance of finger millet from markets. Implemented by the Tamil Nadu Agricultural University, the DHEA Foundation in India, and the University of Toronto, the project "Scaling up small millet production and consumption in India",[11] developed and reintroduced improved higher-yield varieties into a large number of farms in northern India. It also developed more efficient and user-friendly equipment for de-hulling and processing small millets, which increased production and reduced women's drudgery. Increasing equipment manufacturers' capacity to produce and sell more de-hulling machines supported the development of decentralized small millet processing infrastructures in eastern and central India. Over the course of the project, 72,490 people, mostly women, farmers, and schoolchildren, were trained on the health benefits of small millets (Adekunle et al., 2018) and more than 200,000 people learned about the values of small millets through programs aired on community radio and local TV, and through text and audio messages.

A wide range of small millet food products such as cookies, flour, chips and noodles were developed collaboratively with women farmers; the project worked with 66 food enterprises, 152 pushcart millet-porridge vendors, four Farmer Producer Organizations, and 15 non-governmental organizations to expand the market for ready-to-eat small millet products. Private-sector involvement was instrumental in the marketing of these new products and close coordination with local governments facilitated seed production and distribution systems, making it possible for finger millets to be sold, once again, in most markets and supermarkets in northern India. Small millets are being introduced through public food programs in India and there are opportunities to scale up the experience in other countries in South Asia and Africa, where the demand for drought-resistant varieties is growing and where millet is slowly being reintroduced into local diets and avant-garde cuisines in urban centres.

A new time for traditional cereals: Sorghum is a staple food consumed by more than 60 million people in Ethiopia. The crop is also a major source of animal feed, fuel, and building material. However, sorghum production is risky and the increased frequency and severity of droughts compound the challenges faced by Ethiopian farmers, exposing them to food shortages and livestock losses due to a lack of feed. Currently, 70% of sorghum grain is consumed domestically, with women providing most of the labour and trading for the crop. The project "Climate-smart interventions for smallholder farmers in Ethiopia"[12] is trying to revalorize the production and consumption of sorghum through the development and promotion of improved and drought-tolerant sorghum varieties and post-harvest management technologies, and the development of new value chains. The project will improve productivity and climate resilience for 240,000 smallholder sorghum farmers, reduce post-harvest loss through farm-scale grain storage options, and increase economic opportunities for women through

422 Marco Antonio Rondon and Renaud DePlaen

value-added products, small-scale threshers, and improved storage facilities. It is implemented by the Ethiopian Agricultural Research Institute.

An ancestral taste from the Andes going global: Even though quinoa (*Chenopodium quinoa*) is increasingly marketed in Western countries as a "superfood", it has been, for millennia, the main staple food of the Indigenous Peoples of the highland Andes in Bolivia and Peru. It is a robust plant that can grow at high elevations, tolerate extreme temperature fluctuations, and adapt to dry zones and even saline soils. Appetite for this gluten-free cereal with a high nutritional value is growing rapidly among the affluent consumers in Europe, North America, and Asia. Limited research investments to improve varieties and agronomic practices has led to the tripling of yields and quinoa has made rapid strides to become a commercially significant crop in a relatively short time. Quinoa production increased globally ten-fold in one decade and the cereal is now grown in more than 70 countries. Given its remarkable qualities, it constitutes a viable alternative for many regions of the world (Bastidas et al., 2016), especially where water is in short supply and where soils are depleted. The project "Scaling up Quinoa Value Chain to Improve Food and Nutritional Security in rural Morocco", implemented by the International Centre for Biosaline Agriculture, brought experts in quinoa production from the Andes to select and introduce best varieties and agronomic practices in Morocco. Although the project is still ongoing, it has generated strong interest from farmer cooperatives and local governments, which are promoting its integration into local diets and developing the value chains necessary to supply a growing external market.

Following efforts from the Government of Bolivia, supported by IDRC, 2013 was declared by the UN as the "International Year of Quinoa". That helped to bring together different sectors and stakeholders, farmers, governments, scientists, and the private sector and contributed to advancing the crop internationally. The example of quinoa illustrates how underutilized species can evolve from being forgotten and underutilized to becoming important staple crops beyond their homelands.

Towards more robust global food systems

There is a renewed interest worldwide in expanding and diversifying food sources. The body of scientific literature on underutilized and undervalued food crop species is slowly growing and receiving international attention, as illustrated by publications on quinoa, which have multiplied ten-fold in the last five years (Inglese et al., 2017; FAO, 2018). Attention to underutilized species is growing in many countries. Quinoa and lentils have been prioritized in the National Food Security Plans of the Government of Bhutan, and taro and moringa by several countries in South and Southeast Asia (Li and Siddique, 2018).

The hype about "superfoods" and a new generation of food-conscious consumers represent an opportunity to bring back some of the great diversity of colours, shapes, and flavours that humanity once consumed. To move forward, we urgently need to preserve germplasms and select high-yield and high-nutrient-content varieties that can adapt to diverse conditions and locations. We must also change the common misperception that these species are "poor people's crops". Local knowledge is rapidly eroding and urgent efforts are required to recover and document traditional knowledge and practices regarding underutilized species. Production and post-harvest practices need to be modernized, value chains for added value products developed, and resources made available to enhance the capacity of researchers, policymakers, extension agents, farmers and farming cooperatives, and especially the private sector to realize the full potential of underutilized crops.

The resilience of food systems and their capacity to provide sufficient healthy food to a growing population require the inclusion of a greater diversity of crops, many of which are currently undervalued. Many cultures around the world have successfully preserved the culinary knowledge, cultural value, and cultivation and processing techniques for underutilized crops over time; it is now time for science to support their efforts and contribute to the improvement of desired traits and the enhancement of productivity and income-generation opportunities. Underutilized crop species have allowed ancient civilizations to flourish and can, once again, regularly feature in our diets and lead to the development of new food preparations and even new medicines. The IDRC is committed to continuing its efforts to rescue and revitalize underutilized crops and alternative food practices as a key contribution to the development of equitable and resilient food systems.

Notes

1 Agriculture and Food Security Program, https://www.idrc.ca/en/program/agriculture-and-food-security
2 The International Development Research Centre (IDRC), is a crown corporation from the Government of Canada whose mission is to fund and to support research in developing countries to promote growth, reduce poverty, and drive large-scale positive change. https://www.idrc.ca/en
3 The Canadian International Food Security Research Fund (CIFSRF), is a joint initiative between Global Affairs Canada and IDRC to accelerate research and development funding by supporting practical innovations that directly improve the lives of the poor and food insecure. https://www.idrc.ca/en/news/celebrating-results-and-promoting-partnerships
4 Cultivating Africa's Future program is a ten-year partnership between IDRC and the Australian Centre for International Agricultural Research (ACIAR). The partnership aims to improve food and nutrition security, resilience, and gender equality across eastern and southern Africa. CultiAf funds applied research to develop and scale up sustainable, climate-resilient, and gender-responsive innovations for smallholder agricultural producers. https://www.idrc.ca/en/initiative/cultivate-africas-future
5 For additional information about the "Nepal Terrace Farming and Sustainable Agriculture Kits" project, including video products and publications, visit the project website at https://www.saknepal.org/
6 LI-BIRD, "Local Initiatives for Biodiversity, Research and Development," is a non-profit, non-governmental organisation. Li-BIRD is committed to capitalizing on

local resources, innovations, and institutions for sustainable management of natural resources for improving livelihoods of smallholder farmers in Nepal. http://www.libird.org/

7 The project "Microveg" was implemented by a multidisciplinary research team from Université de Parakou in Benin, Osun State University and Obafemi Awolowo University in Nigeria, and the Universities of Saskatchewan and Manitoba in Canada, and the Green Generation Initiative. The project operated in Nigeria and Benin. http:/www.microveg.org

8 Pulse Innovations: Food and Nutrition Security in Southern Ethiopia. A research project implemented by Farm Radio International in collaboration with Hawassa University, the University of Saskatchewan, and local radio stations in Ethiopia including Fana FM Asela, Sidama Community Radio, Wolaitie FM, Woylaita Sodo. https://farmradio.org/publications/final-technical-report-scaling-up-pulse-innovations-for-nutrition-security-in-southern-ethiopia/

9 Scaling of Legume Technologies in Tanzania. Joint research project implemented by CABI international, Farm Radio International Tanzania and The African Fertilizer and Agri-business Partnership (AFAP) https://idl-bnc-idrc.dspacedirect.org/bitstream/handle/10625/58550/IDL%20-%2058550.pdf?sequence=2&isAllowed=y http://africasoilhealth.cabi.org/about-ashc/ashc/silt/

10 Information about the "Pulse crop productivity and sustainability research strategy for the next 10 years" and additional information about pulses in general can be found at the website https://iyp2016.org/

11 Scaling up small millet production and consumption in India https://www.dhan.org/smallmillets2/project-2.html

12 Climate-smart interventions for smallholder farmers in Ethiopia https://www.idrc.ca/en/research-in-action/climate-smart-interventions-smallholder-farmers-ethiopia

Disclaimer:

The views and opinions included in this chapter reflect those of the authors and do not necessarily correspond to the institutional perspectives of the IDRC.

References

Adekunle, A., Lyew, D., Orsat, V. and Vijaya, G. S. (2018) 'Helping Agribusinesses— Small Millets Value Chain—To Grow in India', *Agriculture* 8(3), March 2018. DOI: 10.3390/agriculture8030044

Bastidas, E.G., Roura, R. Rizzolo, D.A., Massanés, T. and Gomis, R. (2016) 'Quinoa (*Chenopodium Quinoa* Willd.), from Nutritional Value to Potential Health Benefits: An Integrative Review', *Journal of Nutrition & Food Sciences* 6(3), March 2016 http://hdl.handle.net/2445/109262

Chapagain, T. and Raizada, M. (2017) 'Agronomic Challenges and Opportunities for Smallholder Terrace Agriculture in Developing Countries', *Frontier Plant Science* 8, 331. DOI: 10.3389/fpls.2017.00331 2017

Christenhusz, M. and Byng, J. (2018) 'The Number of Known Plants Species in the World and Its Annual Increase', *Phytotaxa* 261(3). DOI: 10.11646/phytotaxa.261.3.1

Ebert, A.W. (2014) 'Potential of Underutilized Traditional Vegetables and Legume Crops to Contribute to Food and Nutritional Security, Income and More Sustainable Production Systems', *Sustainability* 6, 319–335. DOI: 10.3390/su6010319

Food and Agriculture Organization FAO (2018) 'More Super Crops with Strong Nutritional Properties'. http://www.fao.org/zhc/detail-events/en/c/356766/

Food and Agriculture Organization FAO (2019) 'The State of the World's Biodiversity for Food and Agriculture', *FAO Commission on Genetic Resources for Food and Agriculture Assessments*. Bélanger, J. and Pilling, D. (eds.), Rome. 572 pp. http://www.fao.org/3/CA3129EN/CA3129EN.pdf

Food and Agriculture Organization FAO, World Health Organization WHO, World Food Program WFP, International Fund for Agriculture Development IFAD. (2019) 'The State of Food Security and Nutrition in the World'. http://www.fao.org/3/i9553en/i9553en.pdf

Inglese, P., Mondragon, C., Nefzaoui, A., Saenz, C., Taguchi, M., Makkar., H. and Louhaichi, M. (2017) *Crop Ecology, Cultivation and Uses of Cactus Pear*. Food and Agriculture Organization (FAO) Rome, Italy. October 2017, http://www.fao.org/3/a-i7012e.pdf

Kabata, A., Henry, C., Moges, D., Kebebu, A., Whiting, S., Regassa, N. and Tyler, R. (2017) 'Determinants and Constraints of Pulse Production and Consumption among Farming Households of Ethiopia', *Journal of Food Research* 6(1). DOI: 10.5539/jfr.v6n1p41

Kansiime, M., Watiti, J., Mchana, A., Jumah, R., Musebe, R. and Rware, H. (2018) 'Achieving Scale of Farmer Reach with Improved Common Bean Technologies: The Role of Village-Based Advisors', *Journal of Agricultural Education and Extension*. DOI: 10.1080/1389224X.2018.1432495

Kanyinsola, O.Y., Oluwasola, O. and Babatunde, A. (2018) 'Information Use for Marketing Efficiency of Underutilized Indigenous Vegetables', *International Journal of Vegetable Science* 25(2), 138–145. DOI: 10.1080/19315260.2018.1487497

Li, X. and Siddique, K.H.M. (Ed) (2018) *Future Smart Food - Rediscovering Hidden Treasures of Neglected and Underutilized Species for Zero Hunger in Asia*. FAO, Bangkok, 242 pp. http://www.fao.org/3/I9136EN/i9136en.pdf

Mayes, S., Massawe, F. J., Alderson, P. G., Roberts, J. A., Azam-Ali, S. N. and Hermann, M. (2012) 'The Potential for Underutilized Crops to Improve Security of Food Production', *Journal of Experimental Botany* 63(3), 1075–1079. DOI: 10.1093/jxb/err396

Moussa, D., Alashi, M.I., Sossa-Vihotogbe, A.M., Akponikpe, C.N.A., Djenontin, P.B.I., Baco, A.J., Akissoé, M.N. and Aluko, R.E. (2019) 'Inhibition of Renin-Angiotensin System Enzymes by Leafy Vegetables Polyphenol Extracts Related to Fertilizer Micro-Dosing and Harvest Time', *Acta Horticulturae* 1238, 73–80. DOI: 10.17660/ActaHortic.2019.1238.9

Nyadanu, D. and Lowor, S.T. (2015), 'Promoting Competitiveness of Neglected and Underutilized Crop Species: Comparative Analysis of Nutritional Composition of Indigenous and Exotic Leafy and Fruit Vegetables in Ghana', *Genetic Resources and Crop Evolution* 62, 131–140. DOI: 10.1007/s10722-014-0162-x

Odunlade, T.V., Famuwagun, A.A., Orekoya, O.O., Taiwo, K.A., Gbadamosi, S.O., Oyedele, D.J., Aluko, R.E. and Adebooye, O.C. (2019) 'Development of Maize Ogi Powder Fortified with Polyphenolic Rich Extract from Fluted Pumpkin Leaves', *Acta Horticulturae* 1238, 93–104. DOI: 10.17660/ActaHortic.2019.1238.11

Olarewaju, O. A., Alashi, A. M. and Aluko, R. E. (2018) 'Antihypertensive Effect of Aqueous Polyphenol Extracts of Amaranthus Viridis and Telfairia Occidentalis Leaves in Spontaneously Hypertensive Rats', *Journal of Food Bioactives* 1(1), 166–173. DOI: 10.31665/JFB.2018.1135

Raizada, M. and Smith, L. (2019) 'Picture Book of Best Practices for Subsistence Farmers', University of Guelph, Canada 2019. http://www.sakbooks.com/, http://hdl.handle.net/10625/57569

Seck, P.A., Diagne, A., Mohanty, S., et al. (2012) 'Crops That Feed the World 7: Rice', *Food Security* 4, 7–24. DOI: 10.1007/s12571-012-0168-1

Willett, J.R., Loken, B., Springmann, M., et al. (2019) 'Food in the Anthropocene: The EAT–Lancet Commission on Healthy Diets from Sustainable Food Systems', *The Lancet Commissions* 393(10170), 447–492. February 02, 2019. DOI: 10.1016/S0140-6736(18)31788-4W

World Health Organization (2018) 'Obesity and Overweight'. www.who.int/news-room/fact-sheets/detail/obesity-and-overweight

INDEX

Note: **Bold** page numbers refer to tables; *italic* page numbers refer to figures and page numbers followed by "n" denote endnotes.

abiotic stresses 7, 21, 76, 227–228, 230, 320
açaí (*Euterpe oleracea*) 105
acerola (*Malpighia glabra*) 105
Acta Horticulturae Journal 419
Action for Social Advancement (ASA), India 406
African basil (*Ocimum gratissimum*) 419
African eggplant (*Solanum aethiopicum*) 78
African eggplant (*Solanum macrocarpon*) 418–419
African food systems 78–79, 85
African Forest Ltd., Kenya 352–354; about work 352; main obstacles in promoting NUS 353; private-sector involvement in value chains for NUS 352–353; some recommended solutions 353–354
African leafy vegetables (ALVs) 23, 81–82, 403, 410
African locust bean (*Parkia biglobosa*) 78
African neglected and underutilized species: characterization 79–81; discovery 79; domestication and improvement efforts 83–84; food system diversification with NUS 81–83; overview 78–79; trees on farms-promoting diversity 83
African Orphan Crops Consortium (AOCC) 84, 396, 412; mandate tree species list for domestication and improvement **84**
African plum (*Prunus africana*) 396

African Seed Trade Association 211
African vegetables: germplasm conserved to support research and healthy diets 209–210; germplasm enhancement for increased nutrition 210–211; germplasm enhancement for increased productivity 210–211; improving diets and livelihoods 208–214; overview 208; reflection on the way forward for 213–214; safeguarding traditional vegetable biodiversity 208–209; sustainable production 211–212; TAV consumption 212–213
Africa Vegetable Breeding Consortium (AVBC) 211, 214
Agenzia Lucana di Sviluppo ed Innovazione in Agricoltura (ALSIA) 383
agrarian system, evolution of 238–240
agricultural biodiversity 3; CBD programme of work on 22; conservation of 28, 397–398; and MSSRF 26
agriculture: European 138, 141–142, 144; green 68; hill and tribal 71; Indian 73; modern 416–417; nutrition-sensitive 43–55, 319; research 309; subsistence 219, 338, 420; 'sustainable intensification' of 14; traditional 379
Agriculture and Food Security Program 423n1
agritourism: long term sustainability of 206; in Santiago de Okola, Bolivia 198–206

428 Index

agrobiodiversity 58; Andean 199, 201, 204; and FGDs 176; household 202; leveraging local 53; loss of 198–199
agroecological farming methods 277
agroecology-based intercropping 273
agronomic practices 266
agronomic traits 229
Agronomy for Sustainable Development (Kahane) 28
Agropecuária Aruanã Ltda., Brazil 360–361; case of Brazil nut 360–361
Ahir, Maganbhai 340
Aichi Biodiversity Target 13 12
Alliance of Bioversity International 395–396
All-India Coordinated Millets Improvement Project (AICMIP) 71; crop improvement efforts during era of 74
All-India Coordinated Research Network on Potential Crops (AICRN on PC), India 412
All-India Coordinated Research Project on Underutilized Plants 21
All-India Coordinated Small Millet Improvement Project (AICSMIP) 228
alternative seed systems 252–253
amaranth (*Amaranthus cruentus*) 13, 419
Amazonian Brazil-nut tree (*Bertholletia excelsa*) 102
ambassadors and champions 330–331
Anamolbiu, Nepal 408–409
Anamol Biu Pvt. Ltd., Nepal 354–356; NUS work 354–356; views about NUS 354
andiroba (*Carapa guianensis*) 105
Anmol Desi Beej Sanvardhan Khedut Sangathan (seed saver farmer association): interventions for effective promotion 341; major obstacles for NUS enhanced use 341; perspective of 340–341; views with regard to NUS 340–341
arbuscular mycorrhizal fungi (AMF) 273, 275–276
Archeologia Arborea, Italy 406
Ark of Taste 217–220
arrowroot (*Tacca leontopetaloides*) 150
Asian Development Bank 22
Asian Vegetable Research and Development Center (AVR DC -The World Vegetable Center) 21
Asia-Pacific Association of Agricultural Research Institutions (APAARI) 31, 413
Asociación de Comunidades Forestales de Petén 104

assessment: baseline 176; of genetic diversity 111; of seasonal food availability 175–176
aupa (*Amaranthus* spp.) 150
Australian Centre for International Agricultural Research (ACIAR) 31, 171n1, 409–410, 417, 423n4

Bai, Indra 339
Bai, Meera 339
Bai, Phool 339
Bai, Ramkali 339
baking 314
Balakrishnan, R. 264
Bambara groundnut (*Vigna subterranea*) 13, 20
BAMNET (International Bambara Groundnut Network) 22
bananas (*Musa* spp.) 88, 150
baobab (*Adansonia digitata*) 13, 78
barnyard millet (*Echinochloa crus-gali*) 238
barriers: to adopting diversity of NUS fruit trees 88–107; in consumer demand and marketability 103; from farm to market 102–103; at production stage 89–102; at value-chain stage 102–103
beans (*Phaseolus vulgaris, P. coccineus, P. acutifolius*) 118
bele/aibika (*Abelmoschus manihot*) 150
beneficial microbes 273–278; bioirrigation facilitated by 275
benefit sharing 122, 344, 384n2
BFN mainstreaming toolkit 163–171; development of 164; importance of NUS for food and nutrition change 166–167; increasing production, marketing and awareness of NUS 168–170; monitoring and evaluating mainstreaming of NUS 171; offering 164–166; overview 163–164; partners and entry points for NUS mainstreaming 167–168
BFN project: indigenous NUS that were targeted by *166*; three-pillar approach adopted by *165*
BFN Turkey: sustainability index developed by **169**
El Bilali, Hamid 374
Bilgi Sistem Yönetimi – İthalat & İhracat, Turkey 357–358; major obstacles for use enhancement of NUS 358; solutions for effective promotion 358; views about NUS 357
Bill and Melinda Gates Foundation, USA 411

Index **429**

biodiversity: and diverse land use 63–64; people's interconnected sense of *64*; traditional vegetable 208–209
Biodiversity for Food and Nutrition platform 401
Biodiversity for Food and Nutrition (BFN) project 163, 401
Biodiversity Mainstreaming for Healthy & Sustainable Food Systems Toolkit 163
biofortified varieties 332–333
bioirrigation: definition and concept 274–275; facilitated by beneficial microbes 275; scheme depicting *276*
bioirrigation-based intercropping systems: bioirrigation (definition and concept) 274–275; facilitated by microbes 274–277; finger millet/pigeon pea intercropping system 275–277; technology adoption and upscaling potential of 277
biotic stresses 7, 21, 76, 229–230, 320, 381
Bioversity International (BI) 53, 54, 80, 186, 199, 373
black gram (*Vigna mungo*) 169
Botanic Gardens Conservation International 400
Brazil: conservation and promotion of NUS in 130–136; final considerations 136; historical context 129–130; NUS traditionally used in **133**; overview 128–129
Brazilian Biodiversity: tastes and flavours (Santiago de Andrade Cardoso and Coradin) 132, 168
Brazilian Regional Foods (Brasil) 135
Brazil nut 360–361
breadfruit (*Artocarpus altilis*) 150
breeding advancements 231
buckwheat: benefit improvement for increasing farmer's incomes 114; biodiversity 109–110; conservation and identification 113–114; current status of R&D in China 110–113; future action plans for buckwheat R&D 114–116; gaps and opportunities for promoting 113–114; landrace diversity 110; national plans and policies related to 113; nutritional and health value for marketing 114; overview 109; promoting 109–116; species diversity 109–110; supporting policies for sustainable development of 114

buckwheat biodiversity: conservation and improvement of 114–115; conservation of 110–111
buckwheat R&D: assessment of genetic diversity 111; conservation of biodiversity 110–111; current status of 110–113; enhancing external environment for 115–116; future action plans for 114–116; improvement of varieties 111; national plans and policies 113; process and value-chain 112–113; processing and value-chain of nutritional and healthy diets 115; production and farming system 112; promoting farming practices 115; strengthening conservation and improvement of biodiversity 114–115
bunya nut (*Araucaria bidwillii*) 155
burití (*Mauritia flexuosa*) 105, *167*
bush banana (*Leichhardtia australia*) 155
Bushfires and Bushtucker: Aboriginal Plant Use in Central Australia (Latz) 155
bush tomato (*Solanum centrale*) 155
Bush Tucker: Australia's Wild Food Harvest (Low) 154
Bussey, C. 149, 157

cacao (*Theobroma cacao*) 88
cagaita (*Eugenia dysenterica*) *167*
camu camu (*Myrciaria dubia*) 102, *167*
Canadian International Food Security Research Fund (CIFSRF) 417, 423n3
cañahua (*Chenopodium pallidicaule*) 198
capacity building 22, 44, 51–52, 104, 132, 247–248, 344
Cascudo, L. C. 129
case studies: alternative seed systems in Koraput, Odisha 252–253; collective marketing in Mandla and Dindori, Madhya Pradesh 256; DK Engineering Co. Ltd., Nairobi, Kenya 194; integrated value-chain development in Kolli Hills, Tamil Nadu 246–251; Kieru Foods Ltd., Embu, Kenya 186–193; value-chain development for small millets in Tamil Nadu 253–255; value-chain development of millet in Meghalaya 256–260
cassava (*Manihot esculenta*) 78
Centre de coopération internationale en recherche agronomique pour le développement (CIRAD) 398
Centre for Agriculture and Bioscience International (CABI) 406

430 Index

Centre for Pacific Crops and Trees (CePaCT) 400
Centro Agronomico Tropical de Investigacion y Ensenanza, Costa Rica 398
CGIAR Consortium 3
CGIAR Research Program on Agriculture for Nutrition and Health 171n1
characterization: African neglected and underutilized species 79–81
charichuelo (*Garcinia madruno*) 105
chaya (*Cnidoscolus aconitifolius*) 13
chefs: promotion through 330; Smart Food Triple Bottom Line approach 330
chilies (*Capsicum annuum*) 118
China: current status of buckwheat R&D in 110–113; national plans and policies related to buckwheat 113; supporting policies for sustainable development of buckwheat in 114
Chinese Academy of Agricultural Sciences (CAAS) 110–111; National Crop Genebank (NCG) at 110–111
CIHEAM-Bari 373–375, 376
CIHEAM Strategic Agenda 2025 374
Clarke, P.A. 155
climate change 4, 7, 12, 46, 48, 51, 64, 66, 76, 199, 217, 246, 248, 256, 260, 352, 361–362, 373, 375–377, 388, 406; accelerating 178; coping strategies 32; dramatic 14; and millets 295–297, 305; pervasive 119, 147; and UN Agenda 2030 31
climate-resilient farming 273–278
cold extruder *288*
cold extrusion technology 285–286, 314
Commission on Genetic Resources for Food and Agriculture (CGRFA) 53
Committee on World Food Security (CFS) *Voluntary Guidelines on Food Systems for Nutrition* 171
commodity crops: comparing benefits from NUS and **10–11**; and holistic value-chain approach 44
common mycorrhizal network (CMN) 273, 275–276
common wheat (*Triticum aestivum*) 141
community-based approaches 81, 399
community-based organizations (CBOs) 344
community-based tourism (CBT) 199–200
community-centred value-chain development of nutri-millets 245–261
Community Seed Banks (CSBs) 354–355, 381, 383–384

Conference of the Parties to the Convention 21
Connecting Global Priorities: Biodiversity and Human Health (WHO/CBD) 54
conservation: of buckwheat biodiversity 110–111; of Indigenous crops by custodian farmers in Uttarakhand 340; millets 242; millets in ecologically sensitive areas 237–243; of NUS in Brazil 130–136; of NUS in Mexico 121–122; of small millets 228; of small millets genetic resources 227–233
Consultative Group on International Agricultural Research (CGIAR) 19, 228, 395
consumer demand, barriers in 103
consumers: as priority 330; up-scaling NUS access to 105–106
Convention on Biological Diversity (CBD) 21, 53, 131–132; Conference of the Parties (COP) 53; Framework for a Cross-Cutting Initiative on Biodiversity for Food and Nutrition 53; programme of work on agricultural biodiversity 22
cookie-cutting machine *289*
Cordoba Declaration 27
Corrêa, Pio 130
COVID-19 pandemic 14, 43, 47, 50, 55, 135, 163, 178, 202, 210, 324, 342, 346, 349, 387; food and nutrition insecurity during 387; and Kieru Foods Ltd., Embu, Kenya 191–192; lockdown period due to 342
cowpea (*Vigna unguiculata*) 78
crop diversification in Europe 138–144
crop improvement: efforts during AICMIP era 74; efforts during early years 73; progress in 73
crop production and protection technologies 75
Crops for the Future (CFF) 27, 396, 401
Crops of the Future Collaborative (COTF), USA 406
CSC/ICAR Delhi International Workshop 20
Cultivating Africa's Future program 417, 423n4
cultural erosion 8
custodian farmers 28, 50, 247, *249, 251,* 256–257, 337–343
custodian farmers of Attappdi, Kerala: major obstacles and suggested solutions 341–342; perspectives on NUS 341–342

custodian farmers of Meghalaya, India: interventions for effective promotion 343; major obstacles for NUS enhanced use 342–343; perspectives from 342–343; wild edible NUS 342

custodians of NUS 337–343

custom hiring centres 266

cycads (*Cycas media*) 155

Damon, Jacqueline 352

Das, Biswa Sankar 339

Dass, A. 274

Davy, D. 149

decentralised processing facilities 265

Department for International Development (DFID) 21

desert kurrajong (*Brachychiton gregorii*) 155

desert quandong (*Santalum acuminatum*) 155

The Deutsche Gesellschaft für Internationale Zusammenarbeit (GIZ), Germany 398–399, 411

Devi, Aiti 340

Devi, Kamla 340

Devi, Sangeeta 340

Devi, Shalini 340

Devi, Usha 340

dextrins 315

DHAN Foundation 253–254

Dictionary of Useful Plants of Brazil and Cultivated Exotic Plants (Corrêa) 130

dietary and lifestyle change: Oceania 148–149

dietary diversity: and diverse land use 63–64; people's interconnected sense of 64

Dietary Guidelines for the Brazilian Population (Brasil) 135

diets: African vegetables improving 208–214; healthy 209–210; seasonal calendars and diversification of 178

diploid einkorn (*Triticum monococcum*) 141

Diplotaxis spp. 9

discovery: African neglected and underutilized species 79

diverse land use: and biodiversity 63–64; and dietary diversity 63–64

diverse seed centres, establishing 266

DIVERSIFOOD project 140, 142, 143–144

diversity: genetic 111; landrace 110; promoting 83; and small millets in India 68–77; species 109–110; trees on farms 83

divisions of labour 391

DK Engineering Co. Ltd., Nairobi, Kenya 184–186, 194, 347–348; obstacles in enhancing use of NUS 347–348; recommendations for enhancing use of NUS 347–348

domestication and improvement efforts: African neglected and underutilized species 83–84

donors 409–412

durum wheat (*Triticum durum*) 141

eastern Madhya Pradesh: area planted to small millets 303; food security role of millets under climate change in 295–307; percentage yield increases 304; regression of yields by precipitation 302–304; varying yields for good and bad years 303

Eat the Weeds 401

ECHO Community, USA 407

ecologically sensitive areas, conserving millets in 237–243

ecosystems: diverse range of 147; diversified 178; dryland 273; natural 237, 242; seasonal calendars and diversification of 178; social 237, 242

edge runner machine *287*

edible flowers and/or fruits of agave and cacti 119–120

einkorn wheat (*Triticum monococcum*) 169

Ejere Farmer Crop Conservation Association 383–384

elite cultivars and germplasm improvement 230–231

empowerment of vulnerable groups 52–53

endangered food products: promoting 216–223; protecting 216–223

endangered products: defining 217; locating 218; promoting 218–220; protecting 218–220

Englberger, L. 149

entrepreneurship 49, 51, 184, 345

Eruca sativa 9

Eser, Serkan 357

ethanol 315

Ethiopian Agricultural Research Institute 422

ethnobotanical surveys 79

ethnobotany 79

eucalyptus plantations 240–241

EU H2020 Framework Programme (2014–2020) 140

Europe: crop diversification in 138–144; diversifying genetic structure of

432 Index

major and minor species 142–143; rediscovering "forgotten" species 141–142; shifting cultivation areas 140–141; underutilized crops in 140; underutilized genetic resources in 138–144
European Commission 22
European Search Catalogue for Plant Genetic Resources (EURISCO) database 142
European Union (EU), Belgium 410
Eurotropics S.A., Guatemala 348–349; major obstacles in use enhancement of NUS 348–349; solutions 349; views about NUS 348
evolution of the agrarian system 238–240
extension work 52–53, 82, 375
extension workers 82, 375

Fagopyrum 109–110
farmer producer organisations 267
farm-facing solutions and Latin American food systems 103–104
farming: climate-resilient 273–278; sustainable 273–278
farming communities 337–343
Farming System for Nutrition (FSN) 319–320; millets under 320–322
Federated States of Micronesia 156–157
fermentation 314
fern (*Diplazium* spp.) 150
fig tree (*Ficus tinctoria*) 150
finger lime (*Citrus australasica*) 155
finger millet (*Eleusine coracana*) 78, 238, 257, 263, 320; integrated value-chain development of 270; procurement of 267
finger millet/pigeon pea intercropping system 275–277
flaking technology 285, 313
fluted pumpkin (*Telfairia occidentalis*) 418
focus group discussions (FGDs) 175–176
fonio (*Digitaria exilis*) 13, 78
Food and Agriculture Organization (FAO) 171n1, 373, 397; FAOSTAT 8; International Conference and Program for Plant Genetic Resources 23; International Treaty for PGRFA 31; International Treaty on Plant Genetic Resources 23; regional meeting held in Bangkok in 2016 32–33; Second Global Plan of Action on PGRFA 12; Second State of the World Report on PGRFA 27; State of the World Report 23; *Voluntary Guidelines for Mainstreaming Biodiversity into Policies, Programmes and*

National and Regional Plans of Action on Nutrition 171
food culture 7, 130, 188, 216, 238, 245, 342, 351, 358, 399–400
food diversity: loss of 148–149; Oceania 148–149
Food Forever Initiative 397
Food Plant Solutions, Australia 402
food security 327; millets and 70; role of millets under climate change in Madhya Pradesh 295–307
'Food System Divide' 328
food systems: African 78–79, 85; and African traditional vegetables 213–214; diversification with NUS 81–83; global 4, 216, 222, 419, 422–423; globalization of 198; Latin American 89–106; nutrition-sensitive agricultural 43; roadmap to using neglected and underutilized species for 163–171; Smart Food Triple Bottom Line approach 327–328; sustainability of 178; sustainable 155–157; traditional 63, 148, 216; and traditional vegetable biodiversity 208–209; transforming 54, 271
Food Tank 399
football fruit (*Pangium edule*) 150
The Forgotten Foods Network, Malaysia 413
"forgotten" species 141–142; rediscovering 141–142
Foundation for Food and Agriculture Research 406
foxtail lily (*Eremurus spectabilis*) 169
foxtail millet (*Setaria italica*) 71; recipes 269
frameworks enhancing mainstreaming of NUS value-chains 53–54
Fruit and Vegetable Research Institute (FAVRI) (Vietnam) 404
fruits: of agave 119–120; of cacti 119–120; of trees and shrubs 120–121
fumbwa (*Gnetum africanum*) 78
Fundación Promocion e Investigacion de Productos Andinos (PROINPA), Bolivia 407
funding 19, 22, 171, 357, 410, 413
"Future Smart Food" (FSF) 31

Galip nut (*Canarium indicum*) 150
gaps and opportunities for promoting buckwheat 113–114
Gaud, William 15n1
GEF-funded Biodiversity for Food and Nutrition Project 53

GEF 'Mainstreaming biodiversity for nutrition and health' project 171n1
gender: agricultural linkages with 319; equality 388; equity 388; identity 389; roles 389
gender relations and social equity 388–389
gene bank 23, 82, 228, 400, 408; *ex situ* 8, 43; ICRISAT 228; USDA 230
genetic diversity 111
genetic erosion 199
genetic structure of major and minor species 142–143
GENMEDA (Network of Mediterranean Plant Conservation Centres) 374
genomic resources 231–233
German Agency for Technical Cooperation GTZ (GIZ) 20
germplasm: collection and conservation of 71–72; conserved to support research and healthy diets 209–210; increased productivity and nutrition 210–211; utilisation of 72
germplasm characterization: and evaluation 228–230; novel technologies and nutritional potential of small millets 309–316; for nutritional quality 311
Ghent University, Belgium 405
Gichangi, Lilian 186
Global Action Plan for Agricultural Diversification (GAPAD) 31
Global Affairs Canada 417, 423n3
global commodities, creating 329–330
Global Crop Diversity Trust (GCDT) 397–398
Global Environment Facility 410–411
Global Facilitation Unit for Underutilized Species (GFU), Rome 27, 396
global food systems 4, 216, 222, 419, 422–423; towards more robust 422–423
Global Forum on Agricultural Research (GFAR) 24
Global Plan of Action (GPA) for PGRFA 23
GoBhaarati Agro Industries and Services Pvt. Ltd., India 358–360; three-pronged approach for promoting NUS 360; views about NUS 358–360
golden thistle (*Scolymus hispanicus*) 169
Google Scholar 8
Gopi, Girigan 341
government-driven value chains 263–271
grain popping 313
grass pea (*Lathyrus sativus*) 13

Green Revolution 3–4, 15n1, 19, 227, 310, 327, 387
guaje (*Leucaena esculenta*) 120–121
Guatemala 176–177
guava (*Psidium guajava*) 102
Guidance on Mainstreaming Biodiversity for Nutrition and Health (WHO) 54
gun puff machine *289*

health 5, 13; benefits 33, 47, 188, 255, 315; burden 149; foods 75, 242, 248; professionals 33; risks 175; and small millets 76–77
healthy diets: germplasm conserved in WorldVeg genebanks and 209–210; safeguarding and promoting NUS in Oceania for 155–157
The Heirloom Rice Project 383
hexaploid spelt (*Triticum spelta*) 141
Highlands pitpit (*Setaria palmifoliai*) 150–151, *151*
high-yielding varieties (HYVs) 3, 72
HIV/AIDS 63
hog plum (*Spondias mombin*) 102
holistic value-chain approach 30, 44–45, *46*
home garden 62, 121–122, 129, 212, 342, 370
horse gram (*Macrotyloma uniflorum*) 169
hot extrusion 314
household level consumption, promoting 265
human health and Oceania 148–149

ICAR-IIMR studies on primary processing to increase efficiency 284
IEDA Confectionary Limited 184, 186
"Iermana" 383
India 176–177; area, production and productivity of millet crops in *70*; challenges and best practices of nutri-millets in 245–261; collection and conservation of germplasm 71–72; crop improvement and AICMIP 74; crop improvement efforts during early years 73; crop production and protection technologies 75; diversity and small millets in 68–77; divisions of labour 391; fall in millet area and production 69–70; Food Security Bill 28; future prospects 75–76; implications for policy 393; improved varieties of small millets **74**; land management 389–391; limitations of minor millet processing in 281–282; millets and food security 70;

434 Index

minor millets processing technologies in 281–291; National Bureau of Plant Genetic Resources (NBPGR) 403; progress in crop improvement 73; Public Distribution System (PDS) 28; seed production 75; small millets diversity in 71; social equity and gender relations 388–389; utilisation of germplasm 72; women's agency and decision-making 391–393

Indian barnyard millet (*Echinochloa frumentacea* Link) 71

Indian Council of Agricultural Research (ICAR) 71

India school feeding study 331

Indigenous communities of Meghalaya: value-chain development of millet among 256–260

Indigenous Peoples 8, 14, 19, 21, 48, 50–53, 129, 148–149, 216, 218, 256, 342, 344, 352–353, 366, 397, 400, 422

Industrial Technology Institute (ITI) (Sri Lanka) 404

INFOODS programme 397

innovation 48, 119, 140, 186, 192, 220, 277, 310, 351, 358, 368–369, 375, 407, 410

in situ and ex situ conservation 28, 47, 110, 114, 122, 143, 199, 201, 204–205, 373, 381

Institut d'Economie Rurale (IER) 403–404

Institute of Biosciences and Bioresources of Research Council (CNR), Bari, Italy 403

Integrated Child Development Scheme (ICDS) 265, 267; and Covid-19 pandemic 324; inclusion of minor millets in 290; millet inclusion in 268

Integrated Child Development Services 30

Integrated Tourism Association of Santiago de Okola (ASITURSO) 200

integrated value-chain development in Kolli Hills, Tamil Nadu 246–251

Intergovernmental Technical Working Group on Plant Genetic Resources for Food and Agriculture 53

International Center for Tropical Agriculture (CIAT) 367, 395–396

International Center for Underutilized Crops (ICUC) 20

International Centre for Advanced Mediterranean Agronomic Studies (CIHEAM): founded in 374; perspectives on NUS 374–376

International Centre for Biosaline Agriculture 422

International Centre for Research in Agroforestry (ICRAF) 396

International Centre for Underutilized Crops (ICUC), Sri Lanka 27, 396

International Conference on Indigenous Vegetables 27

International Conference on "New Crops for Food and Industry" 20

international cooperation 22, 398

International Crop Research Institute for Semi-arid Tropics (ICRISAT), Hyderabad, India 228–229, 396; gene bank 228; grain nutrients content of collections of small millets conserved at **312**

International Development Research Centre (IDRC) 21, 22, 410, 417, 420, 423n2, 423n3, 423n4

International Fund for Agricultural Development (IFAD) 43, 199, 365, 373, 410; Action Plans on Mainstreaming Nutrition 12; championing pro-poor crops of the future 366–369; and genesis of NUS 366–368; going forward 369; multipurpose species of the future 368–369; overview 366; role of NUS in inclusive and sustainable rural transformation 366; Rural Youth Action Plan 2019–2021 12; Strategy and Action Plan on Environment and Climate Change 2019–2025 12

International Horticultural Assessment 25

International Institute of Tropical Agriculture (IITA) 20

International Plant Genetic Resources Institute 53

International Society for Horticultural Science (ISHS) 27, 412

International Symposia on New Crops 20

International Treaty on Plant Genetic Resources for Food and Agriculture (International Treaty) 384n1, 397; adopted in 378; aims of 378; farmers and food security and rural economy 378–384; key solutions considered for NUS promotion 380–382; obstacles for enhanced sustainable use of NUS 380; views with regard to NUS 379–380

Introduction to useful plants (Pl@ntuse) 402

IPGRI 26, 130; and African leafy vegetables 23; strategy on NUS 25;

Index **435**

"The Underutilized Mediterranean Species"(UMS) project 22
Iriventi, Sridhar Murthy 358
irrigated rice (*Oryza sativa*) 78
isaño (*Tropaeolum tuberosum*) 198
Issoufou, Malika 330
Italian Development Cooperation Agency (AICS): advancing agenda of NUS 372–374
Italian Farmers' Seeds Network 408
Italian National Research Council, Naples, Italy 24
Italian or foxtail millet (*Setaria italica*) 238

Jackfruit (*Artocarpus heterophyllus*) 169
Jackson, C. 387
James Cook University, Australia 405
Japan Association for International Collaboration of Agriculture and Forestry (JAICAF) 184, 186, 195
Japanese barnyard millet (*E. esculenta*) 231
JK Paper Mills 240
jujube (*Ziziphus mauritania*) 13
jute mallow (*Corchorus olitorius*) 13

Kakadu plum (*Terminalia ferdinandiana*) 155
Karawita, Rohan 345
Kenya: African Leafy Vegetables in 81–82; Kenya Agricultural and Livestock Research Organization (KALRO) 166, 403; marketing Kieru products in Embu market *190*; pop cereal business in 182–196; simulated business model in **189**
Kenya, Leon 352
Kenya Agricultural and Livestock Research Organization (KALRO) 166, 403
Khiloput, Tukuna Burudi 338
Kieru Foods Ltd., Embu, Kenya 186–193; additional promising technology 192–193; commercialisation 187–190; effects of COVID-19 on business 191–192; networking 190–191; overview 186–187; promotion 190
Kinupp, V. F. 130, 133
Kirori, Daniel 186
kūmara (sweet potato, *Ipomoea batatas*) 152
knowledge-sharing websites 400–401
Kodo (*Paspalum scrobiculatum*) 256
kodo millet (*Paspalum scrobiculatum*) 71, 238
Kolab Farmers Producer Company Limited (KFPCL) 252–253
Kolli Hills, Tamil Nadu: community seed banks 247; empowering marginal hill-dwellers 248; integrated value-chain development in 246–251; village agro-bioresource centres for capacity building 247–248
'Kolli Hills Agro-Bioresource Producer Company Limited' (KHABPCOL) 247–248
Komen, Margaret 346
Koraput, Odisha: alternative seed systems enabling access to quality seeds in 252–253; farmers views with regard to NUS 338; major obstacles for NUS enhanced use 338–339; perspectives of custodian farmers of 338–339
Kriori, Daniel 347
Kumar, Rakesh 340
Kutki (*Panicum sumatrense*) 256

lablab (*Lablab purpureus*) 78
land management 389–391
landrace diversity 110
Latin America: barriers to adoption for Latin American fruit-tree NUS **94–101**; high-potential Latin American fruit-tree NUS **90–93**
Latin American food systems: barriers at production stage 89–102; barriers at value-chain stage 102–103; barriers in consumer demand and marketability 103; barriers to adopting diversity of NUS fruit trees in 88–107; barriers to adoption of NUS fruit-tree species in 89–103; farm-facing solutions 103–104; from farm to market 102–103; removal of barriers to adoption of native fruit-tree species in 103–106; up-scaling NUS access to consumers 105–106; value-chain development 104–105
Latz, Peter 155
Law for Agrobiodiversity, Seeds and Promotion of Sustainable Agriculture (Ecuador) 382
lean periods and NUS consumption 60–62
lean season 13
Leipzig FAO IV Technical Conference on Plant Genetic Resources for Food and Agriculture (PGRFA) 23
LEISA magazine 25
The Lexicon, USA 398
The Lexicon for Sustainability 398
little millet (*Panicum sumatrense*) 71, 238
livelihoods and African vegetables 208–214
Local Initiatives for Biodiversity, Research and Development (LI-BIRD) 407, 423n6

436 Index

locating endangered products 218
Lorenzi, H. 130, 133
Lost Crops of the Incas 20
Low, Tim 154

Mace Foods Ltd., Kenya 346–347; challenges in promoting NUS 347; general considerations 346; solutions adopted 347
Madhya Pradesh: farmers views with regard to NUS 339; interventions for effective promotion 340; major obstacles for enhanced use 340; perspectives from custodian farmers of 339–340
Magrini, M.-B. 277
mainstreaming: African vegetables 208–214; biodiversity 132; CBD programme of work on agricultural biodiversity 22; identifying partners and suitable entry points for NUS 167–168; monitoring and evaluating 171; of NUS value-chains for improved nutrition 53–54; orphan crops 83
Mainstreaming Agrobiodiversity in Sustainable Food Systems report (Bioversity International) 54
maize (*Zea mays*) 78, 118
major species: diversifying genetic structure of 142–143
Makavana, Ramesh 340
Malay apple (*Syzygium malaccense*) 150
Mali 176–177; Institut d'Economie Rurale (IER) 403–404
malting 313–314
Mandla and Dindori, Madhya Pradesh 256
mangaba (*Hancornia speciosa*) *167*
mango (*Mangifera indica*) 78
mangrove (*Bruguiera* spp.) 150
Marino, Mario 378
marketability, barriers in 103
Martinez, Virgilio 106
marula (*Sclerocarya birrea*) 13
Mathur, Shantanu 366
Maya Biosphere Reserve (Guatemala) 104
McKnight Foundation, USA 411
MDM, millet inclusion in 268
media and Smart Food Triple Bottom Line approach 331
The Mediterranean Germplasm Database, Italy 403
Melhores Receitas da Alimentação Escolar 168
Mexico: CONABIO 123; conservation of NUS in 121–122; edible flowers and/or fruits of agave and cacti 119–120;

fruits of trees and shrubs 120–121; needs, challenges, and opportunities 122–123; overview 118–119; overview of NUS in 119–121; *quelites* 119; traditional vegetables 119
microbes and bioirrigation-based intercropping systems 274–277
micro value-chain for the "Iermana" 383
"Microveg" project 418, 424n7
Millennium Development Goals 26
millet-based intercropping systems 273–278
Millet Network of India (MINI), India 412–413
millets: challenges of eucalyptus plantations 240–241; components of OMM 265–271; conservation for 242; conserving, in ecologically sensitive areas 237–243; custodian farmers' perspectives from Rayagada, Odisha 339; evolution of agrarian system 238–240; fall in millet area and production 69–70; finger 267; and food security 70; food security role under climate change in Madhya Pradesh 295–307; under FSN 320–322; inclusion in ICDS 268; inclusion in MDM 268; multi-stakeholder consultation 263–264; nutritional composition of millet grains **69**; nutritional relevance 69; other challenges 241–242; overview 68–69, 263–264; primary processing of 282–284; scale 265; secondary processing of 284–289; small millets 238; in social safety net programmes 322–323; stakeholders 265; in support of nutrition and social safety net programmes 319–325
milling 311–313
minor millet processing: cold extrusion technology 285–286; constraints in existing primary processing machinery 283–284; flaking technology 285; ICAR-IIMR studies to increase efficiency 284; identified suitable machines for different millet grains 284; importance of 282–289; limitations of 281–282; primary processing 282–284; puffing technology (grain expansion technology) 286–287; secondary processing 284–289
minor millets: nutrient content of *295*; in public-funded schemes 290
minor millets processing technologies: ICDS and mid-day meal program 290; importance of processing 282–289; in

Index **437**

India 281–291; limitations of processing in India and world 281–282; overview 281; policy support 290
minor species: diversifying genetic structure of 142–143
Mishra, Sharad 339
moringa (*Moringa oleifera*) 13, 396
Moringa News, France 403
M.S. Swaminathan Research Foundation (MSSRF), India 26, 246–247, 253, 319, 407
Muduli, Sunamani 338
multi-stakeholder consultation 263–264
multi stakeholder platforms 82, 354
murumuru (*Astrocaryum murumuru*) 105
Myanmar 58–65
Myanmar malnutrition and acceptance study 331
Myda system 238

Narayan, Meghana 349
National Agricultural Research Institute (Papua New Guinea) 404
National Agricultural Research Systems (NARS) 19, 44
National Bureau of Plant Genetic Resources (NBPGR), New Delhi 71, 403; National Active Germplasm Sites 71
National Food Promotion Board (NFPB), Sri Lanka 345–346; obstacles in use enhancement of NUS 345; solutions for effective promotion of NUS 345–346; views with regard to NUS 345
National Food Security Act (NFSA) 320, 322, 323
National Gene Bank of Tunisia 373
national governmental research institutions 403–404
native gooseberry (*Cucumis melo*) 155
native millet (*Panicum decompositum*) 155
native plum (*Santalum lanceolatum*) 155
Native Seeds/SEARCH, USA 408
Native species of Brazilian flora with actual or potential economic value: Plants for the Future – South Region 131
Naturally Zimbabwean, Zimbabwe 409
Naure, Loknath 339
Nderitu, Simon 352
neglected and underutilized species (NUS) 43; additional frameworks and policies enhancing mainstreaming of 53–54; beyond 220–222; beyond a definition 4–9; comparing benefits from commodity crops and **10–11**;

consumption during lean periods 60–62; context and conceptual challenges 59–60; cross-cutting issues 51–53; custodians of 337–343; description of 6; diverse land use, biodiversity and dietary diversity 63–64; and food and nutrition change 166–167; food system diversification with 81–83; IFAD and genesis of 366–368; importance of 9–14; increasing production, marketing and awareness of 168–170; key features of 7–8; mainstreaming, for nutrition-sensitive agriculture 43–55; monitoring and evaluating mainstreaming of 171; nutritional outcomes in value chains 46–51; nutrition-sensitive agriculture and value chains 44–46; of Oceania 150–155; people associate nutrition with well-being 62–63; and peoples sense of dietary diversity and well-being 58–66; private sector engaged in use enhancement of 344–364; in Santiago de Okola, Bolivia 198–206; similar terms found in literature 6; Slow Food and 216–223; strategic actions 29–30
Nepal Agricultural Research Council (NARC) (Nepal) 404
"Nepal Terrace Farming and Sustainable Agriculture Kits" project 418, 423n5
networks and consortia 412–413
New Agriculturalist 401–403
New CROP, USA 402
New Crops for Food and Industry 21
Newton, John: *The Oldest Foods on Earth: A History of Australian Native Foods with Recipes* 154–155
Ngugi, Michael 356
NIRMAN, India 407
non-CGIAR international organizations 396–400
Non-conventional Edible Plants (PANC) in Brazil: identification guide, nutritional aspects and illustrated recipes (Kinupp and Lorenzi) 133
Non-conventional vegetables from the Amazon (Cardoso) 130
non-governmental organizations (NGOs) 406–408
Nongrum, Melari Shisha 342
noni (*Morinda citrifolia*) 150
North East Slow Food and Agro biodiversity Society (NESFAS) 256–257, 259, 343
NUS Community platform 400

438 Index

NUS events and key publications 19–35; period: 1970–1980 20; period: 1981–1990 20–21; period: 1991–2000 21–24; period: 2001–2020 24–33
NUS fruit trees 88–107
nutri-millets: alternative seed systems in Koraput, Odisha 252–253; collective marketing in Madhya Pradesh 256; community-centred value-chain development of 245–261; integrated value-chain development in Kolli Hills, Tamil Nadu 246–251; overview 245–246; value-chain development for small millets in Tamil Nadu 253–255; value-chain development of millet among communities of Meghalaya 256–260
nutrition: additional frameworks and policies for improved 53–54; germplasm enhancement for increased 210–211; millets in farming systems in support of 319–325; people associating well-being with 62–63
nutritional and nutraceutical traits 229
nutritional quality, germplasm characterization for 311
'nutrition security' 327
nutrition-sensitive agriculture 44–46; NUS for 43–55; and value chains 44–46

oca (*Oxalis tuberosa*) 198
Oceania 147–148; dietary and lifestyle change 148–149; loss of food diversity and human health 148–149; neglected and underutilized species of 150–155; overview 147–148; safeguarding and promoting NUS in 155–157; sustainable food systems and healthy diets 155–157
Odisha Millets Mission (OMM) **266**; components of 265–271; custom hiring centres 266; establishing diverse seed centres 266; farmer producer organisations 267; implementation in KMS 2018–2019 (year 1) 267, **268**; implementation in KMS 2019–2020 (year 2) 267, **268**; implementation in KMS 2020–2021 (year 3) 268; improved agronomic practices 266; key objectives of 264; millet inclusion in ICDS, MDM 268; procurement of finger millet 267; promoting household level consumption 265; recognition and impact 268–269; setting up decentralised processing facilities 265; status of different community institutions of **267**

Odisha State Seed and Organic Products Certification Agency 253
Odisha State Seed Corporation Ltd. 253
Oenema, Stineke 376
oil palm (*Elaeis guineensis*) 78
okra (*Abelmoschus esculentus*) 78
The Oldest Foods on Earth: A History of Australian Native Foods with Recipes (Newton) 154–155
Opuntia species 120
Orana Foundation 154
Oroverde, Switzerland 409
orphan crops 5, 58, 83–84, 332, 368, 411
Ortega, D. L. 277
Overseas Development Agencies (ODA) for NUS 22
Overseas Development Institute (ODI), UK 254
OXFAM 399

Pacific Agricultural Plant Genetic Resources Network (PAPGREN) 400
Pakistan Agricultural Research Council (PARC) 404
palm oil (*Elaeis guineensis*) 88
pama (*Pseudolmedia macrophylla*) 105
PANC 130, 135–136
pandanus (*Pandanus tectorius*) 150, *154*, 155
papaya (*Carica papaya*) 78
Papua New Guinea: National Agricultural Research Institute 404
parboiling or hydrothermal treatment 313
Parida, Prashant Kumar 338
Participatory Enhancement of Diversity of Genetic Resources in Asia (PEDIGREA) 413
participatory research 175, 260
patauá (*Oenocarpus bataua*) 105
peach palm (*Bactris gasipaes*) 105
pearl millet (*Pennisetum glaucum*) 78
PELUM Association 413
perceived seasonal availability 176
perspectives from custodian farmers of Madhya Pradesh 339–340
Peru 12, 25, 49, 58–65
pigeon pea (*Cajanus cajan*) 13
Pipi, Pietro 372
pitanga (*Eugenia uniflora*) 105, *167*
Pitia, Sunadei 338
pitpit/duruka (*Saccharum edule*) 150
Plant Genetic Resources Centre (PGRC) (Sri Lanka) 404
Plant Genetic Resources for Food and Agriculture (PGRFA) 19

Index **439**

plantgrowth-promoting rhizobacteria (PGPR) 273, 275–276
Plant Resources of South East Asia (PROSEA) 22, 402
Plant Resources of Tropical Africa (PROTA) 24, 401–402
Platzer, Marco 372
pochote (*Ceiba aesculifolia* ssp. *brevifolia*) 120
policy: advocacy 150; Common Agricultural Policy 138; enhancing mainstreaming of NUS value-chains for improved nutrition 53–54; public 323; support 290
Polynesian chestnut (*Inocarpus fagifer*) 150
pop cereal business: aim of intervention 182–184; DK Engineering Co. Ltd., Nairobi, Kenya 194; key outcomes and lessons 185–186; Kieru Foods Ltd., Embu, Kenya 186–193; multi-stakeholder joint venture 184; result and prospects 195–196; results and outcomes 184–185; technical challenges encountered 194–195
pop cereal business in Kenya 182–196
potato (*Solanum tuberosa*) 198
primary processing 282–284
private companies 408–409
private sector: African Forest Ltd., Kenya 352–354; Agropecuária Aruanã Ltda., Brazil 360–361; Anamol Biu Pvt. Ltd., Nepal 354–356; Bilgi Sistem Yönetimi – İthalat & İhracat, Turkey 357–358; D.K. Engineering Ltd., Kenya 347–348; engaged in use enhancement of NUS 344–364; Eurotropics S.A., Guatemala 348–349; GoBhaarati Agro Industries and Services Pvt. Ltd., India 358–360; Mace Foods Ltd., Kenya 346–347; National Food Promotion Board, Sri Lanka 345–346; overview 344–345; SanLak Agro-Industries Pvt. Ltd., India 361–364; Simlaws, Kenya 356–357; Slurrp Farm, India 349–352
procurement of finger millet 267
productivity and germplasm enhancement 210–211
PROINPA Foundation of Bolivia 199–200, 202, 205
promotion: of endangered food products 216–223; of endangered products 218–220; of NUS in Brazil 130–136; through chefs 330
proso millet (*Panicum miliaceum*) 71, 238

PROTA (Plant Resources of Tropical Africa) Network 24, 401–402
protecting: endangered food products 216–223; endangered products 218–220
public awareness 24–25, 33, 51, 82, 188, 214, 315
public-funded schemes: ICDS 290; inclusion of minor millets in 290; mid-day meal program 290
puffing technology (grain expansion technology) 286–287, 313
Pūhā (*Sonchus oleraceus*) 152–153, *153*
Pulse Innovations: Food and Nutrition Security in Southern Ethiopia 424n8
pumpkins (*Cucurbita argyrosperma, C. pepo*) 118

quelites 119
Querejeta, J. I. 275
quinoa (*Chenopodium quinoa*) 141, 198

Rainforest Alliance 104
ramón (*Brosimum alicastrum*) 104, 120
Rana, Ram Bahadur 354
Ranee, Bibiana 342
Rayagada, Odisha: custodian farmers' perspectives from 339; farmers' views with regard to NUS 339; interventions for effective promotion 339; major obstacles for NUS enhanced use 339
research: germplasm conserved in WorldVeg genebanks to support 209–210; participatory 175, 260
Research for Development (R4D) agencies: conserve and promote sustainable use of NUS 365–384; IFAD 366–369; International Centre for Advanced Mediterranean Agronomic Studies (CIHEAM) 374–376; International Treaty 378–384; Italian Development Cooperation Agency (AICS) 372–374; overview 365; Self Help Africa 369–372; UN System Standing Committee on Nutrition (UNSCN) 376–378
resilience: of agricultural production systems 227; agroecosystem 119; climate-change 32, 48, 175, 304; and community seed banks 247; economic 88; farm 85; food system 82; of human communities 216; safeguarding traditional vegetable biodiversity as driver for 208–209
Rete Semi Rurali, Italy 408
roller drying 315
roller flaking machine *288*

440 Index

Rotary Action Group 402
Royal Botanic Gardens (RBG), Kew, UK
  396–397

safeguarding: NUS in Oceania for food
  systems and healthy diets 155–157;
  traditional vegetable biodiversity
  208–209
safou (*Dacryodes edulis*) 78
safou plum (*Dacryodes edulis*) 396
sago (*Metroxylon sagu*) 150
saijan/drumstick (*Moringa oleifera*) 150
Sankaranarayanan, Vikram 361
SanLak Agro-Industries Pvt. Ltd., India
  361–364; general considerations
  361–362; key obstacles and solutions
  362–364; promoting minor millets 362
Santiago de Okola, Bolivia: agritourism in
  198–206; background and context 199–
  200; conservation of native Andean crops
  in 198–206; methods 202; objectives of
  the study 201–202; results 202–204
SAVE Foundation, Switzerland 408
scenario model: overview of *302*; as a series
  of linked equations **298–301**
Schiaffino, Pedro Miguel 106
seasonal calendars: collecting information
  175–176; development of, for sustainable
  diets 174–179; for diversification of
  diets and ecosystems 178; and traditional
  knowledge 178
seasonal food availability calendar in
  Bamanankan *177*
seasonality 174
secondary processing 284–289
Second International Conference on
  Nutrition (ICN2) 12, 45, 54
Second International Conference on
  Nutrition Framework for Action 30
seed production 75
Sekiya, N. 274
Self Help Africa (SHA): agency's views
  with regard to NUS 369–370;
  commercialising neglected crops in
  practice 369–372; key opportunities
  provided by NUS 371; key solutions
  for better NUS promotion 372; major
  obstacles for enhanced sustainable use
  371; promotion at community level 371;
  registration of varieties 371; shortage of
  seeds 371; trading to the EU 371
Sendas Altas 199–200, 202, 205
SF Foundation for Biodiversity 217
SF Presidium project 219–220

Sharma, Mahesh 354
shifting cultivation areas 140–141
shrubs, fruits of 120–121
Shujaaz 420
Simlaws, Kenya 356–357; NUS work 357;
  views about NUS 357
Slow Food (SF) 399–400: beyond NUS
  220–222; defining endangered products
  217; locating endangered products 218;
  and NUS 216–223; overview 216–217;
  protecting and promoting endangered
  products 218–220
Slurrp Farm, India 349–352; lack of
  research and development 351; major
  obstacles for use enhancement of NUS
  350; solutions for effective promotion
  351–352; struggles in post-harvest
  processing 350–351; views about NUS
  349–350
small millets 238; custodian farmers of
  Koraput, Odisha 338–339; diversity in
  India 71; germplasm characterization
  309–316; germplasm characterization for
  nutritional quality 311; list of sequenced
  genomes in *232*; millet processing
  techniques *312*; and novel technologies
  309–316; present status of conservation
  of 228; processing technologies 311–315
small millets genetic resources: breeding
  advancements 231; conclusion and
  future projections 233; conservation of
  227–233; and elite cultivars 230–231;
  genomic resources 231–233; germplasm
  characterization and evaluation 228–230;
  overview 227–228; present status of
  conservation of small millets 228;
  utilization of 227–233
small millets germplasm: abiotic stresses
  230; agronomic traits 229; biotic stresses
  229–230; characterization and evaluation
  228–230; elite cultivars towards
  germplasm improvement 230–231;
  nutritional and nutraceutical
  traits 229
small millets processing technologies
  311–315; baking 314; cold and hot
  extrusion 314; fermentation 314; flaking
  313; malting 313–314; milling 311–313;
  parboiling or hydrothermal treatment
  313; popping and puffing 313; roller
  drying, spray drying and vacuum shelf
  drying 315; SM-based RTE and RTC
  products 315; starch, dextrins and
  ethanol 315

Index **441**

smart foods: for big impacts 328–333; dedicated effort on 329; diversifying staples with 328–333

Smart Food Triple Bottom Line approach: ambassadors and champions 330–331; consumer comes first 330; create global commodities 329–330; dedicated effort for smart foods 329; diversifying staples with smart foods for big impacts 328–333; food system solutions incorporating 327–328; importance of 327–333; India school feeding study 331; international platforms 331; make it delicious, convenient and easy 330; media and social media outreach 331; Myanmar malnutrition and acceptance study 331; not ultra-processed or excessive added ingredients 332; popularize whole grain 331–332; promotion through chefs 330; scientific backing on nutritional benefits 331–333; selecting millets and sorghum first 329; selection of biofortified varieties 332–333; Tanzania school feeding study 331; working with the hidden middle 330

SMART objectives 171

social equity and gender relations 388–389

social media and Smart Food Triple Bottom Line approach 331

social safety net programmes: discussion and recommendations 323–325; millets in 322–323; millets in farming systems in support of 319–325; millets under FSN 320–322

Society for Research and Initiatives for Sustainable Technologies and Institutions (SRISTI), India 408

sorghum (*Sorghum bicolor*) 78

sorghum recipes 269

Southern and East Africa Network on Underutilized Crops (SEANUC) 22

"Special Programme for Promotion of Millets in Tribal Areas (Odisha Millets Mission)" *see* Odisha Millets Mission (OMM)

species: diversity 109–110; "forgotten" 141–142; major 142–143; minor 142–143; prioritization in Atacora, northern Benin **80**

spider plant (*Cleome gynandra*) 78

spray drying 315

Sri Lanka: Industrial Technology Institute (ITI) 404; National Food Promotion

Board (NFPB) 345; Plant Genetic Resources Centre (PGRC) 404

stakeholders and millets 265

staples, diversifying 327–333

starch 315

strawberry tree (*Arbutus* spp.) 13

SUAAHARA Project 355

sub-Saharan Africa (SSA) 208–212, 214

Subsidiary Body on Scientific, Technical and Technological Advice (SBSTTA) 25

subsistence farming 58; survey information of area and people engaged in **61**

Sudhishri, S. 274

Sukia, Loichan 338

Sumaq kausay 63

"superfoods" 422–423

sustainability index developed by BFN Turkey **169**

Sustainable Development Goal 2 (SDG2) 4

Sustainable Development Goals (SGDs) 12, 31, 328, 417

sustainable diets 327; development of seasonal calendars for 174–179

sustainable farming 273–278

sustainable food systems 155–157

sustainable production 211–212

sustainable use 21–23, 25, 53, 66, 119, 123, 130, 150, 157, 163, 239, 344, 365–384

SUSTLIVES (SUSTaining and improving local crop patrimony in Burkina Faso and Niger for better LIVes and EcoSystems) project 376

Swaminathan, M. S. 9, 23, 319

Syiem, Ridian 342

Syngenta Foundation for Sustainable Agriculture (SFSA) 412

Tahuri Whenua Incorporated Society 155–156, 157n3

tamarind (*Tamarindus indica*) 78

Tamil Nadu, India: catalysing value-chain development for small millets in 253–255; crop improvement for small millets 73; as finger millet growing state 69

Tanzania school feeding study 331

taro (*Colocasia esculenta*) 150, 152

tava (*Pometia pinnata*) 150

technology development 49, 75, 111, 192–193, 277

teff (*Eragrostis teff*) 78

tempesquistle (*Sideroxylon palmeri*) 120

tepary bean (*Phaseolus acutifolius*) 13

tetraploid emmer (*Triticum dicoccon*) 141

tomatoes (*Solanum lycopersicum*) 78

442 Index

traditional African vegetables (TAVs) 208–210, 212; consumption 212–213
traditional cowpea (*Vigna unguiculata*) 169
Traditional Food Plants of Kenya (Maundu) 79
traditional foods 7, 12, 147–149, 157, 188, 192, 205, 219, 256, 260, 342, 347, 370–371, 417–418
traditional food systems 63, 148, 216
traditional knowledge: and seasonal calendars 178
traditional vegetable biodiversity 208–209; as a primary driver for resilient food systems 208–209
traditional vegetables 119
trees: on farms 83; fruits of 120–121; promoting diversity 83
Tribal Development Cooperative Corporation Odisha Limited 267
tucumã (*Astrocaryum aculeatum*) 167
Tulimara Specialty Foods, Zimbabwe 409

UN Agenda 2030 31
underutilized genetic resources: and assessing them 143–144; in Europe 138–144
Underutilized Tropical Fruit in Asia Network (UTFANET) 22
underutilized crops: bright future for underutilized plants species 417–422; enhancing use of 182–196; in the European context 140; ten thousand years of decline 416–417; towards more robust global food systems 422–423
underutilized plant genetic resources (UPGR) 26
underutilized plants species: ancestral taste from Andes going global 422; ancient crops, new terrains 418; bright future for 417–422; finger millets, from obscurity to supermarkets 420–421; greens to the plate 418–419; new time for traditional cereals 421–422; pulses once again on the menu 419–420
UNIDO 373
UNILEVER 409
United Nations (UN) 35; Millennium Development Goals 26; Sustainable Development Goals (SGDs) 12, 31, 328, 417
United Nations Educational, Scientific and Cultural Organization (UNESCO) 399
United Nations Environment Programme (UNEP) 171n1, 400
universities 405–406

University of Abomey-Calavi, Benin 405
University of Nottingham, United Kingdom 405
University of Southampton, United Kingdom 405–406
UN System Standing Committee on Nutrition (UNSCN): agency's views with regard to NUS 376–377; key solutions for NUS promotion 378; major obstacles for enhanced sustainable use 377–378; perspective on NUS 376–378
UN World Food Programme (WFP) 373
Upadhyaya, H.D. 231
UPOV (International Union for the Protection Of New Varieties of Plants) 142
USAID 25, 411
USA National Academy of Sciences (NAS) 20
USDA gene bank 230
utilisation: of germplasm 72; of small millets genetic resources 227–233
Uttarakhand: conserving Indigenous crops by custodian farmers in 340; farmers views with regard to NUS 340

vacuum shelf drying 315
value-chain development: Latin American food systems 104–105; of millet among Indigenous communities of Meghalaya 256–260; for small millets in Tamil Nadu 253–255
value chains: additional frameworks and policies for 53–54; capacity building 51–52; consumption 50–51; empowerment of vulnerable groups 52–53; entry points to strengthen nutritional outcomes in 46–51; harvest 48–49; input supply 46–48; marketing 49; nutrition-sensitive agriculture and 44–46; processing 49; production 48; stage barriers at 102–103
varieties, improvement of 111
Vavilov, N.I. 140
vegetable biodiversity, traditional 208–209
Vergueiro, Ana Luiza 360
Vergueiro, Sergio 360
Verticillium wilt disease 212
Vetriventhan, M. 231
Vietnam: Fruit and Vegetable Research Institute (FAVRI) 404
Village Agro-bioresource Centres 247
Vivekananda Parvatiya Krishi Anusandhan Sansthan (VPKAS), India 231

Index **443**

*Voluntary Guidelines for Mainstreaming
Biodiversity into Policies, Programmes and
National and Regional Plans of Action on
Nutrition* (FAO) 54
vulnerable groups, empowerment of 52–53

Wageningen University & Research, The
Netherlands 405
Wagstaff, Paul 369
Wallace, J.G. 232
water lily (*Nymphaea pubescens*) 169
weda (*Saba senegalensis*) 78
well-being: people associating nutrition
with 62–63; people's interconnected
sense of *64*
whole grain, popularizing 331–332
wild edible plant species (WEPs) 80–81
wild fig (*Ficus platypoda*) 155
Wild Food Plants of Australia (Low) 154
winged bean (*Psophocarpus tetragonolobus*) 20
Winrock International, USA 412
women: agency and decision-making
391–393; and ALVs 82; Gikuyu 220;
and KHABCoFED 248; knowledge
of cultivation 389; mixed-cropping
practices led by 241; portrayed as risk-
averse subsistence farmers 387–388;

safeguarding the wealth of NUS 50;
specialised local knowledge of 59; tribal
237, *250*; working-class 346
women empowerment 170, 216, 220,
256, 319
woolybutt (*Eragrostis eriopoda*) 155
World Agroforestry (ICRAF) 83, 396;
Genetic Resources Unit of 83; Priority
Food Tree and Crop Food Composition
Database 83
World Conservation Union (IUCN) 400
World Health Organization (WHO) 208;
*Guidance on Mainstreaming Biodiversity for
Nutrition and Health* 54
World Vegetable Centre (AVRDC)
208–211, 398
WorldVeg genebanks: access to germplasm
conserved in 209–210; supporting
research and healthy diets 209–210
World Wide Fund for Nature (WWF) 400

yams (*Dioscorea* spp.) 78
Yano, K. 274

Zimbabwe 58–59, 62, 65; area and the
people engaged in subsistence farming
61; seed fair with NUS 65

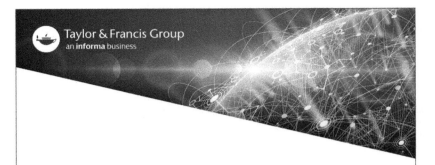

Lightning Source UK Ltd.
Milton Keynes UK
UKHW020628150622
404452UK00015B/259